THE
NEANDERTAL
ENIGMA

Other Avon Books by
James Shreeve

LUCY'S CHILD
(with Donald Johanson)

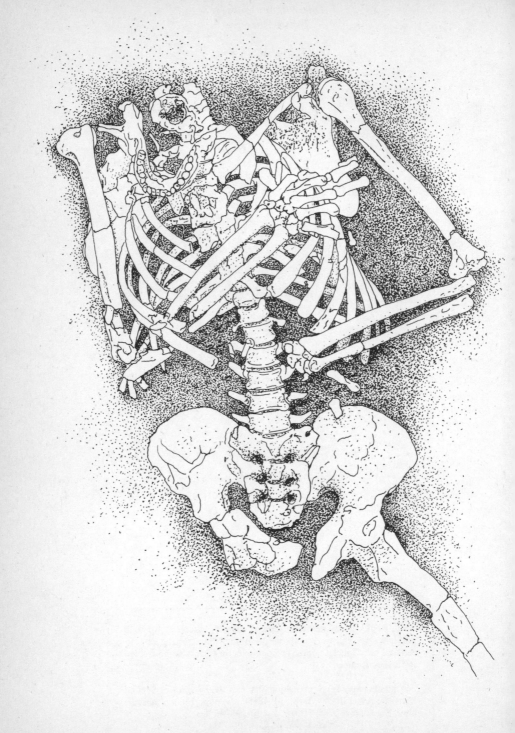

THE
NEANDERTAL
ENIGMA

SOLVING THE MYSTERY OF
MODERN HUMAN ORIGINS

JAMES SHREEVE

AVON BOOKS ◆ NEW YORK

VISIT OUR WEBSITE AT
http://AvonBooks.com

Frontispiece: *Neandertal skeleton from Kebara Cave, Israel,* David Dilworth, redrawn from Bar-Yosef et al. 1986, copyright © University of Chicago Press.

AVON BOOKS
A division of
The Hearst Corporation
1350 Avenue of the Americas
New York, New York 10019

Copyright © 1995 by James Shreeve
Published by arrangement with William Morrow and Company, Inc.
Library of Congress Catalog Card Number: 95-6337
ISBN: 0-380-72881-8

The William Morrow edition contains the following Library of Congress Cataloging in Publication Data:
Shreeve, James.
    The neandertal enigma : solving the mystery of modern human origins / by James Shreeve.
        p.   cm.
Includes bibliographical references and index.
1. Neanderthals.   2. Man—Origin.   3. Human evolution.   4. Behavior evolution.   I. Title.
GN285.S57   1995
573.2—dc20   95-6337   CIP

First Avon Books Trade Printing: October 1996

AVON TRADEMARK REG. U.S. PAT. OFF. AND IN OTHER COUNTRIES, MARCA REGISTRADA, HECHO EN U.S.A.

Printed in the U.S.A.

QPM 10 9 8 7 6 5 4 3 2 1

To John Pfeiffer

Acknowledgments

Science writers enjoy a miraculous privilege. For reasons which I do not fully understand, we are allowed to invade the routines of busy people with phone calls and visits, distract them from their duties with interminable questions, abscond with their best insights and quotes, and later call them back again to make sure we got everything right the first time. For all this trouble, they receive nothing but our gratitude, and even that is often just a hasty footnote attached to the end of a two-hour interview. Among the many anthropologists, archaeologists, and other investigators whose life work forms the basis of this book, I would first like to thank those whom I badgered the most: Alison Brooks, Olga Soffer, Chris Stringer, Erik Trinkaus, Randy White, Milford Wolpoff, and John Yellen. I am especially grateful too to those who hosted me in their homes, tents, and caves during my travels: Ofer Bar-Yosef, Hilary and Janette Deacon, André Debenath, Jean-Philippe Rigaud, Jiří Svoboda, and again Alison Brooks and John Yellen. Olga, Randy, Alison, and John also reviewed chapters in manuscript, as did Clive Gamble, Alex Holmes, Bill Kimbel, Robert Kruszynski, Judith Masters, John Shea, Philip Rightmire, Mark Stoneking, and Linda Vigilant. Any errors are of course my responsibility and not theirs.

In addition to the above, I am in debt to the following scientists for their patience with my many questions: Leslie Aiello, Stanley Ambrose, Baruch Arensburg, Graham Avery, Peter Beaumont, Lewis Binford, Loring Brace, Günter Bräuer, James Brown, Rachel Caspari, Phil Chase, Desmond Clark, Geoffrey Clark, Meg Conkey, Leda Cosmides, Garnis Curtis, Chris Dean, Al Deino, Harold Dibble, Anna Di Rienzo, Pat Draper, Jim Enloe, Robert Foley, Allan Franklin, David Frayer, Clive Gamble, Rob Gargett, Jean-Michel Geneste, Rainer Grün, Michael Hammer, Henry Harpending, David Helgren, Trent Holliday, Ralph Holloway, Clark Howell, Jean-Jacques Hublin, Donald Johanson, Steven Jones, Bill Kimbel, Richard

Klein, Chris Knight, Janusz Kozlowski, Steven Kuhn, Roy Larick, David Lewis-Williams, Gerard Lucotte, Shannon McPherron, Curtis Marion, Anthony Marks, Judith Masters, Michael Mehlman, Paul Mellars, Gifford Miller, Steven Mithen, Masatoshi Nei, Vladimír Novotny, Svante Pääbo, John Parkington, Jacques Pelegrin, Geoffrey Pope, Heather Price, Yoel Rak, Philip Rightmire, Lars Rodseth, Karen Rosenberg, Chris Ruff, Martin Ruhlen, Vince Sarich, Lynne Schepartz, Henry Schwarcz, Jeffrey Schwartz, John Shea, Stephen Sherry, Vitaly Shevoroshkin, Jan Simek, Tal Simmons, Fred Smith, Songy Sohn, James Spuhler, Mary Stiner, Mark Stoneking, Ian Tattersall, Alan Thorne, Anne-Marie Tillier, John Tooby, Hélène Valladas, Bernard Vandermeersch, Pamela Vandiver, Linda Vigilant, James Wainscoat, Douglas Wallace, Bob Walter, Ken Weiss, Robert Whallon, Tim White, the late Allan Wilson, Martin Wobst, and Ezra Zubrow.

Most authors can count themselves lucky if they have a skilled and caring editor; I had two. My sincere gratitude to Maria Guarnaschelli for her passionate support of this project from its birth to late adolescence, and to Anne Freedgood for devoting her acumen and experience to bring it to full maturity. Victoria Pryor gave far more of herself than any agent should have to, and delicately maneuvered the book through some difficult waters. Robert Solomon and Will Schwalbe were instrumental in that effort too. Dave Dilworth, Stephen Nash, and Elizabeth Shreeve supplied illustrations with very little notice, and Marah Stets compiled the bibliography with even less.

For their advice, friendship, and many conversations over the course of the project, my thanks to Josh Fischman, Bill Kimbel, John Pfeiffer, and Jon Weiner. Tom McKain provided some much-needed support at a critical time. My parents, Walton and Phyllis Shreeve, added their enthusiasm and all manner of tangible help, and I am deeply grateful to them. Chris Kuethe Shreeve gave this book the blessing of her spirit, her intelligent eye, and her near-infinite patience. It would not have been possible without her.

Contents

THE

NEANDERTAL

ENIGMA

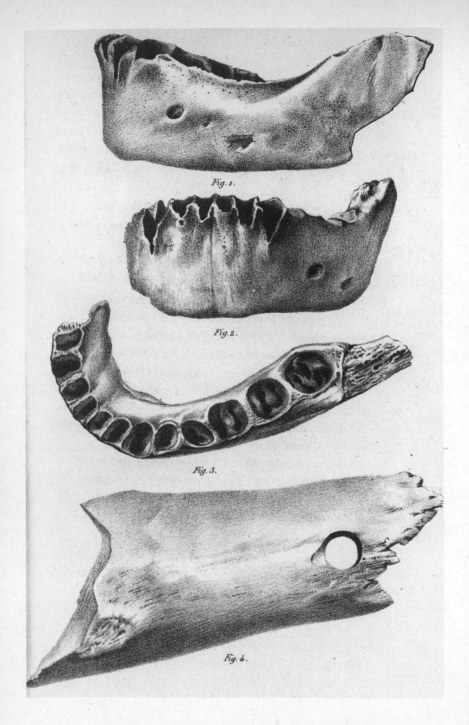

Fig. 1.

Fig. 2.

Fig. 3.

Fig. 4.

Contemporary drawing of a Neandertal jaw found in 1866 in the cave of Le Trou de la Naulette, Belgium, one of the earliest Neandertals discovered. At bottom is an artifically pierced bone from the same cave. FROM E. DUPONT 1866

Chapter One

RAISING THE GARTEL

Non-human animals have life cycles. Humans have biographies.
—GRAHAM RICHARDS

I met my first Neandertal in a café in Paris, just across the street from the Jussieu Metro stop. It was a wet afternoon in May, and I was sitting on a banquette with my back to the window. The café was smoky and charmless. Near the entrance, a couple of students were thumping on a pinball machine called GENESIS!, which beeped approval every time they scored. The place was packed with people of all types—French workers, foreign students, professors, young professionals, Arabs, Africans, and even a couple of Japanese tourists, all thrown together by the rain. Our coffee had just arrived, and I found that if I tucked my elbow down when raising my cup, I could drink it without poking the ribs of a bearded man sitting at the table next to me, who was deep into an argument.

Above the noise of the pinball game and the din of many private conversations taking place in a small public space, a French anthropologist named Jean-Jacques Hublin was telling me about the anatomical unity of man. Anatomically speaking, Hublin himself was compact and muscular, with dark curly hair, a strong chin and mouth, and wide-set brown eyes that never looked away from the person he was addressing. It was he who had brought along the Neandertal. When we came into the café, he had placed an object wrapped in a soft rag on the table and had ignored it ever since. Like anything so carefully neglected, it was beginning to monopolize my attention.

"Perhaps you would be interested in this," he said at last, whisking away the rag. There, amid the clutter of demitasse and empty sugar wrappers, was a large human lower jawbone. The teeth, worn and yellowed by time, were all in place. Around us, I felt the café raise a collective eyebrow. (It is part of the unspoken etiquette of Parisian cafés that while you are free to stare, you may not show too enthusiastic an interest in what you are staring at.) The hubbub of talk sank audibly. Impassive faces turned in our direction. The bearded man next to me stopped in mid-sentence, looked at the jaw, looked at Hublin, and resumed his argument. Hublin gently nudged the fossil to the center of the table and leaned back.

"What is it?" I asked.

"It is a Neandertal from a site called Zafarraya, in the south of Spain," he said. He picked it up, turned it around in his hands, and thoughtfully traced the flow of the jaw's bottom edge with a forefinger. "We have only this mandible and an isolated femur from another individual at the site. But as you can see, the jaw is almost complete. We are not sure yet, but it may be that this fossil is only thirty thousand years old."

"Only thirty thousand years" may seem an odd way of expressing time, but coming from the mouth of a paleoanthropologist, it is rather like saying that a professional basketball player is *only* six feet four. Hominids—members of the exclusively human family tree—have been on the earth for at least four million years. Measured against the age of the earliest members of our lineage, the mineralized piece of bone on the table was a mewling newborn. Even compared to others of its kind, the jaw was astonishingly young. Neandertals were supposed to have disappeared fully five thousand years before this one was born, and I had come to France to find out what might have happened to them. I was definitely interested in the specimen.

The Neandertals are the best known and least understood of all human ancestors. To most people, the name instantly brings to mind the image of a hulking brute, dragging his mate around by her coif. This stereotype, born almost as soon as the first skeleton was found in a German cave in the middle of the last century, has been re-fluffed in comic books, novels, and movies so often that it has successfully passed from cliché to common parlance.

"If I call you a Neandertal, you know right away I am not paying you a compliment," anthropologist David Frayer of the University of Kansas once told me. To professionals like Frayer and Hublin, whose job is to unravel the meaning of fossilized bones, the name represents something much more precise. What makes a Neandertal a Neandertal is not its size or its strength or any measure of its native intelligence, but a suite of exquisitely distinct physical traits, most of them in the face and cranium. Like all Neandertal mandibles, for example, the one on the table lacked the bony protrusion on the rim of the jaw called a mental eminence— better known as a chin. The front teeth were thick but badly worn, almost to stubs. The places on the outside of the jaw where chewing muscles had once been attached were grossly enlarged, indicating tremendous torque in the bite. Between the two last molars and the upward thrust of the rear of the jaw, Hublin pointed to gaps of almost a quarter of an inch, an architectural nicety shifting the business of chewing farther toward the front.

In this, and in several other features, the jaw was uniquely, quintessentially Neandertal; no other member of the human family before or since shows the same pattern. With a little instruction, the Neandertal pattern is recognizable even by a layman like myself, who, until I became involved with them, would have guessed that a "Mental Eminence" was a title given to an unusually erudite cardinal. But to convey what "Neandertal" really means, solely on the basis of physical traits, would be like trying to describe the desire you feel when in love by shouting out your lover's measurements. Neandertals had become, over the past months, a hounding, haunting obsession for me. Unlike Hublin, whose very expertise allowed him to sit there calmly sipping coffee while the jaw of a 30,000-year-old man rested within biting distance of his free hand, I felt like stooping down and paying homage.

Two years before, an issue of *Newsweek* had appeared with an unusually provocative cover. Against a backdrop of shadowy forests and undulating fields of wheat, a man and a woman stood on either side of an apple tree. They were slim and beautiful, with elegantly proportioned bodies and full, vaguely African features. With their polished copper skin and taut physiques, they seemed more like the products of an exclusive health spa than the residents of an eternal paradise. But there was no mistaking this couple. The woman, her naked breasts veiled by her hair, was offering an apple to her partner, her face full of invitation. The man's palm was open to receive the fruit, his lip arched in an innocent smile. Between the two, a fat green serpent coiled around the tree trunk, his yellow eye gleaming.

After a long hiatus, Adam and Eve were back in the news. They had made the cover of *Newsweek* riding on the back of rigorous, quantitative, thoroughly secular science. Based on a comparison of a particular kind of DNA found in cell parts called mitochondria, a team of biochemists in

Berkeley, California, had concluded that all human beings on earth could trace their ancestry back to a single woman who lived in Africa only 200,000 years ago. Every living branch and twig of the human family tree had shot up from this "mitochondrial Eve" and spread like kudzu over the face of the globe, binding all humans in an intimate web of related-ness. Genetically speaking, there was not all that much difference between a New Guinean highlander, a South African !Kung tribeswoman, and a housewife from the Marin County hills across the Bay from Berkeley. Whatever appearances might suggest, they simply hadn't had time enough to diverge.

"The Eve hypothesis," as it came to be known, was a major develop-ment in the science of human evolution. But by putting it on its cover, *Newsweek* was taking a big risk. Articles jammed up with terms like "mi-tochondrial DNA" and "genetic distance" do not typically incite stam-pedes at the checkout counter, even when you warm up the customer's appetite with a couple of naked people on the cover. The news, further-more, was not even new. The Berkeley study had been published in the academic journal *Nature* fully a year before. When the story idea was originally proposed to *Newsweek*'s regular science editor, she turned it down as too old for a topical feature magazine. A senior editor followed a hunch and decided to pursue the story anyway.

His instinct proved right. Later the same year, we watched in amaze-ment as the Iran-Contra affair unfolded and the Iron Curtain rusted away. The AIDS epidemic reached into the heterosexual population, and the hottest summer on record turned the Greenhouse Effect into an in-ternational nightmare. Each globe-shaking development inspired a vivid new cover for *Newsweek*. The American presidential campaign provided the grist for several others. In the end, however, the best-selling cover story of 1988 proved to be the one at the farthest possible remove from current events: how the human race got started in the first place.

To me, the Eve hypothesis sounded almost too good to be true. If all living people can be traced back to a single common ancestor just 200,000 years ago, then the entire human population of the globe really is one grand brother-and-sisterhood, despite the confounding embellish-ments of culture and race. Thus on a May afternoon, a café in Paris could play host to clientele from three or four continents, but the scene still amounted to a sort of *ad hoc* family reunion. If among the Africans and Asians and various Europeans coming in from the rain we could find a table for some Australian aborigines and room at the counter for a couple of Native Americans, it would still be just a slightly larger, better-attended family party. Likewise, if we could bring back to life some cultures driven to the brink of extinction or over it by the overenthusiasm of European colonialists, there might be an Andaman Island woman playing the pinball machine, with her deceased husband's skull strapped on her shoulder for good luck; or an Aztec priest looking dourly out at the Parisian drizzle; or perhaps an Ona Indian from the tip of Tierro del Fuego, scowling from under his ponderous browridges at the trio of high school girls bestowing lavish farewell kisses on one another's cheeks. Family, all.

But Eve bore a darker message too. The Berkeley study suggested that at some point between 100,000 and 50,000 years ago, people from Africa began to disperse across Europe and Asia, and eventually populated the Americas as well. These people, and these alone, became the ancestors of all future human generations. When they arrived in Eurasia, however, there were thousands, perhaps even millions, of other human beings already living there. Among the best known of these were the Neandertals. If the genes of everybody alive trace back to that single African population, what happened to all those *non*-African genes? Eve's answer was cruelly unequivocal. The Neandertals—including the Zafarraya population represented by the jaw on the table—were pushed aside, outcompeted, or otherwise driven extinct by the new arrivals from the south. The genetic replacement was total. The Eve hypothesis strongly implied that the two human types, residents and *arrivistes*, did not interbreed, at least not in a way that produced fertile offspring. Except for those African-bred genes, the human slate was wiped clean, eliminating a million years of human development in a single stroke. Looked at from this point of view, the Eve hypothesis did not endorse the underlying unity of humankind so much as play witness to its universal brutality: "a scientific rendering," as one critic put it, "of the story of Cain."

What fascinates me about the fate of the Neandertals—Abel's side of the story—is the paradox of their promise. Appearing first in Europe about 120,000 years ago, the Neandertals flourished through the increasing cold of an approaching Ice Age, and by 70,000 years ago they had spread throughout Europe and western Asia. As far as appearance goes, the popular stereotype of a muscled thug is not completely off the mark. Thick-boned, barrel-chested, a healthy Neandertal male could lift an average NFL linebacker over his head and throw him through the goalposts. But despite an ingrained reputation for dimwittedness, there is nothing that clearly distinguishes a Neandertal's brain from that of a modern human being except for the fact that, on average, the Neandertal version was slightly *larger*. There is no trace of the thoughts that animated those brains, so we do not know how much they resembled our own. But a big brain is an expensive piece of adaptive equipment. You don't evolve one if you don't use it. Combining enormous physical strength with manifest intelligence, the Neandertals appear to have been outfitted to face any obstacle that the environment could put in their path. They could not lose.

And then, somehow, they lost. Just when the Neandertals reached their most advanced expression, they suddenly vanished from the earth. Their demise coincides suspiciously with the presumed arrival into Western Europe of a new kind of human: taller, thinner, more modern-looking people. I say "presumed" because we do not really have any direct evidence of what these new people looked like. All we have is the residue of their culture: the knapped and polished tools; the pierced beads and pendants they draped on their bodies; the carved animal figurines full of sinuous, sensual expression that appear to mark the arrival not only of a new people but of a new form of consciousness.

The collision of these two human populations—us and the other, the destined parvenu and the doomed caretaker of a continent—is as potent and marvelous a part of the human story as anything that has happened since. If you doubt me, take a quick look at the sales figures for *The Clan of the Cave Bear*, or any of Jean Auel's subsequent romances. The meeting between the first modern humans and the last Neandertals resonates far beyond the actual events in the Paleolithic Age, and novelists have been quick to take advantage of it. This imagined encounter was, after all, the last between ourselves and another intelligent, intelligible form of being. In paleo-fiction, the Neandertals play the same role in the past that extraterrestrials play in the fiction of the future: Their otherness defines our natures, highlights our failures and limitations, more than it disparages their own. The modern humans arrive and prevail, in most cases violently. In William Golding's *The Inheritors*, a Neandertal named Lok struggles to comprehend what is happening as his comrades are systematically exterminated by intruders whose lethal powers so fascinate him that he barely hears his own howls of grief. In Auel's romances, Ayla, the modern human heroine, quickly outstrips the Neandertals who have adopted her; their brains are too crammed with tribal memory to leave room for her kind of innovative thinking. But it is not the triumph of a

superior race that drives the plots of these books. It is the loss of an alternative one.

By itself, the half-jaw on the table in front of me was not nearly so
eloquent as Lok's anguished wonder or as sexy as Ayla's exploits. But it
had its own tale to tell. Hublin had said that the Zafarraya jaw was perhaps as young as 30,000 years old. A few months before, an American
archaeologist named James Bischoff and his colleagues had announced
new dates of their own for some Spanish sites. Applying a new technique
to date some modern human-style artifacts in the caves, they astonished
everyone by declaring them to be 40,000 years old. This was 6,000 years
before there were supposed to have been modern humans in Europe. If
both Bischoff's and Hublin's dates were right, it meant that Neandertals
and modern humans had been sharing Spanish soil for 10,000 years. That
didn't make sense to me.

"At thirty thousand years," I asked Hublin, "wouldn't this jaw be the
last Neandertal known?"

"If we are right about the date, yes," he said. "But our dates are still
very preliminary. There is much work to be done before we can say how
old the jaw is for certain."

"But Bischoff says modern humans were in Spain ten thousand years
before then," I persisted. "I can understand how a population with a
superior technology might come into an area and quickly dominate a less
sophisticated people already there. But ten thousand years doesn't sound
very quick, even in evolutionary terms. How can two kinds of human being
exist side by side for that long without sharing their cultures? Without
sharing their genes?"

Hublin shrugged, in the classically cryptic French manner meaning
either "The answer is obvious" or "How should I know?"

"Don't forget," he said, "those modern human sites Bischoff is talking
about are in the north part of the country. That area may have been part
of a penetration of modern humans into Europe along the coast of the
Mediterranean. But this penetration did not occur in the south. Southern
Spain is very much a *cul-de-sac*."

Hublin outlined his own version of the Neandertal fate, a scenario in
which the old residents of Europe, molded into an increasingly specialized
way of life by centuries of isolation, were done in by a combination of
sudden climatic shifts and new competition. The end result was much
the same as I had heard from other experts: The once-ubiquitous Neandertals were squeezed into ever-shrinking pockets of habitation that even-

tually petered out altogether, like the dying embers of an abandoned hearth. The story made sense, but its validity hung upon the strength of that date.

As he talked, I kept my head down, struggling to keep up with him as I made my notes. When I finally looked around, I was surprised to find that I was not the only listener. The jaw seemed to have broken down the invisible grid that keeps each table in a café locked in its private space. Instead, a circle of common curiosity was spreading out from the fossil: people sitting with their bodies slightly twisted by the effort to keep an eye on the jaw and listen to Hublin. The argument between the bearded fellow next to me and his companion had fallen into a series of jump-started fragments, truncated by bouts of unabashed eavesdropping and fossil-ogling. Even *les mecs* at the pinball machine were staring our way between their turns.

Before Hublin slipped it back in his pocket, I looked once more at the bone. Hinged to an upper jaw, it had once been ribboned in muscle and laced with animating nerve and blood vessels. The jaw had bitten deep into warm flesh and pulled in lungfuls of frosted air as the Zafarraya Neandertal chased game or ran in fear. Though silent now amid the café babble, it too perhaps had once formed words, argued with companions, told stories about the past and made plans for the future. But there would be no future. Somehow, against all logic, another kind of human had come in and stolen posterity from the Neandertals. How that theft was accomplished is the oldest question in human evolutionary science. Lately the familiar answers, the ones I had been brought up believing, had been peppered by an onslaught of new discoveries and quickly, completely, irrevocably discredited. Instead, the Neandertals and their European conquerors were now caught together in a web of mysteries that stretched out of Europe and covered the globe.

What is a modern human being? How did it get that way?

The full and proper name for us is *Homo sapiens sapiens*, a kind of taxonomic stutter meaning "double-wise man." The redundancy drives home the point that our turbocharged intelligence distinguishes us not only from the other animals on earth but also from the other members of the hominid family who came before us. In a nomenclatural sense, "Neandertal Man" is only half as smart, bearing a single *sapiens* in his formal designation, *Homo sapiens neanderthalensis*. Earlier members of our genus such as *Homo erectus* and *Homo habilis*, though cleverer than anything else around at the time, carry no wisdom at all in their names.

The lofty title seems justified, if a bit ironic. Ours is, after all, the species that brought the spearthrower and the bow and arrow onto the planet, as well as the switchblade and the smart bomb; Plato as well as Pol Pot, Moses and Machiavelli too. Along the way, we have also introduced the needle, the button, the bead, the harpoon, fish nets, domesticated crops, art, music, grammar, some five thousand different languages, metaphor, social status, social welfare, religion and religious wars, race and racial wars, civil rights and civil wars, irony, sports, jokes, jurisprudence, stocks and bonds, slash-and-burn agriculture, national parks, national pride, political philosophy, petroleum refinement, the automobile, the traffic jam, the car phone, fast food, slow torture, microchips, mass slaughter, ozone holes, cervical plugs, carbon dioxide blankets, private property, and public relations, to name a few double-wise innovations. Twenty thousand years ago, modern humans were carving Venus figurines out of pieces of ivory; now we sit on the sofa and watch the planet Venus pass by on our television screens.

Though we deserve to be called *Homo sapiens sapiens*, the taxonomic designation is too cold and too clumsy. Until recently, a cozier, more familiar synonym was "Cro-Magnon Man," a name borrowed from a tiny rock shelter in southern France. In 1868, some fossilized skeletons that looked pretty much like skeletons of people today were discovered in the cave. You can probably conjure up an image of Cro-Magnon; you've seen him portrayed at the end of a glossy, full-color time line illustrating the stages of human evolution in some magazine or museum display. Wearing beads and furs and toting a neatly crafted spear, Cro-Magnon Man is shown striding hurriedly off the right-hand side of the page, as if to distance himself as quickly as possible from the progression of stooped and hairy brutes that crowd up behind him in the ancestral queue. Immediately to his left is Neandertal "Man." (Only recently have time lines acknowledged that two sexes are needed to keep them moving along.) Chinless, beetle-browed Neandertal Man has a big head but a dullish sort of look about him, as if he is dumbly wondering where evolution will take him next. In contrast, Cro-Magnon stands tall, white, thin, and relatively hairless, though he often sports a well-trimmed beard. He holds his chin high, his gaze clear and true beneath a worthy vertical brow.

I have always found this Cro-Magnon of the time lines a bit of a bore. With his skins and shorn feet and all his power accessories, he looks too

self-important. This is a hominid who has *arrived*. It was once said (and
often repeated) that, given a shave and a set of clothes, his neighbor
Neandertal Man could travel unnoticed in a subway car at rush hour. If
this is true, it is only because commuters try very hard not to notice
anyone at all. There is no doubt, however, about Cro-Magnon: Strip off
the animal skins, slide him into a Pierre Cardin suit, and he would pass
without comment on any subway, though he'd probably be more comfortable taking a cab. Poised at the very end of a scale of increasing refinement, he is, for all his polish, just a trifle too familiar. You can admire
his spear, his level stare and polished comportment, but then your eyes
drift irresistibly back to the left, to the mysterious half-apes and demi-
humans that portend his coming.

Part of Cro-Magnon's problem is the relative dearth of mystery. Among
all the events and transformations in human evolution, the origins of
modern humans were, until recently, the easiest to account for. Around
35,000 years ago, signs of a new, explosively energetic culture in Europe
marked the beginning of the period known as the Upper Paleolithic. They
included a highly sophisticated variety of tools, made out of bone and
antler as well as stone. Even more important, the people making these
tools had discovered a symbolic plane of existence, evident in their gor-
geously painted caves, carved animal figurines, and the beads and pendants adorning their bodies. The Neandertals who had inhabited Europe
for tens of thousands of years had never produced anything remotely as
elaborate. Coinciding with this cultural explosion were the first signs of
the kind of anatomy that distinguishes modern human beings: a well-
defined chin, a high, vertical forehead lacking pronounced browridges, a
domed braincase, and a relatively slender, lightly built frame, among
other, more esoteric features.

The skeletons in the Cro-Magnon cave, believed to be between 32,000
and 30,000 years old, provided an exquisite microcosm of the joined emergence of culture and anatomy. Five skeletons, including one of an infant,
were found buried in a communal grave, all exhibiting the anatomical
characteristics of modern human beings. Scattered in the grave with them
were hundreds of artificially pierced seashells and animal teeth, clearly the
vestiges of necklaces, bracelets, and other body ornaments. The nearly
simultaneous appearance of modern culture and modern anatomy provided a ready-made explanation for the final step in the human journey.
Since they happened at the same time, the reasoning went, obviously one
had caused the other. It all made good clean Darwinian sense. A more

efficient technology emerged to take over the survival role previously provided by brute strength, relaxing the need for the robust physiques and powerful chewing apparatus of the Neandertals. *Voila!* Suddenly there was clever, slender Cro-Magnon Man. That this first truly modern human should be indigenous to Europe tightened the evolutionary narrative: Modern man first appeared in precisely the region of the world where culture—according to Europeans—later reached its zenith. Prehistory foreshadowed history. The only issue to sort out was whether the Cro-Magnons had entered the continent from somewhere else, or whether the Neandertals had evolved into them.

In these last few years, new fossils, new techniques, and new approaches have completely obliterated this once-comfortable hypothesis. The most damaging blows have come from Africa and the Middle East, where it now appears that fully modern human anatomy—the biological entity known as *Homo sapiens sapiens*—had already emerged as early as 100,000 years ago. According to the "Out of Africa" hypothesis, these earliest modern humans eventually spread out to take over the territory of all other existing hominids. But, so far at least, there is no sign that these hyper-successful moderns were making fancy tools, painting caves, or otherwise *doing* "modern" things. Modern behavior can no longer *explain* modern human form, because by all appearances modern culture didn't even exist for another 60,000 years. Suddenly, the emergence of anatomy and culture have become delaminated in time. You might as well try to account for the origin of the wind by talking about sailboats.

One consequence of this temporal delamination is that the name Cro-Magnon can no longer be used to mean "early modern human." The Cro-Magnons were only one, late example of a much more widely spread phenomenon. In this book, I use the name only when referring to the early moderns of Europe. Otherwise, I am stuck with the rather colorless phrase "modern human" to represent the most astonishing creature that has yet emerged upon the planet Earth. But there is a far deeper consequence to the delamination than just a recasting of names. Where there once was one tidy explanation for our origins, there are now two separate, equally inexplicable enigmas. Why modern human *bodies*, if changes in our ancestors' behavior do not account for them? And why, after a delay of at least 50,000 years, did these modern bodies suddenly start *behaving* like humans? What double-handed power pushed us into being?

In the old days, before a modern human named Charles Darwin was born, there were plenty of answers to go around. In Genesis, God makes man out of clay and woman from the rib of man; they eat a forbidden apple, and the human race is off and running. But this is only one creation story among many. For thousands of years before the Bible was written, other Adams and other Eves had been discovering each other in legends scattered from the Near East to the South Pacific. Everywhere you look in indigenous cultures, our first parents are being crafted from liquid fire or hacked out of melting blocks of ice, or are popping out of the belly, head, thigh, or armpit of some dozing deity. In Eskimo legend, the first man pushes his way out of a pea pod. According to their myths, the fierce Yanomamö of Brazil erupted from drops of blood spilling from a wounded god flying overhead. The Quiche Maya of Guatemala had a deity named Maker, who tried to fashion human beings out of first mud and then wood. When these attempts went astray, he successfully ground humans from yellow and white corn, with a little water added to make fat. If you lived in China around 600 B.C., you accepted on faith that we all began as fleas infesting the pelt of the Great Creator, Phan Ku.

Among world religions, the Judeo-Christian tradition draws an unusually clean line between what is animal and what is human. Many non-Western religions endow animals, birds, and even individual blades of grass with spirit. But in the Bible only human beings have souls and are capable of salvation. For a lion, a tulip, or an HIV virus, original sin is simply not an issue. Evolution, of course, respects no such distinction. In the last century, Darwin proposed a mechanism that could explain how humankind could emerge by descent from more primitive ancestors, just like every other living species. In our own century, the fossil fruits harvested on the African savanna and elsewhere have largely proved him right. But up until just a few years ago, it was possible to believe wholeheartedly in evolution and still keep the division between beasts and human beings clear. You didn't need God's hand to fashion a modern man or woman out of animal clay. You just needed *time*. Lots and lots of it.

I must confess that I am a sucker for time lines. On the far left, at the extreme remove from spiffy Cro-Magnon, the time lines begin with a little apelike thing with a projecting face, walking with bent knees. For the time being we call this first well-established member of the human lineage *Australopithecus afarensis*—the species famously represented by the skeleton Lucy. (In the fall of 1994, scientists working in Ethiopia announced they had found fossil fragments of an even older hominid, believed to have lived very close to the point where the human and ape lineages split off from each other.) To the right of *afarensis*, a handful of hairy intermediates marches awkwardly, inevitably, toward the future, each with its posture a little better and its brain a little bigger than the one just behind it in line. The transformation along the way is mesmerizing. You watch humanity unfold like a flower, each ancestor the ripened promise of the one that came before.

In effect, the single, explosive event told of in the genesis myths has been pulled gossamer-thin and stretched out over millions of years. But the end result is the same. This quaint evolutionary procession is the post-Darwinist, science-validated version of the line drawn between animal existence and human life. The fully animal thing is hidden just off the page to the left—before *afarensis*, before the even more primitive hominid recently discovered in Ethiopia. The contemporary human being—you—is parked just out of sight on the right, the unspoken consequence of Cro-Magnon. Even as the time line demonstrates how we evolved from lower creatures, it keeps us comfortably distanced from that bestial beginning by some *four million years*. Such an enormity of time is literally impossible for the human brain to imagine. You can divide it up into stages and decorate each one with a different hairy, thrust-out face, but no matter how you slice it, four million years is more than enough time to preserve that sense of a separate human destiny, placating the closet creationist holed up inside each of us.

Much as I love gawking at time lines, however, I question whether they are telling the truth. Let's start at the beginning. Ever since Louis and Mary Leakey began to pluck hominids out of Olduvai Gorge in Tanzania in the 1950s, the glamour topic in human evolution has been the quest to identify the point, deepest in the past, when the human lineage diverged from that of the apes. Few scientists use the term "missing link" anymore, but if you were to scratch beneath the skin of a modern-day fossil hunter, you would discover that it is the hope of precisely such a find that motivates their efforts, thrills their followers, and opens the pockets of their donors. Pursued in obscure regions of the world, often under dangerous conditions, the search mixes the excitement of a treasure hunt with the righteousness of a mystical quest.

Consider Donald Johanson, for instance, at the end of his popular account of the discovery of Lucy, looking forward to a return to the fossil-rich deposits of Ethiopia where he found her: "What we find in them could well blow the roof off of everything," he writes, "because science has not known, and does not know today, just how or when the all-important transition from ape to hominid took place. This is the biggest remaining challenge to paleoanthropology."

The key word here is "all-important." *Australopithecus afarensis* is undoubtedly an important character in our evolution. An astonishing abundance of this species' bones was discovered by Johanson and his colleagues in the mid-seventies, and more in this decade, in the remote Afar triangle in Ethiopia. A recently discovered *afarensis* skull is dated to around 3 million years old. It bears resemblances to another skull fragment found earlier in Ethiopia, believed to be around 3.9 million years old, and the rest of the known specimens fall between these dates. Thus the species inhabits a time zone of at least a million years, during which it underwent no meaningful changes in its form.

A number of specific anatomical traits distinguish *afarensis* from the apes, but one feature above all others sets the species apart, right from the start—it walked on two legs. As evidence, we have not only Lucy's undeniably upright skeleton and some other postcranial bones, but the astounding, four-million-year-old footprints, preserved in volcanic ash, found by Mary Leakey and her team at Laetoli in the late 1970s. An upright gait—bipedalism—has been touted as a hallmark of humanity since Darwin's time, one of a handful of early specializations that not only define our particular evolutionary direction, but can be enlisted to *explain* it as well. We have been told over the years that the first hominids stood up on two legs because they were tool users and needed their hands free; or because they became hunters who needed to see over tall savanna grass; or because they were family men who required the use of their arms for carrying food back to the wife and kids. Every explanation for bipedalism is braided into the realization of some grander, uniquely human habit, to be refined and perfected farther on up the family tree. No wonder this transition is declared "all-important." It more or less decides the matter right from the start.

There is no doubt that Lucy's species was hominid and walked upright on the ground. But is there anything about *afarensis* or its style of locomotion that *necessarily* leads to modern humans? Lucy's brain was only slightly larger than that of a modern chimpanzee. Her species did not make stone tools, nor in fact would any other hominid for at least another half a million years. At three and a half feet tall, she hardly seems likely to have gained much advantage over the savanna grass by letting go of the ground with her forelimbs. Preliminary evidence suggests that the recently found, even-more-primitive ancestor of Lucy was in fact already walking upright 800,000 years earlier, even though it had yet to make the move from a forested environment to the open savanna. There is no evidence whatsoever that either *afarensis* or the new hominid hunted game or shared food within a nuclear family unit. Though her bipedal gait on the ground is unquestioned, some anthropologists, such as Randall Susman at the State University of New York at Stony Brook, have pointed to features of *afarensis* limbs that suggest that the species spent a lot of its time *off* the ground, moving apelike through the trees. Other scientists dispute this. But in any case, *afarensis* remained bipedal for at least a million years without exhibiting the slightest tendency to become smarter, more dexterous, or more human.

Bipedalism alone does not augur intelligence, and never did. A dolphin with no legs or an elephant with four are both a lot cleverer than a kangaroo mouse or a sparrow with two. With her confirmed upright stance, Lucy certainly represents the result of *a* transition, but not necessarily an "all-important" one. It is more a shift in the center of gravity than an awakening of the human spirit. The lesson I take from this sounds almost seditious: Our ancestors were *simply animals*—two-legged chimps—long after our lineage had split off from that of the apes. The same would, of course, apply to the carlier, "missing link" ancestor who actually did the splitting off. On its own, without the advantage of evolutionary hindsight, there is nothing about that genealogical divergence that foreshadows, triggers, or even remotely hints at a human being.

In modern versions of the human family tree, there is a fork above *afarensis* about two and a half million years ago. One branch leads toward *Homo*; the other veers off to one side. This side branch bears fascinating fruit, but let us continue for a moment on the main stem. After *afarensis* comes the first species to merit inclusion in our own genus, *Homo habilis*. (In some versions of the tree, another australopithecine squeezes in between them.) You can easily pick this hominid out from the lineup: He's usually the first one carrying a chipped rock in his hand. Discovered by Louis Leakey at Olduvai Gorge in the early 1960s, the species name—"Handy Man"—pays tribute to a creature Leakey saw as the originator of human technology. Since the initial discovery, specimens of *habilis* have been found in several other sites in East and South Africa, dating from between two and one and a half million years ago. Only half as old as Lucy, *habilis* had a somewhat bigger brain. But brain size is at best a very rough indicator of intelligence. *Habilis*'s real claim to inclusion in the same genus as, say, Stephen Hawking, is the manifest ability to manufacture stone tools—crude, chunky things (collectively known as the Oldowan Industry), but tools nonetheless.

In the late 1960s, *habilis*'s toolmaking skills became the centerpiece of a seductive new theory of human origins. What really set the human career in motion, said the theory's proponents, was not bipedalism, toolmaking, or brain growth in itself, but the uniquely human activity that inspired these innovations: the cooperative hunting of large game.

"Human hunting is made possible by tools, but it is far more than a technique or even a variety of techniques," wrote Sherwood Washburn and C. S. Lancaster in *Man the Hunter*, in 1968. "It is a way of life, and the success of this adaptation . . . has dominated the course of evolution for hundreds of thousands of years. In a very real sense our intellect, interests, emotions and basic social life—all are evolutionary products of the success of the hunting adaptation."

As the first toolmaker, *habilis* was naturally seen as the first full-fledged human hunter as well. In the next decade, the "all-importance" of the species was further enhanced by a theory put forth by Glynn Isaac of Berkeley and widely popularized by his colleague in the field, Richard Leakey. The Leakeys' excavations in East Africa had uncovered several sites where smashed-up mammal bones bearing the cut marks of stone tools were found scattered among the tools themselves. Isaac concluded that these accumulations were the remains of hominid "home bases." *Habilis* males were bringing back meat from the hunt to be butchered and shared among members of their social unit. If he was right—and thanks to Leakey's best-sellers, Isaac's food-sharing hypothesis rapidly took on the luster of historical truth—then nearly two million years ago, our ancestors had already developed the social pattern of nuclear families that characterizes virtually all modern human societies.

"The humanness of our ancestors goes considerably further back than Neandertal and probably will go back to the very dawn of culture," Leakey commented.

Fascinating as these theories were, however, they have not stood up well to new discoveries, or to more rigorous scrutiny of old ones. In 1986, Donald Johanson's team at Olduvai Gorge uncovered the first *Homo habilis* skeleton to include both skull and body parts—a combination needed to assess accurately a species' body proportions. (I was lucky enough to be involved in the discovery and wrote a book with Johanson about it.) At three and a half feet, this particular "mighty hunter" was no taller than Lucy herself. With or without tools or cooperative hunting strategies, even the males of this first human species hardly seem to have been a formidable presence on a savanna bristling with lions and saber-toothed tigers. Furthermore, the new specimen's arms were nearly as long as her legs, again resembling Lucy. If functional anatomists like Randall Susman are to be believed, *habilis* too was spending a lot more time hiding in the trees than loping after terrified prey.

Far from confirming a habiline humanity, the skeleton lent solid support to a less romantic portrait of *habilis* put forth in the early 1980s, primarily by Lewis Binford, then at the University of New Mexico. Binford looked at the same evidence Isaac had marshaled for his food-sharing hypothesis, but came back with a radically different interpretation. Assemblages of bones and stone tools, such as those at Olduvai Gorge, were not signs of a hunting hominid, but merely jumbles of material thrown together by natural forces, or collected by nonhuman predators like hyenas. Other archaeologists pointed out that the cut marks made by hominid tools often appeared superimposed *over* the tooth marks of gnawing carnivores. If *habilis* was involved at all in forming these assemblages, its role was hardly a mighty one: After the real predators had left their kill, the hominids scuttled in to scavenge what scraps remained on the carcass.

"The Olduvai toolmaker was no mighty hunter of beasts," Binford concluded. "He was the most marginal of scavengers."

Most archaeologists today would say that Binford went too far in demeaning the abilities of the early hominids. *Habilis* was probably obtaining meat in its diet through a combination of scavenging and opportunistically capturing small animals. In any case, the species was almost certainly making and using stone tools. To add new insult to injury, however, it now appears that *Homo habilis* was not the only Handy Man around.

Go back to the side branch veering off the family tree after Lucy, about two and a half million years ago. There you find a series of lumbering, big-jowled creatures, australopithecines like Lucy but with the species names *africanus*, *robustus*, and *boisei*. It has long been thought that this australopithecine offshoot from the human lineage represents a dimmer-witted alternative to *Homo*, doomed to extinction by its inferior grade of intelligence and an increasingly specialized diet—one that did not include meat.

The robust australopithecines did indeed go extinct, about a million years ago. But not, it now appears, before leaving behind evidence of their ability to make tools. Over the past few years, Robert Brain of the Transvaal Museum in Pretoria, South Africa, has unearthed a number of primitive digging sticks made of bone. These were associated not with *habilis* fossils but with the remains of the supposedly dim-witted *Australopithecus robustus*. Together with Susman, Brain has also examined the fossilized fingertips of the species and found that they bear all the evidence of the manual dexterity needed to make and manipulate stone tools. Farther to the north, there were robust australopithecines running around at Olduvai Gorge and other East African sites at the same time as *habilis*. Who can say for sure now which hominid made the Oldowan tools? Maybe both species were handymen.

A more damning indictment of the "specialness" of our lineage is the presence of these collateral hominids in the first place, rubbing shoulders with *Homo*. Twenty years ago, many anthropologists believed that such a thing was theoretically impossible. No more than one hominid species could exist at the same time, let alone at the same time and place. The Single Species Hypothesis, articulated by Milford Wolpoff and Loring Brace at the University of Michigan, was based on a widely accepted ecological principle known as "competitive exclusion" which maintains that in any given environment, multiple species can coexist only if each carves out a separate "niche" for itself, exploiting its own particular range of resources. If two species compete too closely for the same food and other resources—in other words, try to squeeze into the same ecological niche— one of them will quickly drive the other into extinction.

According to the Single Species Hypothesis, the unique niche hominids carved for themselves was *culture*. Today no other animal uses culture to survive as human beings do, and the same must have been true in the past. If any other species attempted to occupy this niche, it would have been outcompeted by the one hominid filling the niche at any given point in the past. In effect, the hominid lineage was a club so exclusive it could accommodate only one member at a time.

"Because of this hominid-adaptive characteristic implemented by culture, it is unlikely that different hominid species could have been maintained," Wolpoff wrote in 1971.

Thomas Huxley, an early supporter of Charles Darwin, said that the great tragedy of science was "the slaying of a beautiful hypothesis by an ugly fact." Through the 1970s, the beautiful Single Species Hypothesis was picked away at by a steady stream of ugly little facts coming out of the fossil-rich regions of East Africa. The credibility of the theory depended upon seeing all anatomical variation in hominid fossils at any given time in the past as simply the natural differences between the various populations of a single species. With all kinds of new hominid skulls emerging—some with thick, crested crania housing tiny brains, others with larger brains but thin skull bones and far less robust faces—it became increasingly obvious that there were *at least* two species of hominid sharing the African savanna around two million years ago.

In 1975, when Richard Leakey's team discovered an advanced, large-brained skull of *Homo erectus* dating to 1.6 million years ago—a time when the relatively small-brained *Australopithecus boisei* was also surely alive—the single-species hypothesis came crashing down for good. With it went the sense that our lineage was propelled forward in evolution in a kind of enclosed pneumatic tube, removed from the forces affecting other species. Some scholars, like Ian Tattersall at the American Museum of Natural History, argue that there were not two but three or more different branches on the human family bush a million and a half years ago. What could more completely confound the notion of hominid uniqueness than the presence of a multitude of hominid species scurrying over the same landscape?

"This is probably the most far-reaching conclusion of paleoanthropology of the past two decades," says Robert Foley of the University of Cambridge, "for it finally brings home the point that human evolution is like that of other species, and there is no single way of being a hominid."

Among all the contemporary hominids, however, only one would carry forward to the present. After *habilis* in the time line comes *Homo erectus*. Between these two, there was perhaps an "all-important" transition. On average, the brains of *erectus* were about 20 percent larger than those of its predecessor. Their bodies were larger too, probably approaching modern humans in height. Unlike any creature before it, *Homo erectus* apparently knew how to tame fire to its own purpose, and with the appearance of *erectus*, the archaeological record shows the emergence of a much more sophisticated stone-tool industry, known as the Acheulean. Like modern humans, and unlike earlier hominids, *Homo erectus* males were not overwhelmingly bigger than females. This reduced "sexual dimorphism" may indicate that cooperation among males had begun to replace competition in social interactions, a key to the emergence of the complex social organization that characterizes human populations today. To lend support to this notion, archaeologists have pointed to *erectus* sites hundreds of thousands of years old that show the remains of large building structures. Other sites associated with *erectus* have been interpreted as mass elephant and baboon kills, involving dozens of hunters acting together.

Most important, the old geographical aliases for *erectus*—Peking Man and Java Man—bear witness to the fact that this species was the first hominid to be found beyond Africa. The *Homo erectus* migration out of Africa is well documented. Between one million and 700,000 years ago,* the species spread across the warm temperate zones of Mediterranean Europe and Asia as far as Indonesia. By 300,000 years ago, it had colonized the arid steppes and colder woodlands to the north, leaving its traces as far as Britain. This expansion into the Northern Hemisphere is the most treasured testimony to its advanced condition. What could be more human, especially from the Western European point of view that forms the basis of traditional paleoanthropology, than the ability to conquer new and harsher habitats?

Impressive as this first human diaspora was, however, it was not unique. Alan Turner of the University of Liverpool has shown that as *erectus* emerged from Africa, other, nonhuman species were making exactly the same trek, including lions, leopards, hyenas, and wolves, pushed forward by ecological imperatives that have nothing to do with intelligence. Either we have to award these fellow travelers the same merit badge in colonization that we pin on *erectus*, or we must concede that this hominid, like its predecessors, was not governed by any particularly human destiny, but was instead responding, much like other large carnivores, to a shift in environmental conditions. Meanwhile, in the last decade, virtually all the evidence for complex living structures, cooperative hunting, and even the controlled use of fire by *erectus* has been reexamined and called into question, if not outright refuted.

There is also the matter of *erectus*'s tools. The predecessors' stone tools were little more than sharp flakes or lumpish rocks with an edge whacked away. In contrast, *erectus* made a variety of more laboriously produced implements, typified by the symmetrical Acheulean hand ax. These beautiful teardrops of quartz or flint enter the archaeological record as early as a million and a half years ago, along with the first signs of *erectus* skulls and skeletons. The trouble—from an "all-important" standpoint—is that

*New, still controversial dates for some fossils in Java now suggest *Homo erectus* may have left Africa as early as 2 million years ago.

the species then goes right on making the same hand axes and other tool types for the next million years. From this point of view, the whole Acheulean period associated with *erectus* represents not the arrival of a rich and resilient intelligence but, as one archaeologist has called it, a period of "unimaginable monotony."

"It boggles my mind that I can go all over the world in this huge time frame, and all of these guys are making the same damn thing in the same damn way," says Tim White of the University of California, Berkeley. "Nothing happens for hundreds and hundreds of thousands of years. And that's *normal* for the animal. Human behavior isn't like that. One thing you can count on about humans. They change."

In Chasidic ceremony, there is a garment worn by the faithful called a *gartel*, a sash of black silk or wool tied around the waist, meant to remind the worshiper of the distinction between his physical, animal nature below the gartel, and the intellectual, spiritually directed side above. Where along the human lineage can we place the gartel? Which face in the time line is the one that is *our* face, and no longer the face of an animal? The conventional view, that our split from the rest of creation occurred deep in the past, doesn't seem to hold up anymore. In its place is the understanding that our early ancestors lived, died, and evolved like other species: according to the purely biological imperatives of natural selection, and unadorned by the special destiny that we have traditionally, half consciously, bestowed upon them.

Australopithecus afarensis, Homo habilis, Homo erectus—if you look closely, each seems a little less human, a little less noble than we thought. To be sure, these hominids were not *just* animals. But an ape, a lion, or a tree frog is not *just* an animal either. Each species has its own uniqueness, its own dance to do in the planetary ballroom. Given that the price of sitting out a dance is death, the earth dazzles and spins with the splendid singularity of their motions. Through the course of their evolution, our ancestors undoubtedly depended more and more on intelligence to keep the hominid dance going. But that does not set them apart from everything else, and it does not explain *us*. They were animals. They lie below the gartel. What are we?

The simplest answer is to toss off the gartel completely and concede that modern human beings are just animals too. This is no revelation; for centuries scientists from Copernicus to Darwin to Freud have been whittling away at the barrier between humans and the rest of creation. More recently, molecular biologists have revealed that our genes are about 98 percent the same as those of an African ape. Sociobiologists have chased down every aspect of human social behavior from homosexuality to child beating to ice-cream craving and linked it to some hidden evolutionary motive with a counterpart among the lemurs or zebras. Biologically speaking, all that separates an average chimpanzee and an average human being is a membrane so thin they could practically kiss through it.

At the same time, animals have been scratching away at the membrane from *their* side. Given a little training, chimps and gorillas can communicate by using symbols, teach each other sign language, and, by some accounts, even discuss their emotions and ideas of death with their trainers. We now know that even in the wild, vervet monkeys utter different alarm cries depending upon what sort of danger is imminent—perhaps the beginning of language. But it is not just primates who refuse to rest comfortably on their side of the divide. Lions hunt cooperatively, wolves share food, elephants regularly display an emotional depth more profound than, say, some modern human beings working on Wall Street. A while ago, I read about an elephant in the San Diego Zoo who has taken up oil painting. He holds the brush in his trunk. Apparently his paintings are fetching good prices.

This is wonderful stuff to read in the morning paper. But it does not in the least convince me that elephants differ from people only in a matter of degree. I have not had the privilege of viewing the San Diego elephant's *oeuvre*, but I would bet my savings that the best of his work would not sell for $82.5 million, the price recently paid by an anonymous Japanese investor for a van Gogh. I doubt, too, that the elephant could learn to spell *oeuvre*, much less pronounce it correctly. Nor could it invent a computer that can change the typeface of the word to italics with a single stroke, as mine just did.

The fact is, human beings—modern humans, *Homo sapiens sapiens*—are behaviorally far, far away from being "just another animal." The mystery is where, how, and why the change took place. There are no answers to be found in the vast bulk of hominid time on the planet. The gartel has been raised higher. An "all-important transition" *did* occur, but it happened so close to the present moment that we are still reeling from it. Somewhere in the vestibule of history, just before we started keeping records on ourselves, something happened that turned a passably precocious animal into a human being. There must be *some* kind of ancestor struggling toward us, some kind of almost-human thing out there, not in the wilderness of deep time but just beyond the reach of our species' recorded memory.

On the classic time line, there is only one figure left between *Homo erectus* and the fully formed humanity of Cro-Magnon: the ancient enigma, Neandertal Man. People have been wondering about him for a long time.

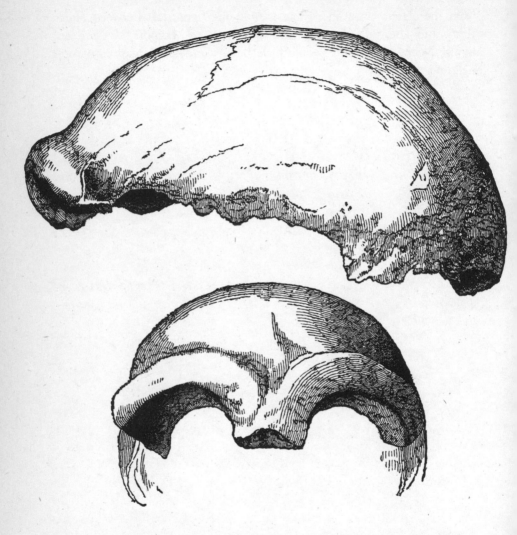

Skullcap of the original "Neanderthal Man," found in a cave in the Neander Valley, near Düsseldorf, Germany, in 1856 FROM THOMAS HUXLEY 1863

Chapter Two

ALL IN THE FAMILY

Hairy or grisly, with a big face like a mask, great brow ridges and no forehead, clutching an enormous flint, and running like a baboon with his head forward and not, like a man, with his head up, he must have been a fearsome creature for our forefathers to come upon. . . . "
                                         —H. G. WELLS ON NEANDERTHAL MAN, *1921*

If [Neanderthal Man] could be reincarnated and placed in a New York subway—provided that he were bathed, shaved, and dressed in modern clothing—it is doubtful whether he would attract any more attention that some of its other denizens.
                                         —WILLIAM STRAUS AND A.J.E. CAVE, 1957

In 1856, some workers quarrying limestone in a secluded valley near Düsseldorf, Germany, hit their shovels against what appeared to be the bones of an ancient bear. The bones were embedded in a thick layer of mud in a cave, sixty feet above the valley floor. The valley was named Neanderthal in honor of an obscure seventeenth-century poet and composer who had lived nearby and wrote under the name Joseph Neander. The quarrymen had never heard of Neander. They were not particularly interested in bones either. They tossed them down the slope along with the other cave fill, where they happened to land at the feet of someone curious enough to pick them up and give them a second look.

If at that moment Neandertal Man* could have foreseen the troubled

*What has happened to the *h* in *Neanderthal?* In old German, *thal* means "valley," pronounced with a hard *t*, rather than with a soft *th* sound. In fact, the German language has no *th* sound. Around the turn of the century, the spelling of many words was changed to bring them in line with their pronunciation, and *thal* became *tal*. The formal spelling of *Neanderthal* likewise changed to *Neandertal*. Though some people prefer the old way, in this book I will use the future in store for him, he would have scrambled back into the cave and reburied himself forever. Ever since, the Neandertals have been tossed back and forth by scientific opinion, now in our ancestry, now out, first too brutish to be seen as ancestral to noble man, later too noble to be indicted by our brutality, only to be brought back for further questioning and slapped with the original charge. At times they have served as an embodiment of a native human virtue uncorrupted by civilization. More often they have been condemned as either too primitive to be associated with us, or too far advanced in their own aberrant direction, or, the worst case, too primitive in some traits and too advanced in others at the same time. Though more "human" than any other ancient hominid, their lumpy primitive features and rudimentary culture have inspired astonishingly nasty reactions, almost as if the specimens themselves were personally responsible for their failure to be more like us.

"Were this the skeleton of the oldest man," declared one prominent German anatomist shortly after the original discovery, "then the oldest man was a freak."

Neandertal Man was off to a bad start.

Among origin myths, one of my favorites belongs to the Blackfoot Indians of the Dakotas. Their Prime Mover is a fellow called "Old Man," a peripatetic divinity who is simply "out there, traveling all over and making things." His roughly northward wanderings leave a wake of creation behind. Brushlands and prairies, birds and animals, rivers and cataracts spring into being with playful largesse. Old Man stops to take a nap, gets up, trips over an ill-placed knoll, and continues on his way, dropping animal species and geographical formations behind him like so many candy wrappers.

One day, he gets it into his head to make some people—a mother and child. Afterward, the trio walks down to a river. While they are staring at the water, the woman turns to Old Man and asks him point-blank: "So how is it going to be for us people? Do we live forever, or do we all have to die?"

---

streamlined spelling unless I am quoting a historical use of the term, as often happens in this chapter. (However you want to spell it, the last syllable is pronounced *tal*, not *thal*.)

While we're talking about names, I use the word "Man," as in "Neandertal Man," when it helps to give the historical flavor of an argument. Otherwise, I prefer to acknowledge that two sexes are needed to make a population grow.

"You know, I really haven't thought about that," he replies. "We'll just have to decide."

Old Man tries to resolve the issue with what amounts to the flip of a coin. He tells the woman that he is going to throw a buffalo chip into the river. If it floats, people live forever, but if it sinks, death will be our fate. Naturally, the dried buffalo turd floats. But then the woman demands a refinement of Old Man's decision-making methods: Instead of a buffalo turd, she will throw a stone, with the same conditions applying. The stone sinks, and so ever after do we.

Picturing Old Man as he ad-libs the world and our place within it, I have to wonder how much easier things might have gone for Neandertal Man if the god of *his* discoverers had been so laid back. It is unimaginable that Yahweh, the god of Genesis, would have neglected to consider so important a question as human mortality. Nothing in the Judeo-Christian version of creation comes about *ad hoc*. Yahweh builds his world like a master stage designer, fixing the fundamentals of time and space before raising the backdrop of mountains and valleys, then filling in the floral props. Animals are introduced sequentially—the great sea monsters and other water beasts, the winged birds, the cattle and creeping things, each brought forth according to its kind. Finally, when all is in its rightful place, Yahweh ushers in his main character and promptly gives him dominion over everything else.

In the Western theological tradition built upon this myth (with a hefty dose of ancient Greek idealism mixed in), plants, animals and humans, like the physical world that serves as their stage, are immutable elements in God's plan. Species are fixed entities whose characteristics have been handed down unchanged through the generations, ever since their "essences" were first established at the moment of creation. The world has no existence, no purpose, no possibility of change, that is not already invested in it from the beginning by the will of the divinity. That nature was created by a Grand Designer is self-evident in the wondrous complexity of its expressions. What purely natural power could account for such diverse and ordered splendor? Every kind of animal, from the lowliest snail upward, is linked in a great Chain of Being, stretching toward the glorifying edifice of Man, and through him, toward God Himself. Man is the critical link in the chain. Among God's creations, only he has a soul that exists independently of his physical body, a life that extends beyond a lifetime. It is he who wears the gartel.

"When we consider the infinite power and wisdom of the Maker," wrote John Locke in 1690, "we have reason to think, that it is suitable to the magnificent harmony of the universe, and the great design and infinite goodness of the architect, that the species of creatures should also, by gentle degrees, ascend upwards from us towards his infinite perfection, as we see they gradually descend from us downwards."

On the surface, it is hard to imagine a scenario more inimical to the notion of evolution by descent. Two fundamental assumptions of Western theology—that life is directed by a purpose that cannot be explained by life itself, and that Man occupies a separate plane from other animals—had to be toppled before a *physical* explanation for life on earth could even be conceived. But, ironically, Charles Darwin's discovery of the physical laws that account for the swarming diversity of nature—and place the human species squarely within it—could never have been made by a society that trusted Old Man's whimsy to invent the world. One must believe that nature was indeed constructed according to some kind of design before the world becomes available to rational inquiry.

Between the Enlightenment and *On the Origin of Species* there is a productive, if increasingly uneasy, tradition of natural science aimed at understanding the wondrous complexity of God's Design which consistently, if unwittingly, laid the groundwork that would undermine His need to exist. The evolutionary principle that every species is adapted to its own environment, for example, was originally understood as another sign of God's benevolence—the work of a sort of divine Quartermaster, outfitting His creatures with the feathers, claws, or other equipment they would need to make a go of it. Similarly, the first searches for ancient traces of man were religiously inspired. In 1726, long before the Neandertals were discovered, the Swiss physician Johann Scheuchzer claimed to have found a fossil representing "A Human Witness of the Deluge and Divine Messenger" on a hill in Franconia. Though Scheuchzer's "Antediluvian Man" turned out to be nothing more divine than a fossilized salamander, the approach he initiated—digging up old bones as evidence of human forerunners—would eventually help to bury the very truths he had hoped to confirm.

By the middle of the nineteenth century, some of the more tenuous notions of the Christian tradition were being seriously challenged. In England, geologist Charles Lyell showed that the earth was far older than the six thousand or so years estimated by biblical commentators. More important, he demonstrated how it could have reached its present state through the action of natural, observable processes—glaciers, volcanoes, earthquakes, and the like—gradually occurring over hundreds of thousands of years. Meanwhile, enough fossils of extinct forms of life had been discovered to make the belief in a static, pre-ordered universe rationally impossible. Clearly, if things existed before that no longer existed, then all could not be as it was on the day of creation.

These testimonies to the mutability of life could, however, still be incorporated in the Great Design if change itself could be seen as merely part of God's overall plan, an expression of the relentless drive toward perfection. The progression of organic forms discernible in the fossil record might, for instance, be seen as a parallel to the increasing complexity of a developing embryo. The *goal* of such development, ordained from the beginning, was still Man, the organism closest to God.

"The object and term of creation is man," wrote the Swiss-American naturalist Louis Agassiz in 1842. "He is announced in nature from the first appearance of organized beings; and each important modification in the whole series of these beings is a step toward the definitive term of the development of organic life."

The historical progression that Agassiz imagined was, like embryological development, a series of incremental steps, each slightly more organized and more complex than the last. The French scientist Georges Cuvier, the most influential paleontologist in Europe in the early nineteenth century, needed no such "modification" of one form into another to explain why the past was littered by so many extinct species. He saw life on earth as having undergone a series of mass extinctions in which one set of organisms was completely destroyed, to be replaced by a fresh set arriving from outside. The fact that extinct animals did not look like modern ones was thus not an argument *for* evolution, but against it. The pattern of change observed in the fossil record was not a natural process of development, but the result of a series of divinely inspired catastrophes, wholly isolating each epoch from the one that came before and the one that would follow. A devout Catholic, Cuvier firmly believed in the biblical age of the earth and confined mankind to only the most recent epoch.

Cuvier's theory of "catastrophism" dominated paleontology, especially in France, long after his death in 1832. But it was not entirely impregnable. In the 1840s, Jacques Boucher de Perthes, a customs surveyor-cum-archaeologist-cum-political economist-cum-romance writer and poet, discovered in the Somme Valley a whole array of stone implements that could have been fashioned only by man, mixed in with sediments also bearing the remains of extinct mammals. The antiquity of man thus seemed to be incontrovertible.

"God is eternal, but Man is very old," Boucher de Perthes concluded, in what seems to me a stunningly anticlimactic fizz of insight. Though he greatly overestimated the age of the artifacts he found, he nevertheless poked a hole through Cuvier's unbreachable time barriers, giving modern man a connection to the past. What was needed was an ancient man to come forward and lie in it—and more important, a theory that could explain *how* a modern human being could arise from something fundamentally different from itself.

Neandertal Man tried to answer the first need. Charles Darwin tackled the second.

From the quarry cave where they had been found, the bones of the original Neandertal made their way to the laboratory of a German anatomist named Hermann Schaafhausen. Unlike most of his contemporaries, Schaafhausen was more impressed by the growing evidence for evolution by common descent than he was by Cuvier's notion of catastrophic change. He pondered the Neandertal skullcap with its heavy browridges, thick walls, and other apelike features, and concluded that this was indeed the remains of a primitive man who had inhabited Europe before the ancient Celts and Germans, "in all probability derived from one of the wild races of northwestern Europe, spoken of by Latin writers."

Other, more powerful pundits were not so quick to believe in the specimen's antiquity. No extinct mammal fauna that might confirm its age had been found with the human bones; indeed, nobody really knew from what level inside the cave the fossils had emerged. All that was available for study was the bones themselves. The braincase, though decidedly odd, was large enough to house a modern-sized brain. The leg bones were curiously bowed and exceptionally robust, but their anomalies could be explained away without resorting to heretical notions involving apelike human ancestors roaming the continent. Perhaps, suggested one anti-evolutionist, the bones were merely the remains of a mental idiot who had suffered from rickets as a child. He had spent his life with his brow tensed against the stress of the disease, so no wonder his browridges were painfully pronounced. The German anatomist A. F. Mayer took one look at the specimen and declared it the remains of a Mongolian Cossack who had deserted from the Russian army chasing Napoleon in 1814.

Given *his* chance with the fossil, the eminent pathologist Rudolf Virchow agreed that rickets in childhood explained much of its weird postcranial architecture. Later in life the poor wretch had been knocked on the head a lot and finally ended his life plagued by arthritis. That such a sorry case should survive into old age, Virchow announced, only pointed out the ridiculousness of believing there was anything primitive about the thing. Obviously the crippled wreck would have needed help to survive, and such aid could have been forthcoming only in a civilized, sedentary, fully human society.

Needless to say, the Neandertal had no chance to voice his own sentiments about his condition, or his views upon the modern humans incessantly fingering his remains and passing judgment on them. But since all of these critics were fervent anti-evolutionists, their dismissal of the Neandertal's ancestral status was nothing personal. It was the idea that humankind had evolved at all that they were attempting to squelch. In 1859, just three years after Neandertal Man made his formal appearance, Charles Darwin published *On the Origin of Species* and gave him a credible means to exist. Darwin is often mistakenly credited or reviled as the inventor of the "theory" of evolution. Evolution is not a theory; it is an observable biological fact. Darwin's real contribution—or heresy, depending on your point of view—was to provide a theory of *how* change could occur in nature and species could develop out of other species without the slightest intervention from a supernatural arbitrator. It is said that shortly after the book was published, a clergyman arched a finger at Darwin on the street and shouted, "There goes the most dangerous man in England!" From a professional point of view, he was absolutely right.

The idea of natural selection is stunningly simple in its basic outlines, proceeding seamlessly from the universally accepted observation that individuals of any species are adapted to their native habitats. Though all members of a given species share common attributes and can breed with one another, there is nevertheless a great deal of individual variation within any given population. This variation tends to be inherited. If my parents happen to have long beaks, for example, then I will probably have a long beak too, while my neighbor with the shorter beak was most likely born of short-beaked folk. This individual variation is the material upon which natural selection can go to work. Among the members of a population, there is always competition for a limited supply of resources, primarily food and mates. The individuals better equipped to extract those resources (and avoid being extracted themselves by predators) are more likely to live long enough to reproduce themselves and pass on to their offspring the traits that allowed them to survive in the first place. If my environmental circumstances are such that a long beak is a more efficient food catcher than a short one, then *on average* long-beaked types like me will probably fare better reproductively than short-beaked types, and the next generation may see a slight increase in average beak length.

This means that biological change *within* a species is a product of competition for resources, which can be either light in times of plenty or harsh when resources are scarce. According to Darwin, species originate when a population is isolated, usually by geographical barriers, from other populations of its kind. Over time, the environment of the isolated population may favor the survival of a different set of characteristics from those that were adapted to the original habitat. Over generations, the effect of this differential in habitat may produce a population so unlike its "parent" population that even when the two are brought together, they are no longer similar enough to mate and produce fertile offspring. Though their similar morphologies and behaviors betray their common ancestry, they can no longer be considered the same species.

Multiplied over millions of generations, this process could account for the entire panoply of life on earth as the various descendants of one original form. Darwin passionately believed that for his theory to have validity, it must apply to humans as well as to other species. Even so, in *On the Origin of Species* he made almost no mention of human evolution, fearful that godless nature and a simian ancestry for humanity would be too much reality for most Victorians to stomach all at once. It was Thomas Huxley, "Darwin's bulldog," who first attempted to bridge the vast gap between man and the animals. In a series of famous public lectures in Britain, and again in *Man's Place in Nature*, published in 1864, Huxley eloquently demonstrated that there was less separating man from his nearest morphological relatives, the chimpanzee and the gorilla, than separated these higher apes from the lower monkeys. If man was thus no more different from the animals than the animals were from one another, why should there be any doubt that man was produced by the same natural forces that account for the rest of nature?

What was needed to support the evolutionary link between man and the apes was a fossil with attributes somewhere between the two. At first glance, the Neandertal skeleton seemed to fill the bill. After careful analysis, however, Huxley conceded that the fossil failed utterly as a "missing link." When viewed in the context of the range of variations in skull shape found among living humans, Huxley concluded that it had to be seen as fully human itself, the "extreme term of a series leading gradually from it to the highest and best developed of human crania."

And whence came those highest and best developed of brains? From Europe, of course. In 1864, the colonialist spirit reigned throughout Europe, and nowhere more than in England. All sorts of primitive peoples were freshly available to feel superior to: ferocious Andaman Islanders, simpleminded Bushmen, and godless Yaghan. The modern specimens deemed closest to the Neandertals were Australian aborigines, their receding foreheads and bulky browridges betraying their morphological primitiveness, just as the simple stone-tool cultures of these "lowest savages" spoke for their behavioral backwardness. To call this attitude racist is stating the obvious, but it is by no means a personal indictment of Huxley. Virtually all Europeans at the time took it on faith that the white race was the apotheosis of the human condition; to most Victorians, believing oneself related to an ape required much less of a leap than believing an aborigine or Negro the equal of oneself. One English naturalist, shocked by the specter of some particularly "hideous" Fuegians gesticulating wildly from a canoe, wrote:

"Viewing such men, one can hardly make one's self believe that they are fellow-creatures, and inhabitants of the same world. We often try to imagine what pleasure in life some of the lower animals can enjoy: how much more reasonably the same questions may be asked concerning these barbarians!"

The naturalist was the young Charles Darwin.

So it was that at the end of the nineteenth century, Neandertal Man was grudgingly accepted into the club of humanity, not so much on his own merits as on the shockingly low admission standards evident in the club itself. If an Australian aborigine or a Hottentot was human, then why not this other thing too? Few Europeans at the time, however, believed that Neandertals were the evolutionary antecedents of *Europeans*. In 1886, two new Neandertal skeletons were discovered in a cave at Spy, in the Belgian province of Namur. Unlike the original skeleton, these fossils wore their antiquity on their sleeve. They were embedded in deposits that also contained the bones of extinct mammals and flaked stone tools. The age and integrity of the Neandertals were no longer in doubt. They could still

be seen, however, as members of an ancient, "inferior" race that had been
swept away by newcomers from Asia, who were the true ancestors of modern Europeans.

The idea of waves of "Aryan" invaders from Asia was popular long
before Darwin's time, and would remain so long afterward. Throughout
the first half of the twentieth century, Asia was still seen as the most likely
birthplace of the human species: Only its harsh, open habitats provided
conditions sufficiently rigorous to forge so stunning an evolutionary product as Man. Like Cuvier's doctrine of catastrophic replacement, the Aryan
ideal could account for a succession of forms in the geological record,
without having to concede that one form—or even one race of modern
human—was biologically linked to another. As late as 1910, the German
anatomist Hermann Klaatsch actually declared that whites were the descendants of Caucasian invaders who could trace their lineage back to the
orangutan in Asia. In contrast, the modern African Negro had evolved
from a Neandertal, which in turn had arisen from an African gorilla.

"We observe that early man was not a forest-living animal," wrote the
American naturalist Henry Fairfield Osborn in 1923, "for in forested lands
the evolution of man is exceedingly slow, in fact there is retrogression, as
plentifully evidenced in forest-living races of today."

In 1891, a young Dutchman named Eugène Dubois had set out to
look for "the missing link" in Asia, and against all odds, his search was
rewarded. Toward the end of the field season, Dubois discovered the skullcap of a humanlike creature in a stream bank in Java. With huge, bulging
browridges, massively thick cranial walls, and a brain capacity well below
the average of a modern human, the skull seemed even more primitive
than Neandertal. Furthermore, it had emerged from a geological layer
containing mammal forms that were extinct long before Neandertal first
appeared. The next year Dubois returned to the site and, among other
fossils, discovered a thighbone near the spot that had held the skull. While
the skull impressed him with its remarkably primitive features, the thighbone seemed very much like that of a modern man. Convinced that he
had indeed found the missing link, Dubois returned to Europe, antici-
pating an instant triumph.

Instead, like the original Neandertal, his *Pithecanthropus erectus*
("Erect-walking Ape Man") was met by curiosity, then doubt, and finally
neglect. Again, some of the criticism was justified. The leg bone and skull
had been found in two separate field seasons, and fifty feet apart, which
undermined Dubois's strident insistence that they had to belong to the

same individual. With the leg bone placed to one side, most anatomists were inclined to view the skull as that of either an extinct ape or, like the Neandertals, a primitive race of human being. Java Man received a cordial reception only in Germany. The great evolutionist Ernst Haeckel, whose theories had prompted Dubois to look in Southeast Asia in the first place, hailed the find as definitive proof that mankind had indeed evolved from lower forms. Inspired by Dubois's discovery, the German anatomist Gustav Schwalbe undertook a thorough reanalysis of the known remains of Neandertals and concluded that contrary to the prevailing wisdom, Neandertal was indeed a separate species from modern humans, one that he called *Homo primigenius*. While it was possible that both Neandertal Man and Java Man were merely offshoots of the main stem of human evolution, Schwalbe preferred another arrangement, placing *Pithecanthropus* as the direct ancestor of Neandertal, which in turn gave rise to modern human beings.

Elsewhere, Dubois's interpretation inspired nothing more than a collective shrug. He took his box of bones to England to show to Arthur Keith, the ambitious rising star of British paleoanthropology, hoping to convince him that he had indeed found the missing link. Keith concluded that it was neither an ape nor a transitional form, but rather just another human being—primitive, but fully human. Deeply disappointed, Dubois snatched his precious fossil out of circulation and refused to let anyone look at it for almost thirty years.

The door was still open for the longed-for arrival of the one *true* ancestor of modern humans, heralded in the scientific community as "Pre-Sapiens Man" to indicate its direct, exclusive link to modern *Homo sapiens*. In time, the blinding passion to locate the only ancestor fit to become human would combine with British national pride to produce one of the greatest scientific debacles of all time. Meanwhile, in France, the obsession with the imagined Pre-Sapiens Man was about to fling heavy, bone-real Neandertal Man out of our lineage altogether, and into the evolutionist's deepest abyss.

There is something in the French soul that loves a nice clean edge. You can see it in their decor and in their politics, in the walls that separate house from house and yard from street, and in the care they take to wrap a package or arrange the food upon their plates. It is the love of edges that makes life in France so aesthetically pleasing. You need edge-love, or at least edge-sense, to distinguish a wine's myriad flavors, each from each, or to delight in the arrangement of fruits and vegetables in a shop into handsome, if wholly inorganic, patterns. Where the gardens of Versailles represent a sort of uninhibited debauch of edge-lust, even a common residential garden in France makes the English equivalent seem under-defined and an American backyard a heathen wilderness. At its best, the French edge is a remarkable clarifier, making sense out of hopelessly mud-dled realities. At times, however, the attention to edges runs away with itself, and the edge seems to become more the point than whatever con-tent it encloses.

Around the turn of the century, the French attitude toward paleo-anthropology might be summarized as a conviction that something is ei-ther human or not human, with a good clean edge between the two. Marcellin Boule, the leading paleontologist in France in the first decades of this century, was firmly convinced that the Neandertals fell on the wrong side of that edge. For Boule, Neandertal Man was "doubly fossil," both harking back to a geological epoch before the present one and rep-resenting a separate species doomed to extinction, most likely at the hands of the people who were the real ancestors of modern humans. Boule's merciless characterization of this "degenerate species" was the first blow in a one-two punch that would send Neandertal Man reeling off the hu-man lineage, creating a paradigm that would dominate the study of hu-man evolution for half a century.

In France, materialist evolutionary theories, especially Darwin's theory of natural selection, were much slower to catch on than in England. The return of monarchist rule and the never-distant influence of the Catholic Church ensured that the anti-evolutionist disciples of the great Cuvier would continue to set the research agenda at major institutions like the Museum of Natural History and the Academy of Sciences. In 1859—the year when Darwin published *On the Origin of Species*—the French gov-ernment allowed the anatomist Paul Broca to establish a Society of An-thropology only on condition that a plainclothes *gendarme* be present at all meetings. His attendance was supposed to ensure that nothing heret-ical, immoral, or politically seditious was said.

In spite of such restrictions, or perhaps because of them, an intensely political, anti-establishment group of young scientists known as the "com-bat anthropologists" arose in the later part of the century. Led by Gabriel de Mortillet, these card-carrying Darwinists based at the École d'Anthropologie looked upon the orthodox naturalists in older institutions as "non-entities," and regarded the Church as a force of moral and intellectual repression. In direct defiance of the belief in the immutability of species, Mortillet and his colleagues held that the human race had evolved directly from the Neandertals, through an edgeless, linear progression. Neither the Neandertals nor any other prehistoric people had actually gone extinct; they had simply stepped up another rung on the evolutionary ladder. The story of human prehistory was progress, driven by the same materialist forces that fueled the evolution of all other species. Meanwhile, progress within science could be achieved only through social change and constant struggle against the Church and its allies.

Partly because their scientific principles were expressed with such political stridency, the combat anthropologists' unilinear evolutionary ideas never took hold in the French scientific community. Sociologist Michael Hammond of the University of Toronto has pointed out that, ironically, the rabid anticlericalism of the École d'Anthropologie deprived it of the very evidence it needed to bolster its position on human evolution. In 1908, some Catholic priests discovered a Neandertal skeleton in a cave near La Chapelle-aux-Saints in France. It was the most complete specimen yet found, including even most of the backbone. Were it not for the École's uncompromising hostility toward the Church, this prize would have been forwarded for analysis to the laboratory of one of Mortillet's followers. Armed with this fossil, the unilinear school in France might have been able to exert more influence, and the whole future of paleo-anthropology would have been quite different. Instead, on the advice of the Abbé Henri Breuil, another cleric who later would himself exert enormous influence on French archaeology, the "Old Man of La Chapelle" was delivered into the hands of Marcellin Boule at the rival—and politically much more conservative—Museum of Natural History.

Boule believed in evolution, but he had little use for Darwinism and detested Mortillet's theories, which he called "a mirage of doctrines." He must have been thrilled at the opportunity the new discovery presented to manifest his own, diametrically opposed views on the course of human prehistory. Contrary to the ladderlike development imagined by the combat anthropologists, he believed that evolution proceeded through a complex pattern of branchings, extinctions, and sudden transformations. Boule had already completed a detailed study of the remains of Grimaldi Man, a human fossil found in Italy which bore a striking resemblance to modern Europeans. What was needed to strengthen his dichotomous position was a type of human who clearly did *not* bear such a resemblance. And Neandertal was right at hand.

Mostly on the basis of La Chapelle, Boule decided that earlier characterizations of the Neandertals as brutish primitives had not gone far enough. Acknowledging that their brains were as large as modern ones, Boule declared that they were nevertheless inferior in "quality," lacking in volume in precisely the areas, such as the frontal lobes, believed to be associated with high mental functions. Not surprisingly, these low-browed brains had been capable of only the most "rudimentary and miserable" tool culture.

"The likely absence of any trace of a preoccupation with an aesthetic or moral order accords well with the brutal aspect of the heavy, vigorous body," Boule wrote, ". . . affirming the predominance of the purely vegetative or bestial functions over the cerebral ones."

Even more damning than the cranial evidence was the postcranial skeleton. Through a detailed analysis covering hundreds of pages, Boule showed that Neandertal Man, far from resembling any human alive, had walked with bent knees and an ape's shambling gait, his big toe splayed chimpanzeelike to the side. The lack of a human curvature to the spine showed that Neandertal would not have been able to stand up straight, so he would instead walk with his head slung forward on a squat neck. Each ratchet turn of Boule's analysis further tightened the distinction between the Neandertals and any other kind of human being alive—including Australians, Eskimos, Fuegians, Bushmen, Polynesians—or with any subsequent offering from the fossil record.

"What a contrast with the humans of the geological and archaeological period that followed," he continued, as his prose approached climax, "with the Men of the Cro-Magnon type who had a more elegant body, a finer head, so much evidence of manual skill, of the resources of an inventive spirit, of artistic and religious preoccupations, of faculties of abstraction— who were the first to merit the glorious title of *Homo sapiens*!"

Contrast, as stark as he could draw it, was what Boule wanted to emphasize. By the time he was finished, the moral and morphological edge between Neandertals and modern humans was absolute. Yet both species, Boule believed, had lived during the cultural period known as the Mousterian. This left far too little time for Neandertals to evolve directly into Cro-Magnons. The only logical alternative was that the Cro-Magnons had existed earlier somewhere else and only later migrated into Europe. The

Neandertals, meanwhile, had gone extinct without issue. He might as well have added, "And good riddance to them."

Given the conservatism of French science, it is not surprising that Boule's Cuvierian-tinged vision of catastrophic replacement in Europe quickly won over Mortillet's radical Darwinism. As in all his writings, Boule had tiptoed carefully around the sensitive implications of evolution, never once discussing what mechanism might explain *how* species evolved, or how one human species could replace another. Rather than antagonize the Church with discussions of theory, he stuck to description, volume upon volume of it. At the time, his scholarship seemed so weighty, his prose so turgid, and his conclusions so inevitable, that no one thought to take another look at the La Chapelle skeleton for almost half a century.

What gave Boule's theory such sustained power was not its authority but its comforting accordance with what most people, not just the Cuvierian "nonentities" in Paris, wanted to believe. It is one thing to accept hypothetically that mankind is directly connected to the apes and, through them, to the lowest mud of creation. It is quite another to see that abstract connection congealed into hard bone, to feel its sharp tug pulling on the deep structures of belief that give oneself a reason to exist. Neandertal Man was—and for many people, still is—a patent insult. If something so different *that had lived so recently* was a direct ancestor, then all that distinguished man from the other animals—art, morality, industry, the splendid expressions of a self-examining soul—was reduced to a hasty concoction of a few evolutionary hours. To move from Neandertal to *Homo sapiens* in so short a time would require, in Boule's words, "a mutation too important and too quick not to be seen as absurd." The Bible may not have been wholly accurate about the means and the timing of our origin, but with Neandertal safely shunted onto a separate lineage, humankind could still have the chronological latitude to emerge with dignity. At the same time that he banished the Neandertals, Boule also declared Dubois's even more apelike *Pithecanthropus* off the human lineage. The hunt for the *real* "pre-Sapiens" ancestor—one old enough, unique enough, human enough—could resume. But the search for primitive man was literally canceling itself out: Anything primitive enough to be ancestral could not be ancestral because it was too primitive.

The Neandertals might still have struggled free of this paleo-Catch-22, were it not for a strange coincidence. A few months after Boule formally expelled Neandertal Man, an astonishing new candidate rose up to claim its place. Near a village common in Sussex, England, an amateur

The Neandertal from La Chapelle in southern France, as rendered in the Illustrated London News *in 1909. Note the bent knees and savage aspect. The "scientific advisor" to the artist was Marcellin Boule.* COURTESY *ILLUSTRATED LONDON NEWS*

fossil hunter named Charles Dawson found parts of a skull and a fragment of a lower jaw, together with some mastodon teeth and other bits of extinct fauna that established their great antiquity. Dawson called on his friend, the respected paleontologist Arthur Smith Woodward, to help him continue the search. Unlike the skulls of Neandertal Man and Java Man, this forehead was vertical and uncorrupted by browridges or other marks of a base beginning. In most respects, it seemed much like the skull of a modern human being, but, amazingly, the jaw was more primitive, more simian than any previous traces of early man. Only the teeth, flat-topped instead of highly cusped, betrayed its nobler lineage. The common where the fossil was found was called Piltdown.

By the beginning of the twentieth century, British evolutionary science had fallen into a decline. Darwin's materialist theories were not nearly as popular as they had been a generation earlier. Influenced by the French, British anatomists like Smith Woodward, Arthur Keith, and Grafton Elliot Smith were abandoning evolutionary ladders and, like their French counterparts, emphasizing multiple lineages and a sharp distinction between *Homo sapiens* and such questionable progenitors as Neandertal. But they had little evidence of their own to argue. Early humans in Europe were all emerging with decidedly Continental names: *Neandertal, La Chapelle-aux-Saints, Le Moustier, La Ferrassie, Heidelberg*. What England needed was its own "First Englishman" deep in time. Much to its ultimate chagrin, Piltdown was it.

No one could have invented a better fossil. Judging from the animal fossils and stone tools found with it, Piltdown seemed to be as old as the early Pleistocene, now dated at about two million years. Yet when Keith reconstructed the cranium from the skull fragments, Piltdown's brain turned out to be very near the average size for living human beings. In this one specimen, three of Britain's leading anthropologists could see their respective theories confirmed. Keith believed in the antiquity of humanity, and here was proof. Elliot Smith advocated the notion that the driving force of human evolution had been brain growth, and Piltdown's brain was much closer to the human condition than other parts of its body. The stark contrast between skull and jaw also confirmed Oxford anthropologist William Sollas's theory of "mosaic evolution," that all aspects of an organism do not evolve together, but each on its own evolutionary trajectory. All in all, the fossil was tailor-made to anchor the human line well before Neandertal or even *Pithecanthropus*; it was essentially like us, but with just a dash of ape thrown in precisely where it would be least insulting to our species' self-respect. This "improbable monster," Sollas wrote, "had, indeed, been long previously anticipated as an almost necessary stage in the course of human development."

And by Jove, the thing had come from British soil. In England, Piltdown was unanimously hailed as "The Earliest Man." Marcellin Boule also granted it ancestral status, if a bit grudgingly. Elsewhere, especially across the Atlantic, Piltdown's reception was more qualified. Aleš Hrdlička and Gerrit Miller, both of the Smithsonian Institution, pointed out that the skull was so human and the jaw so apelike that the two bones could not possibly belong to the same individual. Frustratingly, the hinge between the two that could have decided this issue had not been found. In any case, the American position was seriously undermined when, in 1915, a *second* Piltdown Man was found nearby, once again bearing both manlike and simian traits. Puzzling as Piltdown seemed to be, its reality was hard to deny.

At the time, Aleš Hrdlička was almost the only friend Neandertal Man had. Born in Bohemia in 1869, Hrdlička emigrated to the United States after studying in Paris at the École d'Anthropologie, the hotbed of anticlerical Darwinism. There he had become a committed believer in a lad-derlike progression of stages in human evolution, driven forward by the basic Darwinist engines of adaptation and selection. From Hrdlička's per-spective, there was nothing aberrant or monstrous about Neandertal Man at all—he was simply "Man of the Mousterian Culture," the stage in the biological development of man associated with that level of stone-tool technology. The massive faces and robust physiques of the Western Eur-opean specimens like La Chapelle and La Ferrassie simply reflected ad-aptations to diet and a cold climate. Farther to the east, where conditions during the Pleistocene had been milder, other, less robust Neandertals had been discovered. There was no need to imagine invasions of pre-Sapiens supermen to explain the origins of Cro-Magnon. The ancestors of today's Europeans were already present in the European Neandertals, and when conditions changed, they would change too.

"My conviction that the Neanderthal type is merely one phase in the more or less gradual process of evolution of man to his present form, is steadily growing stronger," Hrdlička wrote in 1916. "I find on the one hand, notwithstanding many lapses, nothing but gradual evolution and involution, and on the other hand no substantial break in the line from at least the earliest Neanderthal specimen to the present day."

In 1927, Hrdlička traveled to London to deliver the Huxley Memorial Lecture, hoping to use that prestigious forum to convince his British col-leagues of the Neandertal claim to human ancestry. He might as well have stayed in Washington. In a subsequent issue of *Nature*, Grafton Elliot Smith expressed a sort of lofty puzzlement over why anyone would even question the status of Neandertal Man, "an issue which most anatomists imagined to have been definitely settled" by Boule and others. At least El-liot Smith deigned to comment. No one else bothered, convinced as they were that if there were any true ancestor to be found in Europe, it was not the low-browed Neandertal, but the First Englishman, Piltdown Man.

As time went on, Piltdown became entangled in its own anomalies, seeming every year less like the missing link and more like a "maze of conflictions and uncertainties," as one investigator put it in 1945. But teamed up with Boule's portrait of Neandertal inadequacies, Piltdown managed to keep the human line clean of any tangible ancestors for nearly half a century. As long as Piltdown existed as data, the unfulfilled promise of the shadowy Pre-Sapiens Man hovered in the air, blocking the accep-tance not only of Neandertals but of a host of otherwise very credible candidates for human ancestry.

Among the most sensational of these would-be missing links was the original *Australopithecus africanus*, or "Ape Man from Africa." In 1925, an unknown anatomist named Raymond Dart announced the find of a juvenile "man-ape" at the Taung limestone quarry in South Africa. Its brain was small, but high and rounded, unlike the flat brain of an ape. Its teeth, too, were surprisingly humanlike for a creature otherwise so primitive. And there was good evidence that it had walked upright. Dart even declared that the creature had used tools.

In spite of Dart's confidence, Keith, Elliot Smith, and the other ruling members of the British anthropological establishment suspected that the Taung Child was merely an extinct form of ape or, at most, another divergent twig off the human family tree. Part of the problem was the

A far more civilized version of Neandertal Man, based on the same La Chapelle skeleton that inspired the drawing on page 40 and published two years later in the Illustrated London News. *Note the use of stone tools, the shell necklace, and the fully human, contemplative demeanor.* COURTESY *ILLUSTRATED LONDON NEWS*

youth of the individual. Juvenile apes often show hominidlike features, so Dart may have been confused by what he saw. The fact that he had found his missing link in the "wrong" continent did not help either—the bracing plateaus of Asia were still thought to be a much more appropriate cradle for Man than the dank darkness of Africa. Most important, Dart's *Australopithecus* was not evolving the way a true human was supposed to evolve. Its brain seemed to be lagging *behind* the much more hominidlike character of the teeth and jaws, rather than forging the way.

In 1931 Dart, like Eugène Dubois before him, journeyed to England with his find, expressly to convince Keith and the others that his little fossil belonged on the road leading to man. But it was no use. "Man is what he is because of his brain," Keith wrote. "The problem of human evolution is a brain problem." The small-brained Taung Child was dismissed.

Meanwhile, a bonanza of new fossils had emerged from a cave near Peking, China, then being excavated by a well-endowed international team headed by the Canadian Davidson Black. With the right continental address, and surrounded by evidence of fire and hints of cannibalism, Black's *Sinanthropus*—popularly known as "Peking Man"—stood a much better chance of winning acceptance as a true human ancestor. In 1924, Dubois had emerged from years of silence himself and presented to the world previously unknown fossils from his *Pithecanthropus erectus* site in Java. The resemblances between Peking Man and Java Man were too obvious to ignore. Another hominid jaw, found near Heidelberg, Germany, back in 1907, also fit the pattern. Eventually, all three collections were merged into the classification *Homo erectus*. Apparently, long before either Neandertals or modern *Homo sapiens* had appeared on earth, another hominid, modestly large-brained and capable of producing a fairly sophisticated culture, had succeeded in occupying much of the Eurasian continent. (*Erectus* was would eventually be found in Africa, too.) But with the hope of Pre-Sapiens Man beckoning, most anthropologists were still content to see *erectus* as merely another "aberrant offshoot" of the family tree.

"Neither Peking Man . . . nor Neanderthal Man has any direct offspring left today in the living world," declared the famous French anthropologist and cleric Pierre Teilhard de Chardin, a close colleague of Boule's, in 1943. "They have been swept away by *Homo sapiens*."

Astonishingly, the entire physical record of hominid evolution covering three continents and two million years of time had been swept away by a spectral ancestor composed almost entirely of preconception, prejudice, and blind faith. By the end of World War II, however, the days of Pre-Sapiens Man were numbered. A new approach to understanding human evolution was emerging, mainly in the United States. Where the pre-Sapiens theory set standards for humanness no mere fossil could reach, the new paradigm welcomed each and all in a happy family embrace.

The 1930s were the "Indiana Jones" decade in paleoanthropology. Adventurers and adventuresses risked heat, hardship, and marauding bandits to push into undiscovered terrain and come back with fossil treasures—as well as malaria and a host of other diseases. In East Africa, Louis Leakey loaded everything he owned onto a dilapidated truck and headed for Olduvai Gorge and beyond, returning with bones that he believed proved the ancient African origin of mankind. The Dutchman G.H.R. von Koenigswald followed Dubois's lead into the jungles of Java. Before disappearing into a Japanese prison camp during the war, he found not only more evidence of *Pithecanthropus* at the site of Sangiran, but more advanced skulls near the Solo River in eastern Java, which, he thought, were a transitional form between *Pithecanthropus* and modern humans.

At the same time, the British were pulling treasures out of the Near East as well as their own backyard. In 1935, a fragment of a skull was found in a gravel pit at Swanscombe, near the Thames, and a year later another piece turned up. Though these amounted to little more than a bit of the rear of the skull, the Swanscombe skull was still enthusiastically enlisted into the camp of Pre-Sapiens Man. Arthur Keith found a place for it after Piltdown on the thick, main stem of human evolution leading to "Basic White." In Palestine, Dorothy Garrod opened up fantastically rich caves on Mount Carmel, sitting under an umbrella at the top of the sunbaked excavations, snapping a whip, and extolling the workers to "Dig faster! Dig faster!" The amateur archaeologist Francis Turville-Petre was also continuing the searches that had earlier brought forth a Neandertal-like specimen in the cave of Zuttiyeh near Galilee. René Neuville, the chief French government official in Jerusalem, was signing all his correspondence "Counsel and Prehistorian," amply justifying the second title with his discovery of human skeletons in the cave of Qafzeh near Nazareth.

Unlike the Swanscombe find, none of the discoveries in the Near East seemed to offer support for a clean, Neandertal-free route to modern humans. In fact, they did the opposite. On Mount Carmel, Garrod pulled what looked to be a Neandertal skeleton from a cave called Tabun. In another cave called Skhul, literally just around the corner from Tabun, she found the skulls and bones of ten more individuals. These people seemed to represent a strange blend of primitive and modern character-istics, and many of the skeletons appeared to have been buried intention-ally; one was even found clasping the jawbone of a pig.

This was precisely what Aleš Hrdlička of the Smithsonian Institution had been searching for: evidence of a clear evolutionary link between Neandertals and modern humans. Hrdlička wanted desperately to get his hands on the actual specimens for a detailed study, and since Garrod's expedition was a joint British-American enterprise, he had reason to hope. Instead, the job of describing the Mount Carmel hominids went to Arthur Keith and a young graduate student from Berkeley, Theodore McCown, who, though he leaned toward Hrdlička's way of thinking, was overruled by the far more powerful Englishman. Keith conceded that the fossils from the two caves represented a single population, but rather than dem-onstrating a transitional form from Neandertals to modern humans, Mount Carmel Man was merely a hybrid—the localized, "bastard" prod-uct of interbreeding between Neandertals and Cro-Magnons. His pre-Sap-iens convictions had perhaps been shaken, but they did not fall.

While Garrod cracked her whip in Palestine and Louis Leakey buried his tires in the mud of western Tanzania, another, somewhat less swash-buckling anthropologist named Franz Weidenreich had taken over the famous Peking Man site of Zhoukoudian. As a Jew in his native Germany, Weidenreich had already escaped from a far greater danger than bone hunting and had accepted a position at the University of Chicago. Be-tween 1935 and 1939, under his leadership, the Chinese excavation pro-duced a dazzling abundance of new evidence. The remains of at least forty-five Peking men—skulls, limbs, jaws, and hundreds of teeth—turned up, as well as tools, hearths, and meals. Unfortunately, the Zhoukoudian enterprise closed abruptly when the Japanese army occupied the nearby village in 1939.

Two years later, fearing for the safety of the fossils, Chinese authorities at the site packed them up and sent them off by train under an American Marine escort, to be transferred to a steamer leaving for the United States. They never reached their destination. Along with other materials, the fos-sils reached the port serving Peking (Beijing) on the day the Japanese invaded Pearl Harbor. The Marines were taken prisoner, and the fossils vanished forever. Fortunately, Weidenreich had made excellent casts.

Weidenreich, like Hrdlička, suspected that human evolution had progressed gradually through a series of stages. What he uncovered in China strengthened his conviction. Working closely with von Koenigswald, he could see that Davidson Black's *Sinanthropus* in China and Dubois's *Pithecanthropus* were regional variants of one of those stages and proposed that they be placed together as *Homo erectus*. Again like Hrdlička, he saw the Neandertals as simply the European version of the next major stage, penultimate to modern *Homo sapiens*. There were no clean breaks, no edges, not even between the stages themselves, and certainly no separate pre-Sapiens phantom awaiting discovery.

"I believe that all primate forms recognized as hominids ... can be regarded as *one species*," he wrote in 1947.

Through the gradual course of human evolution Weidenreich did, however, see telling differences between one geographic region and another. One of the first scholars to rope together the evidence for human evolution on a global scale, he sorted those fossils known at the time into four distinct geographical variants, each of which developed continuously toward a living human racial type: Australian, Mongolian, African, and Eurasian. Modern Australian aborigines bear such distinct Southeast Asian features as pronounced browridges and receding foreheads, which trace back through von Koenigswald's Solo Men all the way to the *erectus* fossils of Java, which are perhaps a million years old. Living Chinese display the same smaller faces, flat cheekbones, and rounder foreheads that Weidenreich found peculiar to the Peking Man remains he was digging out of the Asian earth. Similar lines of "local continuity" could be drawn back in time in the other regions of the world. Nothing about the Neandertals of Western Europe, for instance, prevented them from being considered ancestors to the Cro-Magnons. Binding all these regional types together was a continual mingling of genes *between* the regions. Rather than a bush, Weidenreich's family tree looked like a garden trellis: parallel vertical lines interconnected by a latticework of gene flow.

This notion of a steadily climbing trellis was no more popular in the forties than Hrdlička's similar, ladderlike progression had been twenty years earlier. Nevertheless, it seemed to accommodate the evidence. Garrod's Mount Carmel discoveries practically jumped up and arranged themselves on it. The Tabun Neandertal type had evolved gradually and locally into the transitional population represented by the Skhul people, who had in turn evolved into the modern Cro-Magnon types that had also been found in the Near East. There was only one fossil that did not make any sense at all as an ancestor: Piltdown. Weidenreich took one look at the teeth and jaw of the First Englishman and declared that they belonged to an orangutan. He was completely baffled as to how a tropical ape had come to be living on the banks of the Thames, much less how it happened to be found with other bits of anatomy that were clearly those of a modern human being. But an orang it was.

Weidenreich died just a few years too soon to find out how right he had been. The whole Piltdown business was, of course, a fake. Somebody, probably its "discoverer," Charles Dawson, helped along by a more knowledgeable accomplice,* had mixed fully modern human skull parts with the jaw of a modern orangutan, carefully staining the fragments dark brown to make them seem ancient. The teeth in the jaw had been filed down to make them look a little more like *Homo*. The marks of the file were plainly visible under a low-power microscope. But so neatly did the fossil collage dovetail with the expectations for a more noble ancestor that no one thought to take such a close look at it until the early 1950s. As soon as they did, the forgery became embarrassingly obvious. Piltdown Man, one half of the powerful pre-Sapiens pincers that had put the squeeze on Neandertal Man nearly half a century before, melted into nothing.

The other half was soon to follow. By the late 1940s, enough fossils had emerged to make it extremely difficult to dismiss all of them as "aberrant offshoots" unrelated to living human beings. Even Arthur Keith— by then Sir Arthur—threw in the towel before he died and conceded that Dart's *Australopithecus* as well as *Homo erectus* belonged on the main human stem. Suddenly Neandertal, given its modern-sized brain and relatively sophisticated level of culture, was looking conspicuous as the only outcast. In the 1950s, Henri Vallois, Marcellin Boule's former student and successor at the Museé de l'Homme in Paris, made a final plea for the validity of Pre-Sapiens Man. Vallois believed that new, fragmentary skulls

*Practically every scientist involved in the Piltdown find has at one point or another been fingered as Dawson's mysterious confederate, including Elliot Smith, Smith Woodward, Sollas, and Teilhard de Chardin, who was with Smith Woodward and Dawson when some of the material was uncovered. All had personal and professional motives, but the evidence against them is largely circumstantial. The only British scientist who seemed above incrimination was Arthur Keith. But in 1990, Frank Spencer of Queen's College published compelling new evidence pointing directly at Keith. Following the trail started by Ian Langham, a science historian at the University of Sydney, who died in 1984, Spencer showed that among other anomalies, Keith had written an article about Piltdown before the find had been announced. The case against him is not airtight, however, and the truth will probably never be known.

from a site called Fontéchevade in France, and the Swanscombe skull in England, demonstrated that another human with more modern charac- teristics had been in Europe before the Neandertals. Other scientists, how- ever, pointed out that if anything could be said about such enigmatic fragments, they were more similar to their Neandertal contemporaries than they were to modern humans.

More important, people were at last beginning to take a more gen- erous look at the Neandertals themselves. In 1957, two American anat- omists, William Straus and A.J.E. Cave, reexamined the fossil that had formed the heart of Boule's original low opinion of Neandertal human- ness: La Chapelle-aux-Saints. It turned out that Boule's description of the specimen was flawed from head to toe. Its stooped posture was not testimony to a brutish nature but merely the consequence of advanced arthritis. Boule had noticed the presence of the disease, but his zeal to cast Neandertals in the worst possible light had blinded him to its ef- fects. Indeed, nothing about the skeleton below the waist suggested that the Old Man of La Chapelle or any other Neandertal walked with a di- vergent toe, bent knees, or on the outside of its foot. Straus and Cave reached the stunning conclusion that, morphologically speaking, a Neandertal wasn't all that different from a modern human being. It was they who first suggested that with a shave, a bath, and a new set of clothes, a typical Neandertal could travel incognito on the New York subway system.

Eventually, Neandertal would ride that subway all the way to the thor- oughly human heart of the family. At about the same time Straus and Cave were at work on La Chapelle, a young American paleoanthropologist named Clark Howell was resurrecting a compromise theory of modern human origins known as the Pre-Neanderthal hypothesis. Howell, like the Italian paleontologist Sergio Sergi a decade earlier, did not believe that the recent Western European Neandertals like La Chapelle and La Fer- rassie were part of the modern human lineage. With their fantastically huge faces and bulging noses, these late, "classic" Neandertals were rather the remnants of a population that became extinct. On the other hand, an earlier group of "progressive" Neandertals, discovered mostly in western Asia and the Middle East, was not yet so developed in the Neandertal direction. Specimens like Tabun, along with an enormous cache of fossils found in a cave called Krapina in Yugoslavia, represented the gene pool out of which *both* the later Neandertals of Western Europe *and* modern humans had emerged. Howell did not have to look far from Tabun to point to the earliest truly modern *Homo sapiens*. They were lying next door, in the Skhul cave.

Another American completed the embrace and brought all the Neandertals back into the human family fold. Anthropologist C. Loring Brace of the University of Michigan developed a reputation for radical thinking from the start of his career. In the early 1960s, Brace published a series of articles championing the ancestral status of the Neandertals with a polemic passion rarely seen in the pages of academic journals. Largely because of the blind acceptance of the views of Marcellin Boule, Brace argued, the Neandertals had rarely been given the fair appraisal that they deserved. When they *had* been carefully considered—in the work of Hrdlička, Weidenreich, and the German anatomist Hans Weinert—the studies had fallen on deaf, Boule-blocked ears.

Brace upbraided British scientists like Arthur Keith and fellow Americans like William Howells and Clark Howell. But he saved his most damning invective for the French. Their problem went back far beyond Boule to Cuvier himself. The French simply did not believe in evolution. So thoroughly had Cuvier's influence stifled the seed of evolutionary thought in France that Continental scholars ever since had been engaged in nothing more than a "devious and unproductive delaying action" against Darwinian truth. "Perhaps one of the reasons for the consistently anti-evolutionary position taken by French physical anthropology," Brace wrote in 1964, "is that . . . they have eliminated human adaptation from their thinking, and without an understanding of this, it is of course difficult to interpret the hominid fossil record from an evolutionary point of view."

Brace's own thinking, like that of many American physical anthropologists, had been deeply influenced by the "New Synthesis" of Darwinist principles with modern population genetics. Evolutionary biologists like Theodosius Dobzhansky, Julian Huxley, Ernst Mayr, and George Gaylord Simpson had rejuvenated natural selection as the creative force behind evolution. The spotlight of paleontology had shifted from the *what* of evolutionary change to the *how*. It was no longer enough to slap a label on a fossil, describe its anatomy, and fit it into a scheme relative to other fossils. What was needed was a systematic understanding of how one form had become different from another.

Brace was convinced that Neandertals had evolved directly into Cro-Magnons, and he had an idea of how the change had come about. In spite of Boule's low regard for Neandertal brains and skeletons, there was nothing about either that differed qualitatively from modern human brains and bodies. Where Neandertals and modern humans really parted company was in the face. The heavy browridges, the big noses, and the giant protruding mid-facial regions of the Neandertals were unquestionably distinctive. Most unlike us were the "gross dimensions" of their teeth, especially the front ones. In fact, the various niceties of Neandertal facial architecture amounted to nothing more than a single, integrated system supporting tremendous amounts of loading on the front teeth. All that would have been necessary to change a Neandertal into a modern human was a relaxation of the pressure of natural selection upon that massive chewing apparatus.

What adaptation would have rendered the Neandertals' superjaws obsolete? Like modern hunter-gatherers, the people of the Mousterian were probably chewing a mixture of meat and gathered plant foods. Their diet, however, was not the key issue. "The important thing to look to is not so much the food itself," Brace wrote, "but what was done to it before it was eaten."

The place to look for the cause of the change was the archaeological record. According to Brace, the increasingly elaborate and varied tools that appear in the Mousterian in Europe and continue on into the Upper Paleolithic had taken on the tasks formerly performed by the Neandertals' powerful jaws, such as cutting raw meat and scraping hides. An equally important technological advance was the growing reliance on fire. From the pattern of ashes left behind in Mousterian "hearths," Brace deduced that the Neandertals were earth-baking their food in shallow pits lined with fire-heated rocks and covered with soil. "It seems abundantly clear that the Neanderthals of the Mousterian were cooking their food in the same way that the Polynesians still do. . . . " he recently wrote. "Meat cooked in such a fashion can become quite tender indeed, and in such condition it requires less chewing to render it swallowable than would be the case if it remained uncooked."

An advance in culture, in the form of new tools and fire technology, was all that was needed to make a Neandertal face into a modern human one. No defining biological edge, no catastrophic event, separated the two into different species—only the progressive advance of technology. If one insisted on keeping Neandertal off the line, one would also be forced to say that Victorian gentlemen and ladies were a distinct and separate race from modern Europeans because they had not developed the microwave oven. The mysterious fate of the Neandertals had at last apparently been solved.

"It was the fate of the Neanderthal to give rise to modern man," Brace concluded in 1964, "and, as has frequently happened to members of the older generation in this changing world, to have been perceived in caricature, rejected, and disavowed by their own offspring, Homo sapiens."

Much of what Brace had to propose, Hrdlička had already said half a century before. But this time people listened. Pre-Sapiens Man may have haunted an earlier time, but the Zeitgeist of the 1960s was not so willing to cozy up to an idea based on biological distinctions between one kind of person and another, no matter how far in the past the lines were drawn. Suddenly it seemed as if Neandertals had never been all that different from regular folk. One had only to stop dwelling on their anatomical peculiarities and look instead at the traces of their actual lives. Perhaps their stone tools were not as sophisticated as those of Cro-Magnon, but they were certainly far from primitive. So difficult is the technique they used to manufacture them that only a handful of archaeologists today can emulate their expertise at it. Like the Cro-Magnons, they apparently used these tools to hunt mammoth, aurochs, and other dangerous game.

But the Neandertals' true humanity revealed itself in the actions of their souls. At the 50,000-year-old site of Hortus in southern France, two French archaeologists in 1972 reported the discovery of the articulated bones of the left paw and tail of a leopard. Their arrangement suggested that the fragments were once the remnants of a complete leopard hide worn as a costume. Here, and even more at the Yugoslavian site of Krapina, numerous fragmented human bones hinted at ritualized cannibalism. Ten years earlier, another French archaeologist discovered at the 80,000-year-old site of Regourdou what seemed to have been the scene of a bear cult. The carefully arranged bones of a brown bear had been placed in a stone-lined pit, along with the skeleton of a young adult Neandertal. Perhaps the best evidence of all for Neandertal ritual was what had already been found decades before. In 1939, on Monte Circeo not far from Rome, construction work exposed a cave that had been sealed by debris for 50,000 years. Inside was an isolated Neandertal skull, surrounded by what appeared to be a circle of stones. The base of the skull itself had been smashed, presumably to extract the brains. The rituals of Neandertal Man may have been macabre, but they were also loaded with human meaning and import.

Most important, the very fact that so many Neandertal skeletons had turned up over the years indicated that they had been buried intentionally. At Le Moustier in southern France, a young man's body was found sprinkled with red ocher and buried in a flexed position, as if in sleep. His head rested on a pillow of flints, and burned wild cattle bones were scattered about, as if in offering. At the rock shelter of La Ferrassie nearby, an entire family—including a man, a woman, four small children, and a newborn infant—had all apparently been buried. Far to the east in Soviet Central Asia, another young man had been interred perhaps as long as 100,000 years ago, surrounded by pairs of goat horns placed vertically around his body. There were many others. Tabun was an intentional burial. So was the Belgian site of Spy. The Old Man of La Chapelle, notwithstanding Boule's depiction of him as little more than an animal, had been lowered into a prepared trench by companions, perhaps the same companions who had cared for him long after his arthritis-racked body ceased to be able to find its own sustenance. Hardly an animal's way to die.

The apotheosis of Neandertal spirituality was found in an enormous cave in Iraq's Zagros Mountains. Excavated in the late 1950s by Ralph Solecki of Columbia University, the Shanidar Cave yielded the remains of nine Neandertals, probably around 60,000 years old. One of these had sustained crushing wounds to his head and body during his life, leaving him with a withered arm and blind in one eye. Like the Old Man of La Chapelle, he could have survived only by relying on others. The group had expressed its commitment to another individual after his death. The French archaeologist Arlette Leroi-Gourhan found abundant traces of pollen around the grave of a young Neandertal male in the cave. Her analysis of the pollen indicated that he had been buried on a multicolored bed of flowers, including hollyhocks and others used today for medicinal purposes.

"One may speculate that the [individual] was not only a very important man, a leader, but may have been a kind of medicine man or shaman in his group," Solecki wrote.

The name of his book, published in 1971, in the thick of the Vietnam War, was *Shanidar: The First Flower People*. A little over a century after he had tumbled down from the cave in Joseph Neander's valley, Neandertal Man had pulled off a magnificent reversal. He had rescued his image from that of a brute, driven by violence and instinct, to reemerge as a sixties' flower child, at least as human, and certainly more humane, than the manifestly ruthless species that would inherit his world.

And then, suddenly, it all came crashing down again. In the 1980s, just when it looked as if the Neandertals had secured a permanent place on *our* side of the gartel, their humanness took a precipitous fall. Some anatomists looked at the structure of their vocal tracts and decided that they had lacked fully human speech. A couple of archaeologists gathered together all the indisputable evidence for a symbolic sense among Neandertals, and found they had gathered nothing. People began to question their hunting abilities, their organizational talents, even their habit of burying the dead.

Worst of all, a *true* ancestor of modern humans had once again been identified, one whose very existence shunted the Neandertals into oblivion. This time, it wasn't just a nebulous Pre-Sapiens Man concocted out of prejudice and expectation. It was composed of fresh, hard data. It wasn't a fossil and it wasn't even a man. The new ancestor's name was Eve.

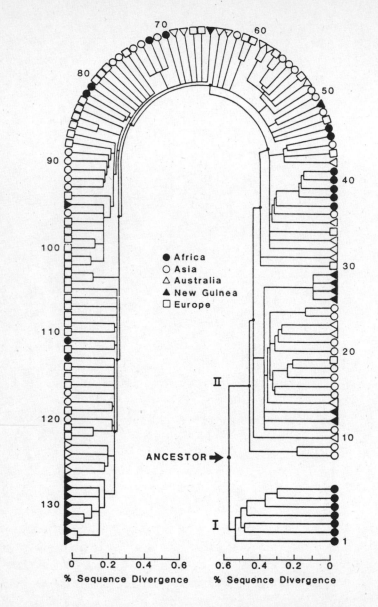

The genealogical tree showing how the mitochondria of all living human beings could trace back to a common African ancestor. When it was first published, one anthropologist said it reminded him of a rose trellis: sweet-smelling from a distance, but covered with thorns. REPRINTED WITH PERMISSION FROM *NATURE*, COPYRIGHT © 1987 BY MACMILLAN MAGAZINES LIMITED

Chapter Three

SPEAK, MITOCHONDRIA

There is all Africa and her prodigies in us. —SIR THOMAS BROWNE

In 1989, she was the queen of science. A year had passed since she made her public splash on the cover of *Newsweek*, two years since she was born in a beaker in Berkeley, and everybody was still talking about the African mother of the human race. Johnny Carson wove her into his opening monologue on *The Tonight Show*. Cartoons appeared in *The New Yorker* and *The Far Side*. She even got a mention in the popular television show *thirtysomething*. In scientific circles, the question of modern human origins dominated every discussion of human evolution, and modern human origins—for a while—were all about Eve.

The story really begins in the 1960s. Back then, if you wanted to find out about human origins, you went out and dug for it. If you were persistent and lucky, you brought home some hard evidence, like a skull or some ancient tools. Then Allan Wilson and another Berkeley scientist named Vincent Sarich got it into their heads to look for clues to the human past not out on the African savanna but among the proteins of living people and primates. They were trying to answer a fundamental question: When did the human line first branch off from our relatives, the apes? They were looking at proteins because they knew that, like beaks, skulls, and other gross anatomical parts, proteins evolve by accumulating mutations. They also knew that, like the beaks and skulls of different organisms, proteins in related species are slightly different from one another because of the mutations that occurred after the species split off from the common ancestor. A human being's albumin protein is similar to, but not exactly like, a chimpanzee's albumin. A human's albumin is much less like that of a rabbit or a frog, which are more distantly related to us. While it is often difficult to measure and compare variations in visible anatomy—does one species have 60 percent more nose than another, or fur four times more brown?—differences in proteins can be quantified.

Nothing about this was controversial until Sarich and Wilson said that the mutations occur across the millennia at a steady rate, like the ticks of a molecular clock. If this is true, the difference in a given protein in any two species would indicate not only how *related* they were, but also how much *time* had elapsed since they shared a common ancestor. The albumin of two primate species that shared an ancestor twenty million years ago should be twice as different, one from the other, as the albumin of two primates who branched apart only ten million years ago. There was nothing inherently threatening about this notion—it had already been put forth a few years earlier by two other biochemists, Linus Pauling and Emile Zuckerkandl. It was only when Sarich and Wilson used their molecular clock to set the time of divergence of one particular higher primate that the trouble began.

Several years before, an American anthropologist named Elwyn Simons had come across a piece of fossilized jawbone belonging to a species called *Ramapithecus* in a Yale museum drawer. He took a careful look at the jaw and decided it was hominid. Dated fourteen million years before the present, it was, in fact, the oldest known member of the human lineage. Though not everyone agreed with Simons, by the time Sarich and Wilson made their announcement, *Ramapithecus* was flaunting its "Earliest Man" status in introductory anthropology texts and museum dioramas around the world and striding confidently through the pages of Time-Life Illustrated books displayed on the coffee tables in thousands of American homes.

Looked at from a biochemical point of view, *Ramapithecus* was a total impostor. Sarich and Wilson's protein analysis suggested that the common ancestor of apes and humans had lived only five million years ago. Even when they factored in all the possible errors in their calculations, the two scientists could not support a date earlier than eight million years ago for the ape-human split. Therefore *Ramapithecus* could not possibly be a hominid, because it had lived at least six million years before hominids existed.

"To put it as bluntly as possible," Sarich later wrote, "one no longer has the option of considering a fossil specimen older than about eight million years as a hominid *no matter what it looks like.*"

What fossils "look like" is, of course, the *sine qua non* of paleoanthropology. Sarich and Wilson were not simply attacking *Ramapithecus*; they were calling into question an entire discipline, granting paleoanthropology about as much scientific legitimacy as telling fortunes with tea leaves. Not surprisingly, fossil experts did not take kindly to their emasculation by a couple of rogue biochemists.

"We were rubbing a lot of nerves raw," says Vince Sarich. "I'd have to paraphrase what they told us to do. It was something like we should go away somewhere in particular."

As it turned out, new fossil evidence eventually showed that *Ramapithecus* was not Earliest Man after all, but some sort of primitive orangutan. Most paleoanthropologists now accept the idea that the human lineage diverged from the apes sometime between five and seven million years ago. The new hominid species announced in 1994 is almost four and a half million years old, and just about as apelike in appearance as a hominid can be and still be called a hominid. So apparently Sarich and Wilson were right. But some say it was for the wrong reasons.

"*Ramapithecus* was a paleoanthropological problem solved by paleoanthropologists doing paleoanthropology," one of their critics—a paleoanthropologist—barked at me when I brought the matter up. "The biochemists had nothing to do with it."

While Sarich and Wilson were finding the right answers for the wrong reasons, other lab scientists were using proteins to reconstruct the much more recent history of living human ethnic groups: how diverse they are, how closely each is related to the others, and ultimately, what pattern of branching formed the human family tree that exists today. The first efforts to create such a genetic history of humanity focused on blood chemistry. Every person alive bears a gene that determines whether his or her blood will be type A, B, O, or AB. Human blood also contains an antigen to destroy invading toxins, which comes in two forms, Rh-positive and Rh-negative. Your blood type and Rh factor are important information if you need a transfusion. But they have a very different use for a population geneticist.

As far back as 1918, it was known that certain blood types could be found more frequently in some ethnic groups than in others. American Indians are virtually all type O. Rh negative blood is found almost exclusively among Europeans and is most prevalent in the Basque people of the western Pyrenees. Since blood groups are determined by specific, known genes and do not seem to be related much to environmental pressures, they are far more useful in judging the relationships of populations than superficial anatomical characteristics like skin color. If you were to judge by skin color, Africans and Australians would appear to be closely related. When you look at their blood chemistry, however, they turn out to be about as far apart as two human populations can be; their dark skins are merely a shared adaptation to living under a tropical sun. There is such a mishmash of overlap among the superficial morphological traits once used to define "race" that the term has scarcely any meaning to a modern geneticist.

In the 1960s, Luigi Cavalli-Sforza, an Italian population geneticist now at Stanford University, and his colleagues began to collect and organize information on the regional frequencies of blood types into a vast human family tree. Their results suggested that Caucasians and Negroids are more closely related to each other than either is to the third major population grouping, the Mongoloids. (Their definition of this group contained Australians and American Indians as well as Asians, because Australia and the Americas were settled relatively recently by migrants from Asia.) Cavalli-Sforza then began to collect the same sort of data on hundreds of other genetic "markers"—human gene types viewed indirectly by comparing the proteins and enzymes that they manufacture. Combining all the data, he suggested that humanity's roots were probably in Asia, with the Negroids of Africa and the Caucasians of Europe branching off at some later time.

By the end of the 1970s, molecular biology had evolved fast enough for scientists to shift their focus from the *products* of genes, such as proteins, enzymes, and blood types, and go straight to the heart of the matter: the genetic code of life itself. Comparing actual genes is a more direct and unambiguous way of measuring differences between populations. Looking at genes instead of proteins is like staring directly at a person, rather than looking at the impression his face would make in a tablet of clay. Not only do genes form the blueprint for the manufacture of proteins, they also mutate faster, so that the differences between one population and another are much more clearly outlined.

In 1986, a year before Eve made her dazzling debut, an Oxford geneticist named James Wainscoat focused on a particular region of the beta-globin gene, the genetic blueprint for the blood protein hemoglobin. Like all other genes, the beta-globin gene is composed of interlocking strands of DNA arranged in precise sequences of joined base pairs like the pearls on the double-spiral necklace of DNA. Among humans the gene exists in a range of different forms called haplotypes, which differ from each other because small, harmless mutations have occurred through the generations along that particular sequence of base pairs. Wainscoat and his colleagues at Oxford examined the haplotypes in eight diverse human populations from Europe, mainland Asia, New Guinea, the South Pacific, and Africa. Their results showed that two of the gene forms were extremely common in Africans but rare or completely absent in humans outside that continent. In Europe and among the genetically related populations of Asia and the South Pacific, three other haplotypes were common instead. Over all, Wainscoat's results hinted that one continent was the origin of all living races—but it was not Asia, as Cavalli-Sforza had maintained.

"Our data are consistent with a scheme," wrote Wainscoat and his colleagues, "in which a founder population migrated from Africa and subsequently gave rise to all non-African populations." In fact, Wainscoat's research suggested that everyone alive today—European, Asian, American Indian, South Sea Islander—ultimately derives from a group of Africans who may have numbered no more than a few hundred people.

Intriguing as this conclusion was, information on a single nuclear gene, like the one for beta-globin, could give only rough information about where evolutionary events took place. And it said next to nothing about *when*. There is another kind of DNA, however, that lives outside the cell nucleus, which is unique in a manner tailor-made for an enterprising molecular evolutionist. It was here that Eve would have her genesis.

On a bright June morning in 1989, I was walking down a street in Berkeley, California, taking in deep drafts of eucalyptus-scented air and feeling blessed that so much research on human origins takes place here, demanding my close personal attention. I was in this paradise searching for Eve, or at least for someone who could tell me more about her. The article in *Nature* that had caused all the stir was authored by three people: graduate students Rebecca Cann and Mark Stoneking, and their famous mentor, Allan Wilson. Unfortunately, Cann, the primary author, had taken a position at the University of Hawaii, and was no longer within easy reach. Wilson was living up to his reputation as a pathologically press-shy hermit, routing my letters to his secretary, who sent me sheaves of reprints but no appointments. That left Stoneking as the only horse's mouth willing and available to be listened to. We had set up an appointment at his house at ten o'clock. I had been told that I must also find Linda Vigilant, a new star in Wilson's stable of graduate students, who was pushing the Eve hypothesis to new extremes of audacity. Conveniently, Stoneking and Vigilant—names that could not help but inspire trust—shared not only a lab but a life, including the West Berkeley bungalow I was trying to locate.

By the time I found the house, Vigilant was on her way out. We shook hands as she stepped around a clutter of camping equipment on the porch, dodged an incoming cat, and agreed to talk another time. Stoneking invited me to sit on the sofa while he made coffee. He was thin, with an appealing face saved from bland, blond handsomeness by a slight off-centeredness. I asked him about his background. He told me he had come to Wilson's lab after getting a master's at Penn State University, where he had been working on the population genetics of brook trout.

"I was hesitant originally to work on humans," he said, hesitantly. "Now I realize that for all the disadvantages, humans offer a number of advantages too."

"Such as?"

"Public interest, for one. I mean, I wouldn't have science writers sitting in my living room if I were still working on brook trout."

From Mark's painfully diffident manner, it was obvious that he did not enjoy publicity for its own sake. Like many scientists, he views the human interest in humans as an instrument, a ladle in which to scoop up gobs of raw data. Because people are more interested in themselves than they are in brook trout or wildebeest, there is a vast amount already known about the characteristics of living human populations. This has the snow-balling effect of making them that much more attractive to study once again. In genetics, virtually all of the accumulating data concerns the genes inside the human nucleus. And so it should, since that is where the vast majority of our genes reside. But a very small amount of DNA— thirty-seven genes in all—is lodged in the cellular structures outside the nucleus called mitochondria. And it was the peculiar nature of this mi-tochondrial DNA that was drawing science writers into Mark's living room, whether he liked it or not.

Mitochondria exist in almost all organisms: algae, geraniums, birch trees, fruit flies, eels, frozen Siberian mammoths, brook trout, science writers—everything alive, except for organisms without nuclei in their cells, like bacteria and viruses. The fact that each mitochondrion carries its own little 16,000-base-pair loop of DNA, separate from the cell's nucleus, has led biologists to think that billions of years ago mitochondria were free-living bacteria that established a symbiotic relationship with their primordial host organisms, providing a key function in return for some reciprocal favor from the host. Eventually, these independent bacteria were integrated into the life cycle of the host's cells and became one with them. A given animal species might shelter as few as a single mitochondrion in each of its cells, like the alga *Microsterias*, or as many as 500,000, like the giant amoeba *Chaos chaos*. Higher primates like ourselves typically harbor a more circumspect 1,700 or so mitochondria in each of our cells. Their function is metabolic. They act like microscopic furnaces, converting carbohydrates into energy to fuel the cells' life processes.

What makes mitochondrial DNA (mtDNA for short) so valuable to geneticists like Mark Stoneking has little to do with its value to his cells. First of all, he told me, mtDNA mutates at about ten times the rate of nuclear DNA, providing a more reliable indicator of change through time. Just as a watch with a second hand can tell time more precisely than one that ticks off only the minutes, a stretch of DNA with ten times as many mutations provides a more accurate molecular clock. A second advantage—the very heart of the whole Eve hypothesis—is its fortuitously simple mode of inheritance. When an egg is fertilized, the nuclear genes carried in the father's sperm mix with those of the mother to form the nuclear genome—or set of chromosomes—of the new individual. Your nuclear genes are a recombined legacy from your two parents, who in turn received it from your four grandparents, eight great-grandparents, and so on. Go back a few more generations, and you will find hundreds of people in each generation contributing to your nuclear genome. It is a patchwork quilt of recombined DNA scraps inherited from all of your ancestors, and it is impossible to sort out precisely who originally gave you which traits— blue eyes or curly hair or a birthmark at the nape of your neck.

Mitochondrial genes, on the other hand, are inherited only from your mother, who received them from *her* mother, who got them from *her* mother, and so on back through a chain of mothers. The mitochondria in your father's sperm were left outside the egg membrane at fertilization and are thus lost to history. If you are male, your direct mitochondrial lineage will die out no matter how many children you have, unless you have a sister who gives birth to a daughter and continues the line into the future. Unperturbed by the genetic reshuffling going on inside the nucleus with every generation, mitochondrial DNA serves as a sort of genetic tracer bullet, lighting up a route to the past.

"If you go back just five generations, you will find thirty-two people all contributing something to your nuclear genes," Mark explained. "But among those thirty-two people, only one contributed your mitochondrial DNA—the female along the maternal line. The same is true if you go back ten generations, or a hundred generations, or a thousand. There will still be only one female carrying all of your mitochondrial heritage. This allows you to track genealogies directly. You can't do that with nuclear genes or blood groups."

In spite of this linear inheritance, my mtDNA is not exactly the same as that of my thousandth-back great-grandmother. In the time between us, new types appeared through mutation. These mutations are not wholesale overhauls of the genetic structure but simply random alterations along the sequence of paired bases that make up the mtDNA genome—perhaps a base pair switched, or one spliced in or subtracted out. According to Stoneking and his colleagues, most of these mutations are "neutral" to natural selection. They had no effect on how the mitochondria functioned, so they did not render any of my maternal great-grandmothers stronger, weaker, more fertile or less.

If natural selection did not favor some of these mutations over others, *then the only reason one person's mitochondrial DNA differs from another's is simply the passage of time.* This is the crux of the Eve hypothesis. You and I might have the same woman in our maternal lines twenty generations back. Since then, however, your mtDNA lineage has been mutating off in one direction, mine in another. If we could measure the genetic distance between us—in other words, count the number of base pairs along our loops of mtDNA that differ from one another—all we would need to know would be the *rate* at which those differences were accumulating. We could then easily calculate how long ago we shared a common female ancestor. Try this with enough people from various ethnic groups around the world, and you come up with a common ancestor for us all: mitochondrial Eve.

You and I cannot measure genetic distances between people, but a well-equipped, well-trained molecular biologist can. The measurement technique first used in studies of human mitochondrial variation is called restriction mapping; it uses enzymes to cut DNA into telltale patterns of long and short fragments to generate a "map" of the particular individual's mitochondrial gene. Any two such maps can then be compared to see how much random mutation has altered them since the individuals shared a common mitochondrial ancestor.

In the late 1970s, Douglas Wallace, a geneticist in Cavalli-Sforza's lab at Stanford University, and his colleagues began to use restriction mapping on mtDNA to construct a human evolutionary tree, collecting mtDNA from a sample of more than two hundred individuals from five ethnic groups around the world (Bantu, Bushman, Oriental, Caucasian, and American Indian). Wallace was the first to prove that human mtDNA was inherited maternally, and soon his work was producing startling results. The first was the evident "shallowness" of the tree's roots. There was very little difference between any two people's mtDNA, suggesting that everyone's common ancestor had lived surprisingly recently. Second, the pattern of variation in mtDNA among the five populations often corresponded to geographic and ethnic origin. Third, the Old World populations sorted out into two distinct groups: African and non-African, with the African mtDNA types showing much more variation than the non-African ones.

Taken together, Wallace's results suggested that there had been a recent, single point of origin for the modern human race. There were two ways of determining *where* that origin had been, and these came in conflict. If the mutations in human mitochondria occur steadily through time, and if more mutations have accumulated in Africa than anywhere else, then Africa should be the homeland whence all populations have sprung. Wallace found, however, that the type of human mitochondria most closely related to that of other primates popped up in Asians more frequently than in other populations, suggesting instead that the point of origin was somewhere in Asia. This second way of determining the home of modern humans agreed nicely with Cavalli-Sforza's study of blood groups. When Wallace published his work in 1983, he opted for an Asian origin but hedged his bets by admitting Africa as a possibility.

"The reason I've been ambiguous is that I don't know the answer," Wallace later told me. "Only God knows."

God, and perhaps Allan Wilson too. In 1979, Rebecca Cann joined Wilson's lab and began to develop some earlier work on human mtDNA. Two years later Mark Stoneking arrived. It was their efforts, under Wilson's ever-present guidance, that eventually put the word "mitochondria" into the pages of *Newsweek*.

Their first task was to collect the human mtDNA itself—no easy feat. "We needed lots of samples," Stoneking recalled. "Blood doesn't have enough mitochondria in it, and livers and hearts weren't appropriate, for obvious reasons. So the best thing to use was placentas."

Many of the placentas were donated by local hospitals. Cann spent much of her time interviewing expectant mothers and persuading them to donate their afterbirths to the cause. Securing a supply of mtDNA representing Europeans and Asians was relatively easy, but the political and financial difficulties of obtaining fresh placentas from African populations proved insurmountable. Cann's solution was to use black Americans to represent people of African descent, a decision that later caused her a great deal of trouble. Meanwhile, Stoneking provided placentas from the South Pacific region, coordinating the collection through colleagues in Australia and New Guinea. By 1986, the two of them had amassed a random sample of 147 different individuals, representing populations from Asia, Europe, Africa, Australia, and New Guinea. The mtDNA from each was extracted and, using twelve different enzymes, cut into fragments. The fragments were then arranged into restriction maps and compared.

These genetic "texts" told a fascinating story. Out of the 147 samples, 14 proved to be identical to at least 1 other. But 133 of the mitochondrial types were unique, and their distinctiveness was curiously patterned. First of all, they were surprisingly alike. A sequence of 100 base pairs on any 2 differed by an average of less than one third of 1 percent. Second, with one important exception, the pairings showed little geographic coherence. A European woman's mtDNA might resemble an African's more closely than that of another European. Both of these observations supported one of Wallace's surprising conclusions: Humans all over the world are amazingly similar, genetically more alike, in fact, than two subspecies of gorilla in Africa that are separated by only a few hundred miles.

Cann fed her data into a computer program designed to determine the simplest way that the 133 human mitochondrial types could be arranged into an evolutionary tree. The most similar mtDNA types clustered together at the tips of the tree's branches, joining their more distant relatives farther down, and so on down in the pattern of branchings. Finally, the "root" of the tree—the hypothetical mitochondrial type in the past that was the ancestor of all 133 living types—was estimated to be midway between the 2 most divergent ones. Rooting the tree this way revealed a strange, lopsided pattern. All but 7 of the mtDNAs linked up into one enormous, radiating branch, bearing multiracial fruit from all five regions in the study. The 7 other types formed a separate group, with long branches running straight down toward the root of the tree, and all of them were of African descent. According to the computer-generated family tree, the root was in Africa. (See illustration on page 56.)

The seven "deep" African lineages showed fully twice as many mutations along their sequences as those in the larger branch. If the number of mutations was a function of time, the all-African group had presumably been around twice as long as the others. The simplest explanation was that all human races had originated in Africa and later some people had migrated out, while others had remained on the home continent. The ultimate point of convergence was the same for everyone: a single African woman whose mitochondria gave rise to all existing human types. There needn't have been anything special about this particular female, and she was certainly not the only one around at the time. But her mitochondrial lineage was the lucky one that persisted to the present, while all the others died out along the way.

If the Berkeley study had stopped there, simply offering high-tech support for the notion that humanity arose in Africa, it might have earned the anthropologists' murmured approvals. What made Eve anathema to them was her comparative youth. Cann, Stoneking, and Wilson figured that since Eve had lived, human mitochondria types had mutated and diverged from each other by nearly six tenths of 1 percent. Previously, they had calculated that human mitochondrial DNA changes at a rate of between 2 and 4 percent every million years. At this pace, it would have taken between 140,000 and 290,000 years of steadily accumulating mutations to account for the number of variations they found among living mitochondrial types. If you could run the molecular clock backward, you could watch the various mtDNA types becoming more and more alike, until they were finally all joined in mitochondrial Eve, about 200,000 years ago.

When a woman reaches the age of 200,000 years, it may seem impertinent to refer to her as astonishingly young. But evolutionary time does not flow so much as trickle, and old is a word reserved for those who can boast of six or seven zeros in their age. When Eve was supposed to have lived 200,000 years ago, the Homo erectus migration out of Africa was already very ancient history—eight times as distant an event to Eve as she is to us now, if the new dates for erectus in Java are correct. The common female ancestor of all non-African people would be younger still. The population Eve represents may have begun spreading out of Africa as early as 135,000 years ago, but it is more likely that it remained there for thousands of years, genetically isolated from other populations. Later, perhaps as recently as 50,000 years ago, this one group of descendants must have left Africa, pushing north and east.

On this point the latent conflict between the geneticists and the anthropologists erupted into open war. Unlike the earlier *erectus* migrants, Eve's descendants would not have arrived in a humanless landscape. By one estimate, some 1.3 million members of our genus were living on earth 50,000 years ago, a good many of them spread throughout Eurasia. These ancient Eurasians were not demi-apes living half the time in trees; they were our trusted European familiars, the Neandertals, along with other large-brained folk in Asia, people who knew how to bend fire to their needs, craft fairly sophisticated stone tools, and otherwise use their wits to survive millennia after millennia in harsh habitats. And, of course, these older Europeans and Asians were passing mitochondrial genes down through the generations, spinning on through their daughters and their daughters' daughters. Then suddenly, *all* of the lines stopped. At least there is no trace of them left in modern people. They have simply vanished from the genetic legacy of the human species. Or so said the Berkeley group.

"If populations representing the earlier inhabitants of Eurasia had contributed to the living human gene pool, we'd expect to find mitochondrial types around today that are five times more variable," Stoneking told me. "They just don't seem to be there."

Like a proper scientist, he outlined the alternative explanations for the lack of older non-African lineages before landing on the one he really believed. With only 147 people in the original study, perhaps they simply hadn't sampled enough people, and sooner or later an mtDNA type is going to turn up in a placenta from New Guinea or New Jersey that is miles away, genetically, from the others outside Africa. But if the 147 people represent a random sample of humanity, the chances of such a new type appearing are very slight, and approaching zero as the work continues. By now, thousands of individuals have donated their mitochondria to the cause of one study or another, with no ancient non-African types turning up.

Another possibility is that the non-African people *did* contribute mitochondria to future generations, just like the migrating Africans, but for some reason all the people carrying these older Eurasian mtDNA types subsequently died out, leaving no children. This is an intriguing idea, but it runs head-on against one of the bulwarks of the whole hypothesis: the assumption that the differences between one human mitochondria type and the next are neutral to selection. If the exact sequence along the mitochondrial genome makes no difference to evolutionary survival, why would all of the old Eurasian types die out, and not the ones from Africa?

"The best explanation for why you don't see these ancient non-African mitochondria is that they are not part of the modern human lineage in the first place," Stoneking concluded. "In other words, the African people replaced the residents without interbreeding with them."

"Wait a minute," I said. "You tell me that the residents didn't contribute mitochondrial genes. But that accounts for just a tiny percentage of our genetic makeup. How do you know they didn't contribute *nuclear* genes?"

"We don't know for sure. But we do know that when a dispersing population with some kind of cultural advantage comes in contact with a resident population, the invading males usually mate with the resident females, and not so much the other way around. Remember, it's the females alone who are contributing the mitochondria. And since we don't see any non-African mitochondria, it looks like there wasn't any mating going on between the resident females and the migrant males—at least none that produced a lasting genetic legacy."

"Why not?" I asked. "Humans aren't known for being choosy about mates. Especially males away from home."

Mark shifted in his seat. "I don't know. Maybe there was too much physical difference between the two groups for mating to take place."

"You mean they were a different species?"

"It's a possibility. Or maybe the indigenous populations were wiped out."

My questions were not quite fair. As revealing as the mitochondria may be, they say nothing about the *how* and *why* of this massive population replacement. On their own, the mitochondria cannot even say much about the *who*. You can stare forever at a restriction map of mtDNA and never learn what Eve looked like, whether she had big browridges or high cheekbones, or whether her descendants were warlike or sat around all day smiling at each other. You cannot even be certain that Eve was a modern human; she might have been a more primitive *Homo sapiens* whose lineage only later evolved into fully modern anatomy. As Stoneking and his colleagues pointed out in their *Nature* article, however, there is intriguing fossil evidence that modern anatomy appeared in Africa earlier than anywhere else, suggesting that the final evolutionary event in our lineage's five-million-year history did, in fact, take place in Africa. The simplest explanation is that either Eve herself, or her purely African descendants, had evolved into *Homo sapiens sapiens* before they left the home continent.

Whatever advantage these African moderns possessed, it allowed them to outcompete, outreproduce, outfight, outthink, and otherwise outsurvive all the other human beings they encountered as they spread over the earth. No wonder the idea upset the anthropologists. If this startling scenario were true, the century-old quarrel over the fate of the Neandertals could be safely tucked into science history. Neandertals could not, in any meaningful way, be ancestral to any living population. Nor could any other human fossil type outside of Africa earlier than the imagined migration. This would include Java Man, Peking Man, and many other museum treasures dear to the hearts of anthropologists. They must all have become extinct, *no matter what they looked like.* As was the case with *Ramapithecus*, the geneticists' argument cleanly and cruelly polarized the issue. If the mitochondria were correctly interpreted, any theory of modern human evolution proposing a non-African origin was flatly, finally, irrevocably wrong. And the memory of their past success was lending the lab-bound scientists a confidence bordering on arrogance.

"Some people don't like our conclusions," sniffed Allan Wilson in *Science* magazine, "but I suspect they will be proved wrong again."

"Some people" in this case happened to include the majority of anthropologists alive. I asked Stoneking whether he was as sure of success as his mentor.

"I don't see why it has to be looked on as a contest," he said. "I'm not trying to prove anybody wrong. I'm simply trying to test the validity of a hypothesis involving variation in mitochondrial types."

"And how valid does it look to be?"

He fidgeted, as if he didn't like to be pinned down. "Becky and I later took into account every source of error we could think of and came up with a time range of fifty thousand to five hundred thousand for Eve. I'm ninety-five percent sure of that. And the latest work makes me more confident, not less."

"But that's a range of four hundred and fifty thousand years," I said. "How useful is it to be positive about something so vague?"

"I've always said that the timing was the softest part of the story."

"But the timing *is* the story, isn't it?"

"Not entirely. And if you want to know what vague means, you should talk to the anthropologists. At one point at a meeting on modern origins in Cambridge, somebody put up a slide of a fossil and went around the room asking, 'Is this a modern human, or not?' They got as many answers as there were people in the room. That's where the paleoanthropologists are at. That's what they have to deal with."

On the other hand, most of those anthropologists were pointing out that believing in Eve leads to an absolute impossibility: the total, global replacement of the population of two continents by invading humans who somehow managed to avoid sex with the equally human inhabitants of those continents for thousands of years. Since this *could* not have happened, there must be something wrong with the geneticists' argument, for all its high-tech appeal that says that it *did*.

I thanked Stoneking and headed back up the hill to the Berkeley campus. A couple of weeks later, at the University of New Mexico, I visited Erik Trinkaus, one of the world's leading authorities on Neandertal fossils, and asked him what he thought of the Berkeley gene work.

"That genealogical tree that they published in *Nature* looks a lot like a rose trellis," Trinkaus said. "Which is really fitting. The mitochondrial evidence smells sweet from distance, but when you get up close, it's nothing but thorns."

I got my first full taste of Eve's prickliness over a burger and fries. My lunch companion was Milford Wolpoff, University of Michigan paleoanthropologist and leading champion of the anti-Eve movement. I remember this lunch for its largeness: great fat burgers on bloated buns, steak fries stacked like lumber on an oversized plate, Wolpoff's big face and grin and the swelling billows of discourse that came pouring over from his side of the booth. Wolpoff may be the most often quoted anthropologist in the world, a position that has fallen to him partly because of the depth and scope of his research, and partly through the vigor, persistence, and sheer volume of his opinions. Lately, he has been quoted more often than ever. The "multiregional continuity" theory he supports stands in direct, hostile opposition to the Out of Africa theory of modern human emergence, so neatly buttressed by the mitochondrial work. When a dissenting opinion is needed for some press report, he is eager and ready to offer a juicy one. It was Wolpoff who came up with the biting phrase "Pleistocene holocaust" to characterize the implications of the Eve hypothesis.

"In one article, Rebecca Cann said that her study proved how all mankind was one great big brotherhood," he told me, chomping into his burger. "Boy, did that tick me off. She adopts this 'more liberal than thou' tone, and ignores the fact that the only way her scenario works is through violence. It amounts to killer Africans with Rambo-like technology sweeping across the world and obliterating everybody they meet. Hardly my idea of universal brotherhood."

"But nobody who supports the Out of Africa model believes that the replacement *must* have been violent," I protested. "In fact, they say that such phrases as 'Pleistocene holocaust' are deliberate misrepresentations of their viewpoint."

"It is *not* a misrepresentation. It's simply taking their argument to its logical extreme. We know our species. This is the species of Pol Pot, of Adolf Hitler. You tell me—how could this total replacement have happened peacefully?"

I assumed the question was merely rhetorical and waited, pencil in left hand and steak fry in right, for him to proceed. But he seemed actually to want an answer. On the plane I had tried to digest one explanation for the replacement, proposed by demographer Ezra Zubrow of the State University of New York at Buffalo. Zubrow had shown statistically how a new population arriving in Europe in the Upper Pleistocene could easily have replaced the Neandertals without physically killing them off. The gist was that the new arrivals had only to live a little longer. If the modern humans' mortality rate was just slightly lower than the Neandertals, the latter would have faded peacefully into mathematical oblivion in a single millennium.

"Well, what about what Zubrow is saying?" I asked, driving home the question with a shake of my fry.

The suggestion was dismissed with the wave of a big hand. "Zubrow has a simple computer model based on the premise that two populations living next to each other didn't interact. Pretty strange behavior, if you ask me."

"But there are plenty of animal populations that replace other populations without violently attacking them," I said. "Look at the squirrels in Europe. When the grey squirrel was introduced in the 1920s, it completely replaced the European red squirrel, and nobody calls that a holocaust."

"If we were talking about just one place, one region, I'd agree with you. But across the globe? Analogies with squirrels can't work on a global level."

"Well, what if the modern humans from Africa were carrying a pathogen of some kind?"

"Show me a case where pathogens carried by an invading population completely wiped out an indigenous population, and I'll buy it. But you can't. The Eskimos were devastated by measles, but they weren't wiped out. They're still up there, interbreeding with the invaders." Wolpoff shook his head. "I'll buy pathogens, aided by bludgeoning. A heavy dose of bludgeoning. Face it. Wilson and his people are stuck with violence to make this thing work. I'm just glad it's them and not me."

Wolpoff's intention was to persuade me not that the Out of Africa migration happened violently, but that it did not happen at all. He passionately believes that the fossil evidence of human evolution clearly demonstrates that the living races of humankind came not from Africa but from ancestors who lived in the same regions as much as a million years ago. Living Asians resemble ancient humans found only in Asia, while aboriginal Australians bear the telltale features of fossil skulls found in Indonesia, whence their ancestors came. Caucasians are the descendants of the Neandertals, who far from going extinct are alive and well and eating burgers in places like Ann Arbor, Michigan. If all of these non-African ancients had been obliterated by migrating hordes of Eve-spawn from Africa, these regional peculiarities in anatomy would have been wiped away. Since, on the contrary, regional traits are clearly evident in modern populations, it follows that the Out of Africa migration did not occur, *no matter what the mitochondria look like.*

That, in a nutshell, is Wolpoff's vigorously promoted view. It is similar to the multiregional continuity models for modern human origins put forth earlier by Franz Weidenreich and by the Harvard anthropologist Carleton Coon. But while Wolpoff is happy to acknowledge his debt to Weidenreich, he grimaced when I mentioned Coon. In his popular book *The Origin of Races*, published in 1962, Coon suggested that the various races of mankind crossed the "*sapiens* threshold" into modern humanity at different points in time, with the Africans crossing last—which accounted for their more primitive cultural state.

"Carleton Coon believed that the races of mankind actually evolved from *erectus* in isolation from each other," Wolpoff explained. "That is not what we are saying at all, and it makes us very angry when people misrender our arguments into sounding like Coon's."

Wolpoff's version of regional continuity maintains that the races were independent, but not entirely isolated. A certain number of genes flowed

across and among the various regional populations of humankind all the
way back to *erectus*, linking the parallel racial branches to some extent.
Together with shared changes in behavior—like the invention of more
sophisticated tools, which would relieve the need for heavy chewing anatomy—this gene flow led to the gradual "sapienization" of all races, even
as the adaptation to local conditions kept their regional characteristics
distinct.

Needless to say, a unifying flow of genes back and forth across regional
boundaries is nothing like the invasion of *people* from a particular point
of origin as envisioned by the Eve hypothesis. Looked at from a broader
perspective, it is, in fact, just the opposite. Among evolutionists, a twenty-
year-old controversy rolls on between those who see evolution progressing
gradually through time, with no clear boundaries between species, and
those who think that long periods of stability are punctuated by sudden
changes from one species to another. Advocates of the "punctuated equilibrium" model, like Stephen Jay Gould at Harvard, welcomed the Eve
hypothesis with open arms. No wonder: The idea of a single origin for
humanity in Africa and a subsequent replacement elsewhere dramatically
fulfilled their theory's predictions. Wolpoff's belief that modern humans
evolved slowly out of a flowing pool of genes worldwide is firmly allied to
the opposing "gradualist" view of evolutionary progress. In his opinion,
the Eve hypothesis fits the punctuated equilibrium model only because
both ideas are equally flimsy.

"Eve is just another bandwagon," he said. "These come and go. Some
of them catch on, some prove to have flat tires."

Wolpoff proceeded to attack the tires of the Eve bandwagon with
gusto. His first point of assault was the pace of the alleged "molecular
clock." In the *Nature* paper, the Wilson group had used a mutation rate
of 2 to 4 percent per million years to calculate Eve's age, a rate based on
a calibration to the dates of colonization for New Guinea, Australia, and
the New World suggested by the fossil record. When the paper was published, the consensus from the fossil evidence held that Australia was
colonized about 40,000 years ago. If that is true, all the different mitochondrial types found among native Australians must have diverged from
one another in the last 40,000 years. The geneticists knew from their
restriction maps how much divergence there was among aborigines, and
dividing that amount by 40,000 gave them the rate. Calculations from
New Guinea and the Americas came up with roughly the same range.

This all sounded fine when Mark Stoneking explained it to me, but Wolpoff treated it with scorn.

"Their rate calculation is littered with assumptions," he said. "We don't really know when New Guinea or Australia was colonized, but it was probably a lot longer ago than forty thousand years, maybe as much as a hundred thousand." (The latest dates for the earliest occupation of Australia hover around 60,000.) "We don't know that everybody arrived there at the same time, or if some people migrated back to the mainland again. All these things could screw up their interpretation of the rate of the clock."

Wolpoff told me another prominent geneticist, Masatoshi Nei of the University of Texas, had worked on the rate of human mitochondrial mutations more carefully. He came up with a much slower rate of change—about three quarters of a percent per million years, instead of an average 3 percent. If mitochondrial evolution in the human lineage is really proceeding at this much slower pace, it would have taken far longer to accumulate the amount of variability the Berkeley group found in their sample, something like 850,000 years. This was a figure Wolpoff could easily abide.

"This means Eve was a *Homo erectus* in Africa," he said. "So what's the big deal? The mitochondria are simply underscoring what everybody already knows: that the human lineage, in the form of *erectus*, migrated out of Africa a million or more years ago. It replaced no one, because there were then no hominids outside of Africa to be replaced, violently or otherwise. If it weren't for Wilson's group, there wouldn't *be* an Eve hypothesis. None of the other geneticists agrees with it."

"None" is a bit of an exaggeration. Before coming to Michigan, I had talked to other geneticists, including Nei himself, and the difference between his rate and the Berkeley one was not quite as large as Wolpoff was suggesting. Nei's rate of about three quarters of a percent was a measurement of the amount of mutation that an average mitochondrial gene sequence will undergo in a million years. The average rate of 3 percent used by the Berkeley group to pin Eve's age at 200,000 years measured the divergence between any two lineages. Thus it made sense that the Berkeley rate was much faster. Nei's rate is like a measurement of the distance traveled by a car going at a fixed speed in one direction. The Wilson lab, in contrast, figured the distance accumulating between that car and another one starting from the same point, but traveling in the opposite direction. To compare them properly, you have to double Nei's rate.

"I've gone over and over this with Wolpoff," Mark Stoneking had told me. "He just refuses to get it."

Doubling Nei's rate makes Eve's age about 430,000 years. This is closer to what Wolpoff would like, but not nearly old enough to place Eve among the *Homo erectus* migrants who left Africa at least half a million years before—and as much as a million and a half, judging from the new *erectus* dates in Java. I pointed out to him that it was within the 50,000- to 500,000-year range that Stoneking and Rebecca Cann are putting forward. He was not impressed.

"A range that big is useless," he remarked. "Would you set your watch to a clock that is accurate to within four hundred and fifty thousand years?"

According to Wolpoff, such arguments over the rate of the clock don't matter anyway, because the clock does not even exist. The clock concept rests on the assumption that mutations occur steadily and are the *only* source of variation among mitochondria. Wolpoff sees at least two wrenches in the clockworks, both of which reduce the amount of variation among mitochondrial genes, producing a common ancestor who appears much younger than she really is.

The first wrench is natural selection, which could favor some mutations and eliminate others. "If for some reason one mitochondrial type is slightly more advantageous than others, it could sweep through the whole population," he says. "This would eliminate other types, so you end up with less variability, making the whole mitochondrial population look young."

The Berkeley people, however, were insisting that the mutations they measured in mitochondria occur in places on the gene that are neutral to selection. They aren't within sequences that form the blueprint for proteins, so they don't affect how an individual adapts to his or her environment. Furthermore, the amount of variation they found in their samples agreed fairly well with the amount that *should* have been there, in a statistical sense, if the mutations were indeed immune to selection. Wolpoff dismissed this.

"There are all sorts of diseases now known to be linked to malfunctions in mitochondria," he said. "They attack your muscle tissue, your heart function, nerves, kidneys, liver. They include a kind of epilepsy, and a condition of the optic nerve that causes blindness in young people. And with all this associated pathology, they say changes in mitochondria can't affect the individual's survivability. Give me a break."

The other wrench Wolpoff threw in the works is the phenomenon called random lineage loss. In each generation, some women will have only male children, or no children at all. When this happens, their mitochondrial lineages will be lost to history. Eventually there *must* come a time when only one lineage will remain. Wolpoff likes to use the analogy of surnames to make this point. Like mitochondrial genes, surnames are inherited through only one parent, in this case, the father.

"Let's say you are an anthropologist sometime in the future," he told me over dessert. "You go to West Chicago and find fifteen hundred families living there, all named Gablinski and all descended from immigrants from Gdansk. It will look like these thousands of Gablinskis were the descendants of a single Gablinski couple who came over from Poland. Through hard work they succeeded, becoming more numerous through the generations and eventually replacing all their neighbors. But what if there had been not just the Gablinskis who came to West Chicago, but thousands of Poles with thousands of different names? Every time one of these families had a generation without sons, you'd lose a family name, right? Eventually, you'd come down to just one name left: Gablinski! If you buy the Out of Gdansk theory like the people who jump on the Out of Africa bandwagon, you'll assume that they all came from that one original couple. But you'll be making a mistake."

Wolpoff's analogy is not just hypothetical. John Avise, a geneticist at the University of Georgia who also works on mtDNA, has pointed to the case of Pitcairn Island in the South Pacific. In 1790, six mutineers from H.M.S. *Bounty* arrived on the tiny island, bringing thirteen Tahitian women with them. Few others have ever gone there to live. Recently, a population of fifty people on the island shared only four surnames, and one of these was that of a whaler who had later settled on Pitcairn. Thus *in only six or seven generations*, 50 percent of the six original surnames had already disappeared. After a few more generations, only one will remain.

In this respect, surname and mitochondrial inheritance are identical. Geneticists like Stoneking freely admit that random loss is operating in their hypothesis—in fact, it is the very process that allows them to trace everyone back to a single parent. But they believe that should make no difference to that mitochondrial parent's age. To Wolpoff, on the other hand, it renders Eve meaningless. To make a mitochondrial clock tell proper time, you have to subtract the shrinking effect of random lineage loss on variability from the multiplying effect of new mutations. But you cannot know how much influence random lineage loss has exerted unless you know the entire population history of the species—how many humans were reproducing at every point in time. Obviously, that kind of information is not available. Thus all that Eve really amounts to is the lucky mitochondrial gene type that happened to survive random lineage extinction and be spread over the world, not by individuals but by the ebbs and tides of gene flow through relentless millennia.

By the time lunch was over, there was little air left in the tires of the Eve bandwagon. Wolpoff also attacked Rebecca Cann's use of African Americans instead of indigenous Africans. "Most African Americans carry a substantial number of European genes," he told me. "By some estimates, as much as twenty percent of their genome is white. How well can they represent true African lineages?" He questioned the Berkeley group's method of rooting the tree because it assumed that all mitochondrial types are mutating at the same steady rate—which may not be true. He questioned the choice of the restriction enzymes Cann used to create the raw data. By the time the check arrived, my faith in Eve was shrinking under the table, as if I were listening to someone recount the sordid infidelities of a woman whom I was just about to take as a bride.

"What I'm saying is there isn't a snowball's chance in hell that the Eve hypothesis is right," Wolpoff said as we got our coats. He gave me a slap on the shoulder and a toothy smile. "If you really want to know where modern humans come from, go look at some fossils."

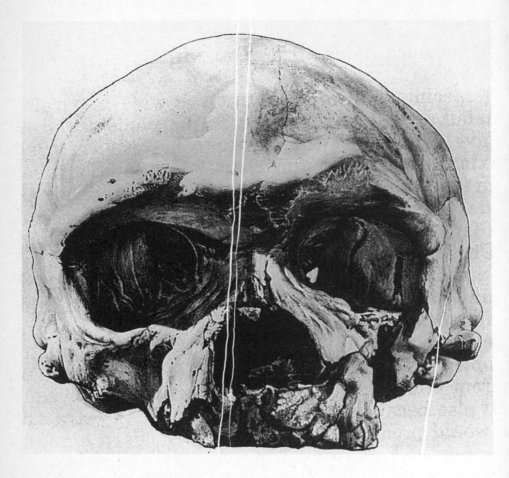

The Dali skull, an archaic Homo sapiens *from China*   DRAWING BY KAI NAI KBANLUNG,
COURTESY GEOFFREY POPE, *YEARBOOK OF PHYSICAL ANTHROPOLOGY*, VOL. 35, 1992, REPRINTED
BY PERMISSION OF WILEY-LISS, INC.

Chapter Four

ARGUMENTS OVER A WOMAN

You know my method. It is founded upon the observance of trifles.
—SHERLOCK HOLMES, IN "THE BOSCOMBE VALLEY MYSTERY"

Go look at some fossils! It sounded so simple. But even if the dry bones of the dead do hold the answer to the mystery of modern human origins, prying the secret out of them would be no easy matter. If I could place Wolpoff's limpid imperative in some kind of syntactical centrifuge and give it a whirl, three messy fractions would quickly parse out. First, consider the object of the sentence: some fossils. Which ones? The fossil record for modern human origins is much more diverse than any we have from earlier periods of human evolution, embracing dozens of skulls scattered from the tip of southern Africa to northern China and Australia. Any responsible answer to the mystery of modern origins has to make sense out of the whole global lot.

Everyone agrees that we evolved from more primitive ancestors represented by fossils and traditionally called *Homo erectus*. The disagreement erupts over where, when, and how. From the critical time period, roughly 400,000 to 100,000 years ago, there is an ambiguous clutter of bones that people have reluctantly agreed to group together and call "archaic *Homo sapiens*," because they don't know what else to call them. It is very unusual in taxonomy for a formal Latin binomial to need an ordinary adjective attached in front of it like a snowplow to do its work, which is to gather together and define a given group of fossils. These fossils show some of *erectus*'s primitive features, such as long, low skulls, thick cranial bones, broad skull bases and heavy browridges. They also share a general trend toward more modern human form. Most important, the braincases are larger. A number of more esoteric cranial and facial features here and there may also signal modernity.

Unfortunately, the mosaic of primitive and modern features is so bafflingly distributed among the various specimens and regions that it is next to impossible to treat "archaic *Homo sapiens*" as a real taxon—a biologically meaningful grouping with some of those nice clean French edges around it. It is more like a bushel basket into which you throw everything that is neither clearly *erectus*, nor obviously modern *Homo sapiens*. The convention of tossing all the Neandertals into the basket as well, calling them "late European archaics," even though they have distinct features that by rights should separate them from the other archaics scattered around the Old World, further confuses the matter. Some scientists also keep a separate bin for the more us-like fossils in the group, producing the exquisite oxymoron "archaic moderns."

So much for the simple object of Wolpoff's imperative. Now look at the verb: go look. You and I can examine a couple of skulls and quickly distinguish differences between them. One may have big browridges, another, more modest ones. Or perhaps the brows differ in shape, forming a straight thick bar across the forehead on one specimen, shadowing the eye sockets in two heavy arches on another. A good deal of human paleontology amounts to eyeballing various features of a dead hominid and comparing them to those on other dead hominids. But what do the differences mean? Are two skulls different because they represent different species of hominid, or were they simply a male and a female, or a couple of individuals from the same population who happened not to look much like each other? Human beings today come in all shapes and sizes. How do we know when a different shape in the past meant a different kind of human entirely?

To try to sort through these questions, experts seek guidance from a variety of techniques and theoretical approaches aimed at revealing the evolutionary logic that put together a skull in the first place. These techniques bring into relief the patterns running between individual features of the specimen, giving a clue to the specimen's ancestry and its relationship to other fossils from the same time period. Not surprisingly, the method of looking influences what the expert sees. Some tend to underscore differences, so that the viewer sees more variation among individual fossils, more species, and cleaner edges separating them. Other techniques emphasize similarities, giving the impression that one form of hominid flows gradually through time and space into another. Traditionally, scientists who look at things this way are called "lumpers," while those who draw clean lines are called "splitters." The ultimate lumper would put everything from Lucy to your own grandmother in a single hominid species. The splitter of all splitters, on the other hand, would have a separate species name for every fossil known.

The messiest part of Wolpoff's imperative is the subject: the people doing the looking. When he said, "Go look at some fossils," the implied subject was myself. But what he really meant was "Let's you and me go look at some fossils." Which is exactly what we did in his spacious laboratory, where shelves full of fossils and casts climbed to the ceiling, industrious grad students bent to their work, and computers purred at the ready, to help with all the looking. But as should be clear by now from the history of paleoanthropology, the same piece of ancient bone can reveal quite separate truths to two different, equally qualified experts. As it happened, I had already met some of the fossils that figure in the modern human origins mystery. From Wolpoff's point of view, I could not have picked a worse person to show them to me.

In the summer of 1971, when Eve was just another Bible story, a twenty-three-year-old East Londoner named Christopher Stringer set off for the Continent in his Morris Minor. Stringer had decided he would go look at some fossils. In his pocket he had three hundred and fifty pounds, hardly enough to keep him in hotels for the four months that he would be gone. But he was more than willing to sleep in youth hostels, under park benches, or curled up on the backseat of the car. He ate little, saving his money for the gas needed to carry the Morris over three thousand miles, crisscrossing Europe on both sides of the Iron Curtain. It was the expedition of his dreams.

When he was nine years old, a BBC radio broadcast on "How Things Began" had planted a fascination with Neandertal "Cave Men" in his imagination. Soon he was tagging along with a teacher on local fossil-hunting excursions and drawing skulls. His parents, working people, thought this a bit strange, but they knew he was bright and were sure he would eventually direct himself to more serious things.

"By sixteen, I was on track for medical school," Stringer says. "And then somebody handed me a college catalog, and I saw courses listed under 'Anthropology.' Until that moment, I had no idea that you could actually do this for a living." To his parents' shock, he dropped his medical-school plans and applied immediately to the Anthropology Department at University College, London. Six years later, he was holding the original Neandertal skullcap in his hands.

Stringer's plan for the trip was the model of a graduate student's ambition. Backed by a newly fashionable measurement technique called multivariate analysis, he planned to examine as much of the evidence for modern human origins in Europe as he could get his calipers around, and to return to London with the answer to the oldest mystery in human evolutionary science. Were Neandertals the immediate ancestors of Europeans, as Loring Brace was eloquently arguing? Or was there a Pre-Sapiens Man still lurking undiscovered in museum cabinets, something that would prove that the modern human lineage was rooted separately, deeper in the past? Or was the truth somewhere in between? Given the Solomon-like judgments of multivariate analysis, Stringer was sure a solution was finally at hand. "All I had to do was gather the data, feed it into a computer when I got home, and the truth would emerge," he remembers.

Bearing some well-fingered letters of introduction, Stringer traveled from Brussels to Bonn, to Brno in Czechoslovakia, to Vienna, down to Zagreb in the former Yugoslavia, then across the Pindos Mountains of Greece to Italy. In Rome, a thief broke into the car and took his clothes and his glasses, plus a fresh human skull he carried around for comparisons. But the thief disdained his accumulating pile of data, so Stringer was able to keep on going. His long hair, beard, and uncrisp toilette did not always make the best impression, especially in Eastern Europe. But most curators received him with at least cordial indifference. With a little pressing they unlocked their storage cabinets to give him a look, a touch, or best of all, a chance to measure.

Like every liberal-minded anthropology student at the time, Stringer believed the Neandertals had been given a bum rap. "Loring Brace's articles were a powerful influence, and at first I believed totally in the Neandertal ancestry idea," he says. "But I wanted to keep an open mind about it. And the more fossils I looked at, the more I could see that there was, in fact, a huge metric difference between the Neandertals and the modern humans of Europe. After a while, my open mind started to close."

Whatever fossils he examined, from Spy in Belgium to the abundant bones of Krapina Cave in Yugoslavia to the skull in the circle of stones of Monte Circeo near Rome, Stringer could find no clear support for the idea of a Neandertal ancestral role in the evolution of modern Europeans.

But neither could he see signs of a Pre-Sapiens Man. In France, Henri Vallois, Marcellin Boule's protégé, was holding up skulls from Swanscombe and Fontéchevade as evidence of such a separate, non-Neandertal link to the past. But in the cold light of the new analytical technique, those fossils appeared either Neandertal-like or merely generalized primitive types that could not decide the issue one way or the other. Exhausted, Stringer arrived in Paris, his last stop, only to discover that some of the fossils he most needed to see were beyond his reach. Every day he went to the Musée de l'Homme, and every day he was told by the curator that the fossil he wanted most to examine was being cast, cleaned, reconstructed, or used by another investigator. He worked all day with what scraps he was given and slept in a park. Winter was approaching, and he was ready to give up.

Among the specimens on loan to the Musée de l'Homme was an enigmatic skull recently found in a cave called Jebel Irhoud, on the Moroccan coast. Other investigators had referred to it as an "African Neandertal," so Stringer wanted a look at it. But the curator of the collection told him the skull had been returned to Morocco. Then one morning, while the curator was out of the building, the French paleoanthropologist Yves Coppens approached Stringer's worktable and set the Irhoud skull down in front of him. It had been in a safe in the next room the whole time.

"Here," whispered Coppens, "you have two hours."

It was a decisive encounter.

"I could see right away that this thing wasn't anything at all like a Neandertal, at least not in the face," Stringer says. "The face was very broad and flat, with a short nose."

In other words, the face returning Stringer's stare of astonishment was much closer to that of a modern human being than any other Neandertal he had ever seen. When he turned the skull around, however, the cranial vault looked neither modern nor Neandertal-like, but more primitive than either. Jebel Irhoud had not yet been well dated, so it was hard to know what to make of this particular blend of the new and the old. It was years before Stringer recognized the paleontological hot potato he held in his hands. But the seed was planted.

"If it weren't for Coppens showing me that skull, I might never have formed the view that Africa held the key to understanding the origins of modern humans."

* * *

By the time I met Chris Stringer, almost twenty years later, he was long past begging a look at fossils. Still bearded, but with his hair trimmed short, he was a senior paleontologist at what was then the British Museum (Natural History)—now called the Natural History Museum—overseeing one of the finest collections of human fossils in the world. Today, if you want to study a hominid in the British Museum collection, Stringer is the man you must appeal to, a circumstance assuring him a warm welcome at every other museum in the Western world. Though he is now well known and respected, his views on modern human origins were once considered near the fringe. Then the Heavens looked kindly on him and blessed him with Eve. When I first encountered him, he was about to give a seminar at New York University. Surrounded by a clutch of admiring students, he was helping the staff identify a boxful of obscure casts that had turned up in a closet, unlabeled. I could almost hear the mental oohs and ahs as he wiped the dust from the casts and, in his gentle Cockney lilt, gave each back its lost name: "Now this one 'ere is what we call Egbert, which is to say, the boy from Ksar' Akil."

Stringer insists, justifiably, that the Out of Africa scenario he has been pushing for years stands up well even without the Eve hypothesis. Nevertheless, it did his theory no harm to have it bolstered, out of the blue, by evidence from a totally independent, supposedly more objective science, put forth by a MacArthur "genius" grant winner and splashed across the cover of *Newsweek*. The year before we met, he and his British Museum colleague Peter Andrews had published a highly influential article in *Science*. Reviewing both the genetic and the fossil evidence, the article established the Out of Africa model as the new paradigm for modern human origins. If Milford Wolpoff was the most frequently quoted paleoanthropologist in the world, Chris Stringer was suddenly not far behind, usually in the next paragraph. And when Stringer spoke, Wolpoff bristled.

"Milford doesn't usually like my reading of the fossil record," he explained to me with typical British understatement.

"It's been acrimonious between me and Chris for years" was Wolpoff's way of putting it. "And either he is going to start backing off, or things are going to get a lot more acrimonious."

The starting point for the feud is a common battlefield in paleoanthropology: Both investigators ask the same provocative question, dip into the same pool of evidence, and come back with absolutely opposing answers. Stringer's Out of Africa scenario is the latest rendering of what Harvard anthropologist William Howells called the Noah's Ark model for modern human origins. *Homo sapiens* arose in only one place—in Stringer's version, Africa—later spreading out to replace all other archaic human populations. According to this theory, the racial differences apparent in living populations appeared very recently, probably within the last twenty thousand years. If Stringer is right, the general pattern of modern human anatomy should show up first in Africa, and in other parts of the world only after the alleged Out of Africa migration took place, beginning perhaps 50,000 years ago. Intermediate types, linking modern humans with their archaic past, will likewise be found only in Africa. Elsewhere, the fossil record should show an abrupt shift, registering the extinction of the local inhabitants and the appearance of the unrelated, African-derived people into their area. Thus, in Europe, it was Neandertals out, Cro-Magnons in, all at once, with no fossil forms showing a transition between them.

Wolpoff's model, a version of what Howells called the Candelabra theory, is contrary in virtually every respect. Modern humans arose not in one place but ubiquitously throughout the Old World; there was no need for Africans or anyone else to move in and import modernity to a given region, because it was already there, gradually developing out of the archaic stock of that particular region. Thus, intermediate forms between the archaics and moderns should be found everywhere, not just in Africa. Nobody is replaced, and nothing "migrates" but the flow of genes and culture, maintaining the unity of the "great human family" in spite of racial differences between regions, which are traceable all the way back to *Homo erectus*.

What might seem like the makings of a lively debate for most scientists has evolved for these two into something harsh and personal. Wolpoff thinks the trouble may have started in the early 1980s, when he was asked to review one of Stringer's papers before publication.

"At one point in my review, I meant to refer to 'Stringer's disparate analysis'," he told me. "But in the text it came out 'Stringer's desperate analysis.' It was just a typo. I really did mean disparate! But Chris was all upset about it."

An exchange of vitriolic letters followed. Wolpoff launched a more public attack at a conference on modern human origins held in Zagreb in 1988, shortly after the *Science* article by Stringer and Andrews had

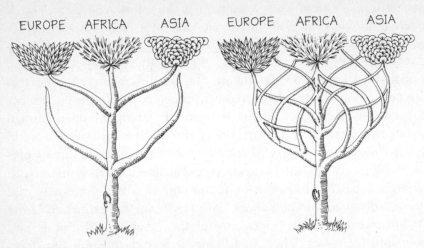

EUROPE   AFRICA   ASIA   EUROPE   AFRICA   ASIA

*Two contrasting views of modern human origins. On the left, the "Out of Africa"
replacement theory, suggesting that modern human populations all derive from a
relatively recent migration of people from Africa. Earlier non-African populations like
the Neandertals died out, like the dead branches in the figure. In contrast, the
"multiregional evolution" tree on the right suggests that people in all parts of the Old
World trace their ancestry back within their own region at least a million years, the
evolution of the species kept in sync by the flow of genes from one population to
another.* DAVID DILWORTH

appeared. Stringer was scheduled to give the last talk of the day, following
Wolpoff.

"Milford was furious that I had been put last on the program, where
I could get the final word," Stringer told me in New York. "So one of
the conference organizers asked me if I would mind if we switched our
talks so that his would follow mine. I said no problem, if it means that
much to him. I'll never do that again. It was all so he could do his par-
lor trick."

In his talk, Stringer presented new arguments for his replacement theory and then joined the audience. Wolpoff mounted the podium. One of
his opening slides showed a picture of Piltdown Man, the fake that had
fooled the English anthropological establishment for forty years.

"The British Museum has a long history of contributions to paleoanthropology," he said, his voice heavy with sarcasm. "Next slide please."

"I thought that was a really low blow," Stringer remembered. But Wolpoff insists it was all meant as a joke, or at most a gentle warning. In the
Science article, Stringer and Andrews claimed that the regional continuity

model depended on "parallel evolution" to account for the emergence of modern humans independently in different geographic regions, neglecting to emphasize the importance of gene flow in Wolpoff's scenario. Parallel evolution was the mechanism Carleton Coon had used to explain how the modern human races had evolved at separate and distinctly unequal rates. The use of the term embittered Wolpoff and his colleagues, who thought they had taken great pains to divorce their theory from Coon's debunked, racist notions.

"The Piltdown thing was simply a signal to Chris that things had gone too far on his side, and he should stop misrendering our hypothesis," says Wolpoff. Stringer denies that he "misrendered" anything at all. Wolpoff simply has to face the implications of his own position, he says, one of which is a certain amount of parallel evolution.

For all the acrimony, the relationship has not been one long mutual smear. Later in the Zagreb conference, there was a point where Stringer appeared to be on the brink of extinction himself. He was delivering a paper on a famous archaic *sapiens* skull called Petralona, found in Greece, which a Greek anthropologist had previously dated to more than 700,000 years ago. Stringer, who is lightly built, had new evidence to present, suggesting that the skull was in fact less than half that old. The Greek anthropologist was in the audience, growing more livid with every word he was hearing, until he suddenly started yelling and charged the podium. A bearlike figure intercepted him. It was Wolpoff.

"I guess you could say that Zagreb represented the two extremes of my relationship with Chris," he told me.

Most of the time, however, the relationship has stuck on only one of those extremes—angry letters, ruined dinners, and shouting matches, leaving both men exhausted and further entrenched in their own positions. Why? What matters so much? Perhaps it is only unfortunate personal chemistry, or just another academic debate feeding on its own momentum. Or perhaps the rivalry has less to do with the rivals, and more to do with the issue that yokes them together, the mystery they have both latched onto. On the surface, they are simply offering interpretations of the fossil record for modern humans. In the process, they seem to have independently grabbed the tail of a larger, meaner beast that is whipping them back and forth, almost against their will. In the beginning, I could sense the beast only by the angry lashing of its tail. But surely the argument between continuity and replacement, stasis and change, is not just a quarrel over old bones. Somewhere beneath the surface, it connects to a deeper struggle between two conflicting approaches to the very nature of existence. Witness the way Wolpoff and a colleague, Australian anthropologist Alan Thorne, express the conflict in *Scientific American:*

"According to multiregional evolution, the pattern of modern human origins is like several individuals paddling in separate corners of a pool. Although they maintain their individuality over time, they influence one another with the spreading ripples they raise (which are the equivalent of genes flowing between populations). In contrast, the total replacement requirement of the Eve theory dictates that a new swimmer must jump into the pool with such a splash that it drowns all the other swimmers. One of these two views of our origin must be incorrect."

Are we bathers in a slow swirl of genes and time, or the products of some cannonballing catastrophe?

Let's go look at some fossils.

I met Chris Stringer for the second time in his office in London. He was in a good mood. He had recently survived a purge at the British Museum (Natural History) and been promoted to head of the Human Origins Group at the renamed Natural History Museum, one of the most powerful positions in anthropology in England. He had also just returned from a conference in Rome, where revelations were brought to light about the cannibalized Neandertal of Monte Circeo, the one found in 1939 surrounded by an alleged circle of stones. New evidence announced in Rome showed that Monte Circeo was not the haunt of cannibals after all. Damage on the base of the skull, long thought to have occurred when the unfortunate fellow had his brains removed to be eaten, turned out to have been made by the teeth of hungry hyenas, who probably dug up the skull from a shallow grave nearby and schlepped it home to their den. As for the "circle of stones," someone else in Rome pointed out that when you remove one item from a dense scatter of objects—let's say, a skull from a random collection of rocks on a cave floor—you are inevitably left with an empty space that roughly defines the surrounding clutter as a circle.

"And so one more piece of evidence for Neandertal modernness is laid to rest," Stringer said, a trifle smugly. The demotion of Monte Circeo from human sacrament to hyena fodder was indeed only the latest in a string of recent Neandertal behavioral deflations. New looks at old evidence were suggesting that the Neandertals left no unambiguous trace of ritual activities, circular or otherwise. They spoke crudely (if they spoke at all) and lacked foresight, organizational ability, efficient fire techniques, hunting prowess, and emotional depth. If Rob Gargett, a young archaeologist in Berkeley, was to be believed, they may not even have buried their dead. Right or wrong, all this Neandertal-bashing was bolstering Stringer's replacement scenario, by characterizing the Neandertals as an eminently replaceable bunch in the first place. He is not out to prove they were less capable, imaginative, or emotional than was thought. But he does want to prove that they were not our ancestors. To that end, the news he had received from the Middle East was even more satisfying. Some new dating techniques on modern human fossils in the caves of Qafzeh and Skhul in Israel had come up with ages of around 100,000 years, far older than some Neandertal fossils found in the same area. If the moderns arrived in the Middle East *before* the Neandertals, they obviously could not be the Neandertals' descendants, any more than my grandparents can be my children.

"Morphologically, Neandertals and modern humans are very, very different," Stringer told me. "I don't think you can get from one to the other, in an evolutionary sense. I'll show you what I mean."

We entered a workroom lined with neatly labeled boxes holding part of the museum's immense collection of casts. He pulled down a few and opened the lids. "Milford is always accusing me of using extreme examples to make my point," he said, "so let's use a pretty mild Neandertal specimen." He set a mottled green and brown cast on the table. "This is the adult male from La Ferrassie in southwestern France. The cranium's brain capacity is over sixteen hundred milliliters, which is more than a lot of living people's. But the shape of the cranium is very different. Viewed from the back, the skull has this characteristic globular shape, rather than the parallel-sided, loaf shape of modern skulls."

For comparison, he brought out a well-known early modern fossil, a cast of the Old Man of Cro-Magnon, and placed it on the table next to the Neandertal. I got down on my knees, putting the skulls at eye level, and the two differing shapes became obvious: What the French call *en bombe* in the Neandertal and the loaf-shaped outline of Cro-Magnon.

"Where you really pick up the difference is in the face," he said. "You can see how the face of La Ferrassie is pulled out in the middle, as if somebody grabbed the nose and gave it a yank. The cheeks are swept back, not horizontal as in moderns, and there's no hollowing of the cheek-bone, what we call a canine fossa. The mandible is long and narrow, with this characteristic gap behind the last molar. And the front teeth are huge and heavily worn down. In short, a classic Neandertal."

Stringer helped me place the skull—about 70,000 years old, dating from the Neandertal *belle époque*—into a larger evolutionary scheme. If the continuity argument is to hold water in Europe, he said, one should see a gradual change in the fossils from one age to the next, forming a chain from the earliest specimens to living Europeans. The earliest European fossil known is an enigmatic jaw from near Heidelberg, Germany, believed to be about 500,000 years old.* After that come some much more complete fossils of early archaic *Homo sapiens*, typified by the skull from Petralona in Greece, another from the French site of Arago, and just recently, by a magnificent new collection of skulls and other bones from Atapuerca in northern Spain. According to Stringer and others, these date to about 300,000 years ago.

Stringer brought out a cast of the Petralona skull. "There are a few features here that might be called Neandertal," he said, "but overall it is much too primitive for that name."

Today, most investigators prefer to call these fossils "pre-Neandertals," indicating their future evolutionary direction. They link up nicely with a group of early, true Neandertals beginning some 130,000 years ago, and from there to a whole host of later ones scattered over the continent. So there does indeed seem to be a gradual trend through time in the fossil record of Europe, but it does not appear to be headed toward modern Europeans. Instead, more time and more evolution only deepen the Neandertal stamp on the fossils, with the most recent ones waving their Neandertal colors—the pulled-out face, big nose, and so on—the most stridently of all. In Western Europe, nothing looks much like people today until Cro-Magnon Man makes an abrupt, audaciously modern appearance around 30,000 years ago. Stringer concedes, however, that the evidence for replacement is not so dramatically carved into the record in Eastern Europe, where the browridges of some fossils from Czechoslovakia and Yugoslavia show a possible evolutionary link, a gradual change, from Neandertals to moderns.

"There's some comfort there for those who go with regional evolution,"

---

*In 1994, Stringer and others announced they had found a leg bone in England, dating to around the same age as the Heidelberg jaw.

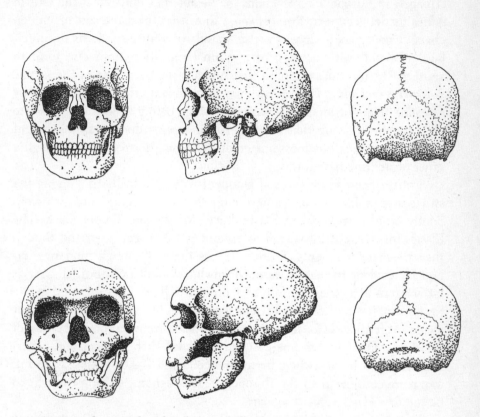

A *comparison of a typical modern human skull, (above), with that of a Neandertal, (below). The Neandertal skull shows the characteristic forward projection in the middle of the face, the voluminous nose, rounded eye orbits, double-arched brow ridges, and the lack of a chin. Viewed from the side, the skull is low and long. The rear view reveals the globular outline of the brain case, compared to the "bread loaf" shape of the modern human skull.* DAVID DILWORTH

he said. But even if the fossils in question show interbreeding between
the two groups, in his view that particular local lineage died out.

The Eastern European fossils notwithstanding, practically everyone
agrees that the case for replacement of one population by another is
strongest in Europe: Cro-Magnons in, Neandertals out. But if the Out of
Africa model is correct, Stringer must show that the same sort of sudden
change takes place everywhere else outside of Africa: moderns in, everybody else out, with no sign of a gradual transition from one form to
another. He must also establish that Africa is the homeland of the people
doing the replacing. Fossils can help that second effort in two ways: by
revealing traces of modern anatomy in Africa before it pops up anywhere
else; and by showing that modern Homo sapiens doesn't spring out of
nowhere in Africa but blossoms gradually, from Homo erectus through a
series of intermediate forms.

Stringer believes he can call eloquent witnesses for both. Among the
strongest candidates for "earliest modern human being" are two from
South African caves called Klasies River Mouth and Border Cave. The
Klasies fossils are firmly dated to around 100,000 years ago, and though
they are mere fragments, Stringer thinks there is enough modern morphology evident to make a case. Meanwhile a skull from Border Cave is
"so modern it boggles the mind." Unfortunately, it was excavated in the
forties, not by professional archaeologists but by workers shoveling bat
guano for fertilizer. Consequently the Border Cave mind-boggler is rather
shakily dated. If it did emerge from the level that Stringer and others
think it did, it is somewhere between 80,000 and 100,000 years old. So it,
too, is much older than the European Cro-Magnons, so long thought to
be the first Homo sapiens sapiens.

Stringer is even more confident about the evidence for intermediate
kinds of hominids in Africa, a chain of fossils tracing the path from erectus
to true moderns. "The record for a transition in Africa is clear," he told
me. "Not even the continuity people dispute that."

Anchoring the African chain are two strikingly similar skulls, Broken
Hill and Bodo, the first from Zambia (formerly part of Rhodesia), the
second from Ethiopia. Stringer happened to have the original Broken Hill
fossil in his office, and he carefully took the skull out of its box and placed
it alongside the casts on the table. Back in 1921 when it was found,
everything primitive was called a Neandertal, and everybody knew Neandertals hadn't evolved far enough to stand up straight. So Broken Hill was

given the name "Stooping Man of Rhodesia," in spite of the fact that the limb bones bore no hint whatsoever of a stoop. Later, Carleton Coon used it to support his misbegotten theory that evolution proceeded more slowly toward *Homo sapiens* in Africa than anywhere else. Coon thought Broken Hill was the contemporary of the fully modern Cro-Magnons of Europe, but with its hulking browridges and other primitive traits, it looked much more like *H. erectus*. Thus while evolution in Europe had spurted ahead, Africa was evidently still home to "a tired-looking peripheral survivor of an ancient and vigorous race of very primitive men."

Laid alongside the shiny cast of Cro-Magnon on the table, Broken Hill Man did look more primitive. But the comparison is unfair. The fossil turned out to be some ten times the age that Coon had thought; its true European contemporary was not Cro-Magnon, but the Petralona skull from Greece. The same is true of the Bodo skull in Ethiopia. If these African fossils seem *erectus*-like, it is because they are almost as old as *erectus*, not because Africa lagged behind in evolution. Both crania housed larger brains, however, and, viewed from behind, show the parallel sides so typical of later modern humans, along with other traces of modern anatomy. In short, they are a credible bridge between *erectus* and the next link in the African chain toward modern humans.

Stringer pulled down another box and put into my hands a salmon-colored cast with the face down. "Now this one here is Jebel Irhoud," he said. "Viewed from the back like this, it seems pretty primitive. But look at the front."

I turned the skull around, revealing the face that had launched Stringer on the road out of Africa twenty years ago in the Musée de l'Homme. "The face still has some robustness to it, and it projects a bit in the lower region. But it's broad and flat, with this hollowing of the cheekbone. You can see much the same pattern in the Ngaloba skull from Tanzania," Stringer went on, rummaging in another stack of boxes to find another cast. "To my eyes, this is the embodiment of a transitional fossil. It's still primitive, but if you were trying to produce a modern human, you would want this level of morphology to start with. Look at it next to La Ferrassie. The Neandertal has gone off in another anatomical direction completely."

Irhoud and Ngaloba joined the fossils on the table, which was getting too crowded for comfort. We moved the casts to the floor and squatted down among them. More skulls and faces appeared as Stringer needed them to make a point, and soon we were awash in bones. Though he did not have Wolpoff's bullish intensity, Stringer was matching him in raw stubborn confidence stroke for stroke, throwing up a cascade of bones to dazzle away any doubt I might be harboring. I got the feeling that if I didn't accept the Out of Africa hypothesis on the evidence already strewn around my knees, Stringer could always open another box, going shelf by shelf through the museum until the whole history of mankind was summoned.

A crucial question is the age of the African "late archaics." For fossils like Irhoud and Ngaloba to serve as an evolutionary link between the *erectus*-like Broken Hill group and full-fledged modern human beings, their ages should be in-between as well. Irhoud was originally guessed to be around 40,000 years old, much too recent to make any sense of the fossils as precursors, since moderns were already present much earlier in the Middle East and South Africa. But new dating tentatively fixes the Irhoud remains at about 90,000 years, and possibly much older, and Ngaloba at around 130,000. Another, even more modern skull from a site called Omo in Ethiopia appears to be about the same age. If these fossils are that old, they are positioned in time to be the true immediate ancestors of modern human beings. Stringer and other replacement believers see clear anatomical connections between the nearly-us group from North and East Africa and the frankly modern skeletons from Qafzeh and Skhul in Israel.

With this last link, the African chain of evidence takes on the fullness of a narrative, albeit a narrative peppered with unanswered motives. Something happened in Africa (what?) somewhere around 130,000 years ago that favored the evolution of more modern-looking people (why that anatomy and not something else?). Between 130,000 and 100,000 years ago, these true human ancestors spread north and east from their homeland into the Levant (why?), with perhaps another, separate population pushing south to establish the lineage of Klasies River Mouth and Border Cave. Whatever their reasons for moving, when the northern-bound populations reached the Middle East, they appear to stop dead in their tracks (why?). After a delay of some 40,000 years, the moderns pushed out again, eventually replacing archaic populations everywhere else as well (by what power, to what advantage?). As they established their dominion in various parts of the world, they adapted to local conditions, giving rise to the human races we see today.

And that, in the barest of outlines, is the Out of Africa scenario, neatly buttressed by the genetic evidence, but cooked up wholly independent of it. To its critics, the hanging questions are like open wounds, bleeding the life out of the theory. But they don't trouble Stringer much.

"But why does it happen?" I asked, as we quit for a tea break in the common room. "How can one group of people just pick up and take over the world?"

"Well, that's the big mystery of course, and you'll find a lot of people banging their heads up against it," he told me. "But on another level, it needn't really figure in our hypothesis. I'm getting a bit tired of people like Milford saying we have to explain how the replacement occurred before they will accept that it happened. I mean, we've known for a century that the dinosaurs went extinct all of a sudden. But we've only begun to figure out why."

I mulled that over while Stringer poured tea. At that moment Peter Andrews, Stringer's colleague and co-author on the *Science* article, wandered in. I saw a chance to tweak the Big Mystery.

"Some of your critics refer to the Out of Africa scenario as a story of 'making war, not love,'" I said to him, after we were introduced. "How does one kind of people just pick up and walk over all other kinds of people, if not through violence?"

"Why do Americans seem to have such a problem with the very idea of replacement?" asked Andrews. "Species replace other species all the time in the paleontological record without slaughtering each other. Why should there be a different standard for humans?"

"Then you think Neandertals were a different species?"

"Possibly," he replied. "But the Out of Africa scenario doesn't depend on it. They could have been interbreeding, although the genetic evidence doesn't show it. In any case, they didn't have to be killed off in order to be replaced. Perhaps the Neandertals were simply pushed into a less desirable region by a superior population. This sort of thing happens all the time among animals."

"But it's the notion that one group of humans is innately superior to another that a lot of people in America don't like," I pointed out. "It sounds a lot like Marcellin Boule, doesn't it? The inferior brute Neandertals pushed aside by a higher race of man?"

"The Out of Africa scenario hasn't got anything to do with Boule," Stringer answered. "He was a Eurocentric racist. We have the origins in Africa. Boule would never have said that."

Stringer and I went back to his workroom, where a lot of little mysteries were still hanging. If the replacement holds true, Stringer has to demonstrate that *all* the non-African archaics were supplanted by the new breed from Africa, not just the Neandertals of Europe. Outside Europe, his job is more difficult. Stringer has yet to convince everyone that the Chinese fossil record, for example, speaks for replacement. The skeptics include virtually all of the anthropologists in China itself. "The ancestor worship in China is literal," he told me. "It's in their cosmology. If you say these Chinese *erectus* things aren't their ancestors, they get offended."

The traditional arguments for continuity in Asia derive from the abundantly bone-rich cave of near Beijing Zhoukoudian, the home of the original Peking Man. Over the decades, the cave has yielded skulls, artifacts, ancient meals, traces of fire and perhaps cannibalism, altogether some 250,000 years of human occupation in a single site. Franz Weidenreich believed the Peking Man *erectus* fossils from the lowest levels of Zhoukoudian, although half a million years old, already bore distinct "Mongoloid" features still evident in Asians today: small, delicate faces; flatter, flaring cheeks; and less prominent noses, flattened on top. Modern Asians often also possess a peculiar, scooped-out appearance in back of the incisor teeth, called shoveling. (So do the American Indian populations, who probably descended from Asians by way of migration across the Bering Strait.) If the regional traits appearing on Peking Man some 500,000 years ago can also be found on people crowding the streets of Beijing today, it seems to rule out the notion that Africans or anyone else cannonballed in and drowned out the ancient lineage. Weidenreich certainly believed so. But he lamented the paucity of fossil evidence after Peking Man— transitional forms that could link it with living Chinese. He had found three early modern skulls from another part of Zhoukoudian, but none of these bore the regional traits he was looking for.

New bones have been found since Weidenreich's time, and while the Chinese keep access to their fossils highly restricted, a few Western scientists have had the chance to look at them carefully. One is Geoffrey Pope, at the University of Illinois in Urbana when I met him, and now at William Paterson College in New Jersey. Growing up as an embassy brat in Taiwan, Pope is one of the few Western paleoanthropologists who speaks fluent Chinese, which helps him navigate the bureaucratic swamps that lead to the original fossils in Chinese museums. According to Pope, there is little evidence for replacement in East Asia.

"There are a lot of transitional fossils in China, which bothers Chris quite a bit," he told me. The best of these is a haunting enigma from a site called Dali, found in 1978, along with an unusually complete but equally enigmatic skeleton from Jinniu Shan (Golden Ox Mountain), found in 1984. In addition to the traditional arguments for Asian continuity, the fossils show a host of newly defined regional traits, mostly in the face. Or so Pope believes.

"A good example is a little notch in the cheekbone called the incisura malaris," he told me, pointing to the feature on a cast of Peking Man. "You find this in Dali, in the Zhoukoudian early moderns, and right on through to contemporary Asians."

He also showed me how the cheekbone in the Asian line runs horizontally, while both European and African specimens dip down in a more fluid curve. Meanwhile, the maxilla (the upper jawbone) in the Chinese line is shorter, top to bottom, than those from other regions. Not every Asian skull shows every Asian feature, but they appear more often there than elsewhere, which is enough to convince Pope.

"I can walk into a room and tell you if a skull on a table is Asian or not," he said. "I don't know how I do it, but I know I can."

In London, when Stringer and I got back to his workroom, I read him some of my notes from my earlier talk with Pope. He fidgeted and scowled, and before I was finished he was off again, flinging open boxes and digging into file cabinets for photos until he had his rebuttal laid out on the floor. Pope had offered notched, horizontal cheekbones and short jawbones to define "Asian-ness," from Peking Man to living Mongoloids. Stringer's counterargument was to show that these same traits are well represented outside of Asia too.

"Geoff is grounding everything on this jaw from Zhoukoudian," Stringer said, picking up his own cast of Peking Man. "Here's the incisura he's talking about, and the short maxilla, and you can see it in the early moderns from the Upper Cave too. But look over here at Broken Hill. It's got the incisura and short max as well, and it's from southern Africa. That's about as far as you can get from Asia. Jebel Irhoud shares some of the supposed Mongoloid cheek morphology, and so do some modern Australians. How can you call these Asian traits, if they appear all over the place?"

As Stringer explained, there are other reasons besides recent common ancestry why two fossils may share anatomy. For example, the shared character might be what's called a "primitive retention": a feature inherited from a more distant common ancestor of *all* modern humans, rather than one marking that region alone. Even Geoff Pope concedes that shovel-shaped incisors, long a hallmark of Asians, have turned up on an important *Homo erectus* skeleton in Kenya.

There is a more difficult challenge to the Out of Africa hypothesis farther to the south. Somewhere around 60,000 years ago, the first human beings reached Australia. At that time, sea levels were much lower than they are today. Australia, Tasmania, and New Guinea were all part of a larger land mass called Sahul. Most investigators think that humans reached Sahul from the Asian mainland by way of Java, hopscotching over eight small islands, then crossing a stretch of some fifty miles of open ocean. Such a crossing would require a vessel that could stay afloat for a few days, perhaps some kind of bamboo raft. The rafts, of course, are gone now, long ago rotted into the earth. But traces of a stone-tool culture remain, as well as a baffling cache of fossils.

Nowhere else is the record for modern human origins so deeply, utterly entangled as it is in Australasia. At the heart of the knot is a weird duality in the fossil record itself. While some of the early modern humans from Australia look much like people today, others bear all the markings of a more robust kind of human, with thick skull bones, swollen browridges, and huge teeth, even bigger than those of *Homo erectus* in some specimens. All would be well for Stringer's replacement theory if these primitive-looking ones came first, only to be swept aside later by arrivals via the southeast Asian mainland, bearing modern features from Africa. But the fossils don't seem to tell that story.

"I have to admit that Australia poses special problems for the Out of Africa scenario," he told me.

"You bet it does," said Milford Wolpoff, when I entered his lair in Ann Arbor a couple of months later. "If Chris would actually look at what he's talking about, he'd see that the Australian evidence completely blows apart his hypothesis. Maybe that tells you why he never looks at it in the first place."

In 1978, seven years after Chris Stringer took off to look for Neandertals, Milford Wolpoff set out on a journey of his own. Like Stringer, Wolpoff had come to a crossroads in his career, and he believed that only original fossils could tell him which way to turn. He did not need letters of introduction to get into museums, and he never once slept in his car. At thirty-six, he was already one of the most influential paleoanthropologists in the world, famous for the vigor of his opinions, if not for his tact in expressing them. His trip was backed up by a four-year, $250,000 grant from the National Science Foundation, one of the largest travel grants ever awarded to an individual in his field. It was awarded for a comprehensive study of hominid teeth and jaws, but Wolpoff had something even more ambitious in mind.

"I took advantage of the money to look at everything," he told me. "In four years, I looked at every bloody hominid fossil that existed at the time, every single one."

Out of the thousand-odd fossils Wolpoff examined on the trip, one alone changed the course of his professional life. Along with Loring Brace, Wolpoff had been the driving force behind the "single species hypothesis," which declared that no more than one hominid species could exist at any one time on the earth. After guiding the direction of human evolutionary thought for a decade, the idea faded under the glare of new discoveries, the most important being Richard Leakey's find at Koobi Fora of a 1.5-million-year-old *Homo erectus* who must have lived at the same time as *Australopithecus*. Wolpoff was shaken by the evidence but still convinced that the human lineage had evolved gradually through a series of uniform stages, from the australopithecines to early *Homo*, and thence through *Homo erectus* to the archaic *Homo sapiens* forms like the Neandertals, and on to modern humans. The whole fabric of this evolving lineage was stitched together by the homogenizing flow of culture and genes. Thus at any given point in time the hominids of one region would look pretty much like those anywhere else in the world.

Two years before, Wolpoff had met the Australian Alan Thorne, who was pushing quite a different view. Rather than look for resemblances horizontally through time, with fossils of the same period matching up, Thorne insisted that the important connections were vertical. As evolution went on, *Homo* in each region of the earth preserved an exclusive set of characteristics that defined that region alone, just as Franz Weidenreich had believed. At the time, Wolpoff would have none of it. Any variation between, say, an African *Homo erectus* from Koobi Fora and a Peking Man *erectus* from Zhoukoudian was mere window dressing, the result of insignificant local adaptations that had little to do with the grand themes of evolution.

"If you've seen one *erectus*," he told Thorne, "you've seen them all."

On his trip, Wolpoff pored over the museum drawers of four continents and found nothing to alter his faith until one day in a laboratory
in Java he held in his hands a recent discovery from the site of Sangiran,
where the Dutchman G.H.R. von Koenigswald had worked forty years
before. Called Sangiran 17, the new find was the most complete *Homo
erectus* skull yet discovered. But in spite of its importance, a long-standing
feud between two Javanese investigators—both of them famous for guard-
ing their fossils like jealous lovers—had kept the skull largely hidden from
sight. Given a privileged peek at it, Wolpoff was surprised to see that the
face had been attached to the skull with rubber bands, so that it swung
in and out, as if it were hinged onto the cranium at the forehead. Fossils
are hard enough to interpret when their features are fixed in place; a
specimen with movable anatomy is impossible to pin down.

"I saw this thing, as badly reconstructed as it was," Wolpoff remembered, "and I said, 'Look, can I take a crack at redoing this?' "

Instead of using rubber bands Wolpoff built a framework of toothpicks,
and in a couple of hours he had the face glued in place. He let the glue
set for another half hour and only then picked up the specimen and
turned it in profile. "I nearly dropped dead. Instead of being just another
erectus, here was this great big, hyper-robust Australian aborigine. I knew
at that moment that Thorne was right, and I was wrong."

What astonished Wolpoff was the fossil's face, especially the way it
projected out from the skull. Once he had completed the toothpick reconstruction, he could see that the jutting face was unlike anything he'd
seen in *erectus* specimens from Africa or among the Peking Man casts in
Beijing, where he had been just days before. Though some 700,000 years
old, the face eerily resembled those of far younger, modern human fossils
from Australia. The "robust" Australian *sapiens* were as modern in brain
size as any in the world, but they showed the same facial projection—big
browridges, thick bones, sloping foreheads, and heavy molars—that Wolpoff saw in Sangiran 17. Many living Australian aborigines carry the same
traits today.

Decades before, Weidenreich had suggested a connection between the
erectus fossils of Java and modern Australian aborigines. But Weidenreich
had only skullcaps to work from, like those of Solo Man from Ngandong.
In the face of Sangiran, Wolpoff saw the missing anchor to Weidenreich's
Australian lineage. Another researcher had suggested that the anatomy of
Australian aborigines bore "the mark of ancient Java." For Wolpoff, San-

giran was stunning proof. In their arguments, Alan Thorne had been trying to convince Wolpoff that regional features would appear first at the remote edges of the hominid range, farthest away from the African birthplace of the earliest hominids. And here sat Sangiran 17, three quarters of a million years old and about as far from the African "center" as one could get—but already a full-fledged native Australasian.

From Java, Wolpoff continued east, carrying his excitement and his load of crow to eat in front of Thorne at the Australian National University in Canberra. Over the next few weeks, the two scientists sat down and took a look at the fossils. The modern human bones in Australia fell into two distinct groups. Some were very robust, clearly bearing "the mark of ancient Java." Typical of this group were the skulls from a site called Kow Swamp in Victoria, dated to only around 10,000 years ago, and a new find called Willandra Lakes 50, a massive skull made of pure white, opalized bone. These robust skulls, Thorne and Wolpoff believed, were clearly descendants of H. erectus from Java. Linking them to the deep past of Sangiran was the collection of skullcaps of Solo Man. These faceless fossils, ensconced on the cusp between erectus and archaic sapiens, had bigger braincases than Sangiran, but the same pattern of regional features as Kow Swamp and the other robust Australians. The latest dating put them at about 100,000 years old.

Wolpoff and Thorne also had to contend with another set of moderns in Australia, ones with much more gracile or slenderly built faces and cranial bones. Among them was a cremated human skeleton from Lake Mungo, several hundred miles into the bush north of Kow Swamp. Such lightweight skull and face architecture is usually thought of as a trademark of advanced sapiens. But these fossils were older than Kow Swamp and most of the other robusts. At 30,000 years, Lake Mungo was the oldest well-dated fossil on the continent, and the earliest known human cremation.

Thorne already had an explanation for this strange dichotomy, one that slid neatly into the regional framework to which both men were now committed. There had not been just one migration onto the continent, he reasoned, but two at the least. The gracile fossils like Lake Mungo were not part of the Java lineage at all, but the legacy of another migration from northern Asia, bringing "the stamp of ancient China" onto the continent with them, by way of the Philippine Islands. Once these two populations were established, they interbred and mixed their genes with other, later arrivals from the west. Out of this complex human gene soup came the modern Australian aborigine.

With the duality of the Australian fossil record accounted for, Wolpoff and Thorne thought that they had the best evidence yet for regionality in human evolution. In a major paper, published in 1981 with Wu Xinzhi of the Institute of Vertebrate Paleontology in Beijing, they marshaled the arguments for continuity from other areas of the world, folded in Wolpoff's ideas about gene flow, and added Thorne's hypothesis that regional features first appear at the edges of the hominid range. Out of this blend came the multiregional continuity theory, the most plausible, most appealing explanation of modern human origins around—until the discovery of Eve. In the early 1980s, Chris Stringer's African replacement scenario was merely a dim candle flickering on the distant extreme.

"Nothing he said or did made any impression," said Wolpoff. Even today, Wolpoff is convinced that when the Out of Africa idea reaches Australia, it withers up and dies like an African violet planted in the outback. There is no trace of Africa, he believes, anywhere in the Australian fossils. "If Chris is right, then this whole regional pattern of tremendous browridges and facial prognathism and large teeth that you see in Sangiran must have been wiped out in Australia by the African arrivals, and then magically evolved all over again."

Wolpoff revealed all this to me in the back room of his lab, thrusting a cast of Sangiran 17 or some other fossil into my hands, then whisking it away and moving on to another specimen, another argument, another nail in Eve's coffin. After Australia, we ran through the Chinese fossils again, then the European record, where Wolpoff showed me the Eastern European skulls that he believes speak for a blending of Neandertals and modern humans. If anything, the performance was even more spectacular than Stringer's in London. I did my best to poke at some of the weak chinks in his armor, but the points I made were not so much rebutted as ridden out of town by a posse of counterpoints.

"Everybody who has taken the time to look at the original fossils, they all agree," he concluded. As it happened, Wolpoff had organized a seminar for the American Association for the Advancement of Science meeting in New Orleans, coming up in a couple of months. Besides himself, all the people who had "looked at the fossils" would be on hand—more than enough expertise to bury Eve under an avalanche of hard bone evidence, with Chris Stringer by her side.

"Come to New Orleans," Wolpoff urged me. "I guarantee it will be worth the trip."

I promised to be there. But I was beginning to wonder where all the drumming on the bones was leading me. I had already seen too many, and understood too little.

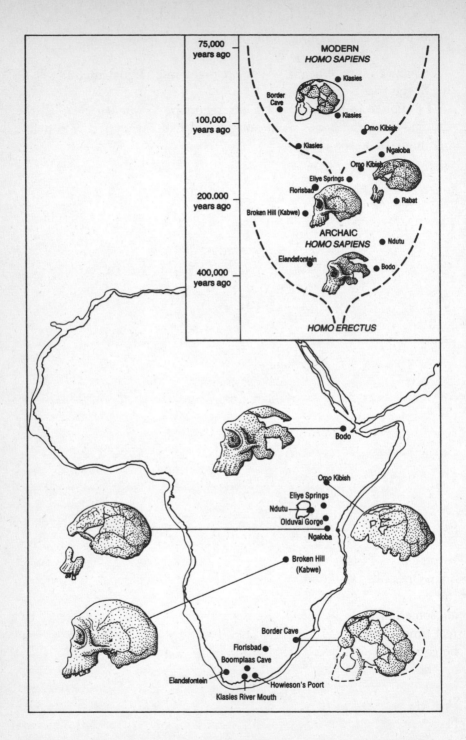

The fossil evidence for an African origin of modern humankind. Above, a diagram
shows how these key specimens might reveal the evolutionary transition from archaic to
modern Homo sapiens. COURTESY CHRIS STRINGER

Chapter Five

DOWN TO NEW ORLEANS

So, maybe reality is untestable. —ERIK TRINKAUS

As I flew out of Ann Arbor that evening, I was staring absently at the in-flight magazine. Suddenly a thought fixed me rigid in my seat: *What if Wolpoff was right?* It had not occurred to me until that moment that I had been hoping to escape Ann Arbor with my faith in Eve still intact. My motive was depressingly obvious: I wanted a good story. In Chris Stringer's eyes, all the fossil evidence traces back to a single place and a single time. His views dovetailed neatly with the genetic evidence for an African mother. All that was left to be explained was the delicious mystery of how African-bred *Homo sapiens* ended up triumphant everywhere else. But switch to Wolpoff's lens, and the harmony between bones and genes fades away. The fossils speak of continuity, not change. It would be harder to write a book on modern human origins if modern humans had no definite origin, if the face of humanity evolved with languid inevitability, its progress kept in global sync by a flow of genes lapping across regional boundaries like the tidy turbulence in a public swimming pool. I wanted the big splash. But I also wanted the truth. Something did *happen*, after all, something that fashioned a human being out of a different animal, in real time, in bone and blood and flesh. At most, only one theory could be correct.

One more time, I ran through the arguments in my head. Whether or not modern human anatomy first appeared in Africa, most experts agree that the fossil record there reveals a fairly clear trail of transitional fossils. Not even continuity advocates dispute that; in their view, modern Africans should trace their ancestry back within their own region, just like everyone else. The real strength of Stringer's replacement model is the one-two punch of this African continuity, combined with the quick change in the fossils of Western Europe—from Neandertal to Cro-Magnon in a matter of a few thousand years or less. There doesn't seem to have been enough time for one to have evolved into the other. The fossil record is most complete in Europe, so Stringer looks best where the evidence is best. In Asia and Australia, there is no clear evidence of a sudden change. Stringer would say that he cannot make a more positive case for replacement outside Europe because there aren't enough fossils. Wolpoff would counter that he has no case to make. If Eve's descendants left Africa to repopulate the planet, then the earliest modern *H. sapiens* everywhere should look like Africans. But they don't.

"If there are African features in the early moderns of Europe and Asia, I wish Chris would show them to me," Wolpoff complained to me. "Where are they? Where's the beef?"

"That's ridiculous, and Milford knows it," Stringer replied when I asked for his response. "The *Homo sapiens* that stayed in Africa have had a hundred thousand years to evolve the traits that distinguish them today, just like other populations. We have to get away from the *Newsweek* syndrome, picturing the common ancestor looking like a contemporary African."

Instead, he suggests, we should look for more generalized traits in the earliest modern humans, both within Africa and in other regions: a sort of *sapiens* blank slate. These features would include a high vertical forehead, a rounded cranium, reduced browridges, a short, tucked-in face, a chin, and lightly built skeletons. In other words, modern human anatomy *is* the African pattern, and its appearance later, outside Africa, is proof positive of the replacement scenario. Wolpoff counters by pointing out that his moderns from China and Australia resemble archaics from their own region more closely than they do archaics from Africa. And so it goes, the whole controversy hunkered down into a feature-by-feature trench war. The replacement advocates gain ground when they can show that an alleged regional trait—the shovel-shaped incisor, for example—crops up frequently outside of the region it is supposed to define. But Wolpoff and his supporters come out ahead when Stringer has to work harder to keep his hypothesis moving forward, as in Australia.

When Stringer's momentum slows down, his hypothesis gathers messy implications. Witness the trouble he got into when he tried to offer a working definition of a modern human being and, according to Wolpoff, ended up excluding thousands of living aborigines from the species. Wolpoff says Stringer fell into similar trouble in his *Science* article in 1988, when he and Peter Andrews tried to force-fit the Out of Africa model onto the complex Australian fossil record. One way to explain the lingering presence of archaic traits like low foreheads and browridges, they suggested, is to imagine a whole pattern of "apparent evolutionary reversals" back to the primitive condition, after modern anatomy had already been established on the continent thousands of years earlier.

"We do not agree with the description of any Australian aborigines or their immediate ancestors as cases of 'apparent evolutionary reversals,' " Wolpoff and his colleagues wrote in a reply to *Science*, "and such statements and their implications are unfortunate." This amounted to a veiled accusation that the British scientists were characterizing a living human race as evolutionarily backward. Stringer had a seething response to the charge when I saw him in London.

"If you want to say somebody is racist, it would have to be Milford," he told me. On the floor in front of us was his cast of Sangiran, the hulking *erectus* from Java, looking all the more brutish for its dark color. He picked it up and held it over his head. "Here they are, Milford's modern Australians! It's the mark of ancient Java!"

Neither man accuses the other outright of bigotry. But each insists the other must reckon with the full implications of his theory, including those with a sinister potential for misuse. No doubt, a budding new Hitler would happily embrace a theory of total replacement, implying that one group of humans in the past held some kind of intrinsic advantage over all the others. "There is a lot of unfortunate baggage in this idea," one archaeologist observed. "It reinforces the old Western notion of the tree of life, with some people important and others marginal."

Multiregionalists like Wolpoff have their own "unfortunate baggage" to lug around, most of it inherited from Carlton Coon. His idea that human races are *biologically* separated from one another by a million years of evolution would certainly give aid and comfort to a racist. "My model says there is a historical element to human variation," Wolpoff explained to me. "Call it race if you like. But at the same time, the species as a whole has been connecting and interbreeding and cooperating and evolving into one great family for hundreds of thousands of years. I wouldn't do research on this issue if I thought it would be useful to a racist. I'd quit and go work on australopithecines."

Wolpoff believes that human racial distinctions extend deep into the past, but not racial isolation. In his view, the idea of genes flowing be-

tween regions keeps modern multiregional theory free of Coonian connotations. Gene flow could mean the movement of one people into the
territory of another, with interbreeding between the two groups; or it
could be simply the exchange of mates between neighboring tribes along
their common border. Eventually, the genes brought in from mate-
exchange on one end of a population's range would trickle through to the
other side, to be passed on to yet another population, and so on through
the great human fabric billowing through time—call it *erectus* or archaic
sapiens or modern humans or anything you like. Take away our nervous
need to give it a name—to put edges around a shifting, edgeless entity—
and it sounds racially unbiased, politically neutral, and, most important,
plausible.

At least it sounds plausible so long as gene flow can really keep the
human fabric in one piece. According to some population geneticists,
genes could simply not have flowed from region to region in the way that
Wolpoff suggests.

"What shoots down the theory are the logical difficulties," Steven
Jones of University College, London, told me. He believes there simply
would not have been enough time for enough genes to flow through the
vast range of *Homo erectus* to affect the changes in anatomy that define
a modern human being. Jones's colleague Shahin Rouhani worked out a
theoretical model for calculating gene flow among *Homo erectus* populations, assuming that their population dynamics were much like those of
hunter-gatherer people today. He assumed a total population of one million *erectus* individuals, the accepted estimate arrived at by other workers.
If the sizes and densities of *erectus* tribes were the same as those of modern hunter-gatherers, there would have been about 10,000 tribes of *erectus*
worldwide, each with about 500 individuals. Again assuming that interbreeding took place between tribes at the same rate as today, about 5
percent of each tribe's genes would be exchanged with those of its neighbors in each generation. At this rate, it would take 20,000 generations—
or about 400,000 years—for a single advantageous gene to "flow" throughout the *erectus* range, even under ideal circumstances. Cultural and geographical barriers would slow it down much more. Furthermore, a single
gene, or two or three, or even ten, are not enough to account for the
massive morphological differences between *erectus* and *sapiens*.

"If you plug in the whole gene pool from South Africa to Asia, you
find that it simply couldn't happen," Jones told me. But Wolpoff dismissed this.

"Jones and Rouhani just don't understand multiregional theory," he explained. According to him, they were basing their estimates on the flow of new genetic mutations through populations. Instead, fluctuations across borders in the frequency with which existing genes appear would have been enough to keep the *erectus* populations evolving in sync.

And so it went on in my head, 30,000 feet above Lake Ontario, bouncing back and forth between two models for our species birth—a total change or a measureless constancy—with both starting to look equally unacceptable. Suddenly my thoughts converged on a new revelation: *Of course; Stringer and Wolpoff are both wrong.* If the fossils cannot unequivocally support either one, why not look for the truth somewhere in the middle?

Many paleoanthropologists would gladly settle for a scenario of modern origins somewhere between the stringently Stringeresque and the wholly Wolpoffian. For more than a decade, German paleoanthropologist Günter Bräuer has offered a milder version of the Out of Africa scenario, which he calls the African Hybridization and Replacement Model. Like Stringer, Bräuer believes that Africa played the central role in the origin of all humans alive today. The modern human anatomical pattern arose there, and the genes that determined it were carried by immigrants who left Africa and replaced everybody else. But the replacement Bräuer imagines was not as cut and dried as Stringer's version. Pushed out by the desiccation of the African continent, the migrating moderns entered the territories of tribes in Europe and Asia. Though they were morphologically somewhat different, they were still the same species as the people they encountered, so they were able to interbreed with them to varying degrees. These matings account for the traces of mixing found in the fossils outside Africa, as well as the bits and pieces of regional anatomy that pop up in some populations today. Modern Europeans, for example, often have a bony protrusion on the back of the skull called an occipital bun. Many Neandertal fossils show this feature too, and Bräuer believes that they passed it on to us.

Even so, Bräuer's vision is nothing like the continuity imagined by the multiregional theory. He believes that the basic pattern of modern human form all over the world still derives from one place only—Africa. It has merely picked up a little regional flavor here and there as the newcomers interbred with the resident populations they were replacing.

"Evolving a modern human from an archaic one is a very complex process," Bräuer told me when I met him in London. "How could this happen in separate regions over and over again, and all of them end up in the same place—us? It just isn't possible."

Bräuer's counterpart on the multiregional side is Fred Smith, who was at the University of Tennessee when I first talked to him and has since moved to Northern Illinois University. A former student of Wolpoff, Smith has peeled away the more extreme expressions of his mentor's theory. He believes that the genetic change triggering modern human anatomy must have had a single point of origin. He accepts the idea that the fossils point to Africa as that starting point. But his "assimilation theory" does not buy the notion that this new anatomy was carried out of Africa on the backs of migrating *people*, who systematically replaced the Neandertals and others they found along the way. Though some population migration must have occurred, the real Out of Africa story is the movement of the *genes* for modern anatomy, which were quickly sucked into populations outside Africa as soon as they bestowed a key survival advantage.

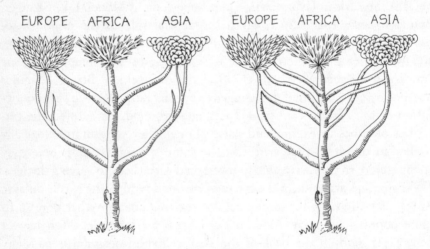

Two more modern interpretations of modern human genealogy. On the left, Günter Bräuer's "hybridization and replacement" model: Modern humans originated in Africa and migrated out, replacing archaic populations in Europe and Asia, but not without interbreeding with them. On the right, Fred Smith's "assimilation" theory also suggests that the genetic change that produced modern human anatomy happened first in Africa. But this development later spread to other parts of the Old World through gene flow among populations, not through the replacement of one kind of human by another. Compare these trees to the more extreme versions on page 88.
DAVID DILWORTH

Paradoxically, the advantage may have been that the modern humans were weaker. Archaic anatomy is more robust, especially in the skeletons and the facial architecture that supports the chewing muscles. Those massive Neandertal physiques evolved because they were *needed* to negotiate the perils of living in a colder, harsher climate. But all that extra bone and muscle was energetically expensive, requiring more food, and more work to find it. If new behavioral innovations came along that could relieve the need for robust builds, then any genes for *lighter* anatomy flowing through the population would quickly rise and spread, because lightly built people could survive on less food. Where and why these genes originated is almost beside the point; what matters is that they were incorporated into the continuing lineages in Europe and Asia as soon as they became advantageous. Smith's assimilation theory is thus the inverse of Bräuer's hypothesis. The main story is continuity—with a pinch of gracility added to the regional lineages when it became adaptive to mix it in. This might explain why the Neandertals were the last archaics to disappear.

"The modern human pattern could not have been established in Europe until the Neandertals had better cultural means of adapting to the extreme cold," Smith told me. "We know that Upper Paleolithic people made better hearths, more clothing, and better tools. I don't know for sure, but maybe there's a link."

His theory positions Africa as the single source of modern anatomical form without having to imagine some intrinsic superiority about the early African people that compelled them to leave home for a rampage of replacement everywhere else. It accounts for the possible traces of continuity between moderns and Neandertals in Eastern Europe without denying that the modern human form was the tangible result of a real genetic shift rather than a nebulous "state of mind," as Wolpoff has maintained, that somehow drifted into being all over the earth. It cannot explain why this genetic shift occurred, and why in Africa—but then, no one else's theory can either.

"I don't know the answer," Smith acknowledged. "But if you're going to tie me down, I'd say the switch to lighter skeletons might have been a change in some regulatory gene controlling bone growth that just came along randomly."

By the time my flight from Ann Arbor landed, I had decided that Fred Smith's assimilation theory was the most compelling answer around. Then I remembered Eve. The Berkeley geneticists' mitochondrial revelations had started me on this search, and if they had any validity at all, then Fred Smith's continuity argument *must* be false. And so, for that matter, must any theory suggesting that fertile matings were commonly going on between modern humans from Africa and the people they encountered— including Günter Bräuer's "hybridization" notions. Indeed, even Chris Stringer's work bordered on irrelevancy. From the geneticists' point of view, *no* argument based on bones could escape from the pitiable interpretive subjectivity that has plagued human paleontology from its beginning. No wonder the fossil experts were upset by Eve. She put them in danger of becoming fossils themselves.

"Our real quarrel is not with Chris Stringer," said Wolpoff. "Our public enemy number one is Allan Wilson."

When I got home, there was a fat envelope on my desk from the Berkeley Biochemistry Department. I hoped it contained an answer to my request for an interview with Wilson, but when I tore it open, I found nothing inside but a batch of reprints and a note from his secretary suggesting that I talk to "Dr. Wilson's associates." With Wolpoff's Eve-bashing still ringing in my ears, I wrote to Wilson that I *had* talked to some of his associates, and would talk to more, but as he was the guiding intellect behind these investigations, wouldn't it be a shame if his views went completely unrepresented? Especially since I was being inundated by arguments from the opposite side. My letter came back with a note scrawled in red ink in one corner: "I am desperately trying to do science and keep this operation afloat midst the slings and arrows, but since you insist I will try to make a few minutes available for you."

There was ample reason to go back to Berkeley besides a few minutes with anthropology's Public Enemy Number One. The pace of research in Wilson's lab had suddenly spurted forward as a result of a wildly productive new lab technique.

"They're going so fast, they're down to looking at mitochondria from single human hairs now," Chris Stringer told me on the phone.

Those hairs were telling an astonishing story: "Based on the new data, Eve lived in Africa 142,000 years ago," Wilson told the *Los Angeles Times*. "We are all derivative of one of the !Kung lineages."

This was a double shocker. First, it meant that Eve was some 60,000 years *younger* than the already heretical 200,000 years originally proposed. And she was a !Kung Bushman to boot. Widely known through the burlesque portrayal of their innocence in the movie *The Gods Must Be Crazy*, the !Kung are the most intensely scrutinized group of hunter-gatherers on earth. As such, their behavior has come to be used—and misused—as a sort of generalized model for that of our primitive ancestors. Now it appeared that they were carrying around the oldest mitochondria as well. Such a suspiciously tidy rapport between traditional ethnography and trailblazing molecular genetics seemed almost too fantastic to believe.

Judging by the reprints in the envelope, Wilson's group was moving faster than ever. The engine powering this rejuvenated juggernaut was the polymerase chain reaction, or PCR for short. Invented in 1983 by Kary Mullis of Cetus Corporation, PCR is a laboratory method to isolate whatever fleck of DNA one has a mind to examine and to produce a truckload of identical flecks. Once a stretch of DNA is "amplified" in this way, a researcher can read the literal, word-for-word genetic code it contains. Because it can quickly provide a genetic fingerprint of an individual, PCR has been adopted by forensic specialists to implicate or exonerate rape and murder suspects. (The prosecution in the O. J. Simpson case, for instance, used the technique on blood samples found at the murder scene and on Simpson's property.) It can also test for AIDS and a variety of other diseases, ferret out infectious organisms in city water supplies, even speed up cancer research. Ten years after its discovery, it earned Mullis a Nobel Prize.

The people in Allan Wilson's lab were using PCR for more esoteric reasons, but it was proving just as effective. My first stop in Berkeley was a visit with Svante Pääbo, a young Swedish investigator in Wilson's lab skilled in this new biochemical art. "PCR is basically a genetic Xerox machine," Pääbo explained. He was dressed in a white lab coat, and his narrow, inquisitive face and steel-rimmed glasses made him look like some kind of natural human probe. "You can take any bit of DNA and produce thousands of exact replicas. It's like cloning, only much faster and much more reliable."

The copying is achieved, Pääbo told me, with the help of a "polymerase enzyme," a naturally occurring chemical whose normal function is to rebuild damaged DNA in a cell. You also need a couple of DNA "primers," bits of single-stranded DNA whose sequence of base pairs is known. These ingredients are added to a test tube containing the double strands of the unknown "target" DNA, which break apart when the solution is heated. The two primers bind to the two separated strands of the target at specific points, bracketing the stretch of target DNA to be copied. The binding of these primers triggers the polymerase enzyme, which reproduces the complementary bases of the target DNA in between the two primers. Where you once had one double-stranded piece of DNA, you now have two. Heat the solution again, and these two double strands break apart. Again the primer DNAs latch on, again the polymerase enzyme goes to work, and you now have four copies. The increase is exponential each time. Reheat the solution twenty times, and you end up with a million copies of your original mystery DNA, which is fast becoming not so mysterious anymore.

Once a researcher has "amplified" DNA in this way a million times or more, it is a relatively simple matter to read the code of bases inscribed on it, using standard techniques. Because PCR can target even tiny bits of preserved DNA among a mass of fragmentary or damaged material, it is particularly useful for studying the vestiges of DNA found in ancient tissue. Pääbo himself had already used PCR to squeeze mitochondrial DNA from the brain of a 7,000-year-old Paleoindian preserved in a Florida peat bog, as well as from the preserved hide of an extinct marsupial wolf. And his work was just beginning. More recently, Pääbo has used PCR to probe the genes of the "Ice Man," the 5,000-year-old corpse found frozen in the Italian Alps.

While Pääbo was probing into these ancient molecules, others in the Wilson lab had adopted PCR as the "method of choice" for comparing living human mitochondrial DNAs. Among them was Linda Vigilant, Mark Stoneking's wife and co-worker. Vigilant's research was the source of Wilson's comment about a !Kung common ancestor for humankind.

"We're just about certain now about an African origin," she told me, back in the bungalow where I had met Stoneking several weeks before. "And we're more sure than ever that the origin was very recent. A lot of the criticisms just don't apply anymore."

The most thoroughly squelched objection was the attack on Rebecca Cann's use of African-American individuals as substitutes for native Africans. While even Cann's most diligent attempts brought her only two African placentas to work with, PCR had enabled Vigilant and her colleagues to forgo placenta-hunting altogether. She got all the bona fide African mitochondria she needed—including samples from !Kung Bushmen, Herero cattle herders, Western and Eastern Pygmies, Hadza hunter-gatherers from Tanzania, and Yorubans from Nigeria—from single plucked hairs. Her study also included New Guineans, Asians, American blacks, Europeans, and one Australian aborigine. Just as Rebecca Cann had suspected, the reliance on native Africans rather than their American descendants made no difference: The oldest mitochondrial lineages in Vigilant's ongoing studies were all of African descent.

The ease of harvesting mitochondria from hair instead of placentas was only one advantage of working with PCR. Because it produces actual base pair by base pair sequences, Vigilant could focus on two tiny fragments of mitochondrial DNA, each only about four hundred base pairs long, and glean more information from them than Cann's restriction-mapping technique gathered by looking at the whole genetic load inside the mitochondria. Vigilant's fragments were part of a loop of mtDNA called the control region.

"We chose the control region because we knew it doesn't code for any proteins or other products," she explained. "Mutations there don't affect the functioning of the molecule, so they can accumulate much faster."

That faster mutation rate translated into a more accurate measurement of how much one person's mtDNA type differs from another's. When all the mtDNA types were compared, the global portrait that emerged showed a phenomenal geographic cohesion. Among the individuals sampled, identical mitochondrial types appeared only within the same ethnic populations. Even the Eastern and Western Pygmies of Central Africa, separated by fifteen hundred kilometers of jungle yet presumably very closely related, shared not one mitochondrial type between them. The closest relative of every Eastern Pygmy in the study always turned out to be another Eastern Pygmy, and the same was true for the Western group. Similar but not quite identical mtDNA types still tended to cluster together among the same population or geographic area.

Thanks to PCR, human mitochondrial DNA could also be compared directly with that of a chimpanzee, which had previously been impossible. Adding a chimp to the analysis parried at least two other attacks on the original Eve hypothesis. The first involved the way Rebecca Cann rooted the human mitochondrial tree in Africa. Her "midpoint rooting" method assumed that all human mitochondrial lineages mutated at the same rate. Certainly the most diverse types were all African. But Doug Wallace, who at the time was pushing for an Asian origin for modern humans, pointed out that the African lineages might be more diverse than others simply because they had mutated faster.

Using a chimp as a common outside reference point buried this objection. By definition, the chimp mtDNA in Vigilant's study *had* to be equally related to every human type, African and non-African alike, because all human lineages trace back to the same point of departure from the apes. With this indisputable fact as anchor, Vigilant ran the data through the computer program designed to look for the most "parsimonious" way to arrange them on an evolutionary tree. Once again, the computer produced a tree with clearly African roots. In fact, the pattern of branches on the first go-around through the computer led straight back to the !Kung—hence Wilson's claim in the *Los Angeles Times*.

Even Vigilant, however, wasn't sure about a specifically !Kung origin. When she spread a printout of her latest genealogical tree on the coffee table, all the different mtDNA types were arranged along the right-hand side, with their branches tailing off to the left. The length of the branches corresponded to the degree of difference between each type and the branch next to it, the differences arising from the mutations that had occurred since the two shared an ancestor. If the number of mutations was indeed a function of time, then the long-branched, very different mtDNA types were the oldest.

"We reanalyzed the data with a more powerful computer program," Vigilant explained. "In this new tree, the longest branches led to Pygmies. Then came some !Kung."

"Does that mean we came from a Pygmy ancestor instead of a !Kung?" I asked.

"Not necessarily. The computer can take the mtDNA types we collected and come up with a whole mess of different trees that aren't meaningfully different. This one is the simplest explanation of how all the lineages relate to one another. But we can ask the computer to develop a tree with one of the !Kung samples as the common ancestor, and it comes out almost as parsimonious."

"What happens if you put an Asian or a New Guinean at the root instead?" I wanted to know.

"We asked the computer to root the tree with non-African samples too. But it looks as if the parsimonious trees have African samples at the root. Which is what you would expect if the mitochondrial mother came from Africa."

"Even if you're sure that Eve is African," I said, "what about the timing?"

"Ah, the timing. The chimp comes to our rescue there too."

Having a monkey in the works, she explained, helped to figure out the rate of the mitochondrial clock—and hence the all-important age of Eve—in a different and more reliable way. Cann, Stoneking, and Wilson had calibrated the rate of the clock to the times of colonization of Australia, New Guinea, and the Americas. As Milford Wolpoff had pointed out to me, such a calibration method was fraught with assumptions. But because PCR had the power to compare the DNA sequences of humans and chimps directly, Vigilant could figure the rate of mutation based on the time of the split between apes and hominids. Almost everyone agrees that the apes split off from the hominid line between five and seven million years ago. Knowing the number of mutations that have accumulated over that vast stretch of time, Vigilant obtained a rate of change in mtDNA that she could then apply to splits within the human lineage itself. This gave her a date for the common human ancestor.

"Two hundred and eight thousand!" she exclaimed, raising a fist in mock triumph. "That's my new number!"

In essence, Vigilant's PCR-boosted research had examined more evidence, in greater detail, from a different perspective—and come up with the same starting point and practically the same age for Eve. Meanwhile, another graduate student in Wilson's lab, Anna Di Rienzo, had produced a study that put Eve even closer to the present—the 142,000-year date that Wilson had mentioned. Vigilant was nonchalant about the difference.

"Dating is still the weakest part of the whole procedure," she said. "We need more data from nuclear genes to really solve the problem."

It was time for my rendezvous with Public Enemy Number One. I thanked Vigilant and walked on back up the hill. At the time, neither of us realized that there was a still more critical weak point in the procedure, one that would eventually bring the Eve hypothesis close to extinction itself.

Allan Wilson had always been described to me in superlatives, such as "one of the real geniuses in science," or "the most arrogant guy I know." When I knocked on his office door, I was thus surprised to find it answered by a fragile man with wispy shoulder-length white hair. While his chief critic Wolpoff exudes size and solidity, Wilson seemed physically at the opposite extreme, a milkweed seed about to blow off in the wind. But the frailness was only corporeal. He apologized for putting me off for so long and bluntly explained that the reason he had done so was that he did not trust me.

"I am concerned about the role of science writers in this business," he said. "The anthropological perspective on evolution is no longer valid; it has been overthrown. And yet the science writers who insist on talking to me come drenched in an anthropological perspective, and there is really no point in talking to them. Frankly, I suspect that will probably be a problem with you too."

"What's wrong with the anthropologist's perspective?" I asked, unsure whether I was drenched in it or not.

"It is paralytic. It prevents you from asking certain questions, and it forces you to ask others. The whole discipline invites you *not* to investigate."

As Wilson defined it, the anthropologist's perspective is paralyzed especially by "the uniqueness problem"—the tradition of explaining the evolution of humans by behavioral and cultural processes that apply only to ourselves. Wilson's own lifework had been devoted to placing the evolution of our species firmly within the context of molecular biological principles that apply to any living thing. To this end, he had authored or co-authored well over two hundred molecular evolutionary studies, using a menagerie of nonhuman species, including geese, fruit flies, salamanders, extinct marsupials and extinct horses, hagfish, mice, frogs, pheasants, bacteria, and pine trees. Humans figured very heavily in his CV too, but they were pointedly *not* the be-all and end-all of existence. These studies had opened a door on the molecular basis of evolution, providing Wilson with solid funding for his lab over the years and a MacArthur "genius" grant for himself, deep respect from his colleagues and students, and persistent badgering by science writers and journalists.

This last group was being further encouraged by his tendency to push his ideas to extremes, often in highly visible forums. A few months before my visit, Wilson had announced at the annual meeting of the American Association for the Advancement of Science—surely the single most writer-rich science meeting in the world—that the Neandertals were replaced because they could not speak. In itself, this was not an original or extreme notion. If modern humans migrating into Europe squeezed the Neandertals into oblivion, it stands to reason that they possessed a powerful adaptive edge, and language is an often-cited candidate. But in his lecture, Wilson had taken the idea a giant step further, suggesting that a particular gene for language might have been carried in the mitochondria themselves. Since invading males would have been more likely to mate with resident females than the other way around, the offspring of sexual contact between the two groups would be "linguistically deaf-mute," like their Neandertal mothers. Thus disadvantaged, these "village idiots" would face the same fate as the mothers: extinction. Only the language-endowed African lineage would continue.

The language gene idea, and especially the unfortunate term "village idiots," elicited hoots of derision from the anti-Eve camp, and gave no joy to Wilson's colleagues.

"It isn't something we like to talk about," Mark Stoneking told me when I asked him for comment.

"Yes, everyone said this idea was outrageous," Wilson told me. "People were scathing about it. They said it was an embarrassment to the University of California."

For a moment he stared out the window, then he turned to me and explained. The idea that a single key mutation might lay the foundation for new nerve networks associated with language is not new. It was assumed that such a mutation would occur in the nuclear genome, rather than along the comparatively minuscule stretch of DNA in the mitochondria. But in spite of its diminutive size, the mitochondrial genome is known to code for proteins affecting functions of the nervous system—for instance, the eye disease and epilepsy Wolpoff mentioned. There was thus nothing inherently outrageous about the notion that a mutation in mtDNA could also have the positive adaptive effect of laying the foundation for complex language, which is, after all, a function of the nervous system.

"If such a thing were maternally inherited, this could explain why one population could expand at the expense of another even if there *was* interbreeding between them," Wilson said. "Cavalli-Sforza pointed out recently that until a hundred years or so ago, deaf-mute people rarely reproduced, because they were socially ostracized. If language was a function of the mitochondrial genome, then hybrids between a Neandertal and a modern might have suffered the same fate. Anyway, it was just a hypothesis. People called it 'bad science,' but I still don't think it's such a ludicrous idea. Already it has led us to look for coding regions on the mitochondrial gene that might be unique to humans, and it turns out that there are some. So even if the hypothesis is wrong, it has already proved useful. Which is good science, not bad."

But if mutations in mitochondrial genes can produce such an enormously valuable adaptation as language itself, how could these same mutations be considered *neutral* to selection when it came time to construct a molecular clock? Wilson sighed when I asked this, a shiver seeming to pass through his thin frame.

"We never said the mitochondrial genome as a whole was neutral to selection," he replied. "That would be silly. But the vast majority of the mutations there seem to occur in regions of the DNA that do not code for proteins, so they have no effect on the survivability of the organism. If you concentrate on noncoding regions, as Linda has, then of course you are even more secure. But it doesn't really matter. Wolpoff says we are assuming neutrality. This is bull. The steady rate of the clock is empirical. We observe it, we don't assume it. Does he want to know why we think as we do, or does he just want to win an argument?"

"But the mitochondrial clock even within your own lab now looks as if it is telling two different times for the common ancestor," I protested. "How old *is* Eve?"

"The uncertainty as to time continues, but in a direction that I am afraid will cause even more flak. The lower limit may not be one hundred forty thousand, but one hundred thirty or even one hundred twenty thousand years. We are bending over backward not to overstate our case and alienate the anthropologists. But I don't see how we can avoid it."

Listening to Wilson, and later to his student Anna Di Rienzo, it was clear that there would be more trouble ahead. The new, younger dates Di Rienzo was producing were even further from the million-year-old age for Eve that anthropologists like Wolpoff and Fred Smith would love to see. Her results were even more volatile in another respect. In her study, she analyzed mitochondrial DNA from the placentas and hair of 117 Caucasians from the island of Sardinia and from the Middle East. Her goal was not so much to try to find an age for the common mother as to pin a date on the second significant event in the Eve hypothesis: the time her descendants first left Africa to populate the rest of the world. When Di Rienzo and Wilson organized the Caucasian mitochondrial types into a genealogical tree, they noticed a sudden rush of new branchings taking place about 60,000 years ago. There could be several explanations for this explosion of new mitochondrial types, but the most likely was a rapid increase in population. This is precisely what one would expect to find if a population was expanding into new territory. In other words, the study may have discovered inside the mitochondria what archaeologists have not been able to find in the ground: a physical trace of the migration of moderns out of Africa.

"So the origins of Caucasians seem to correspond with the demise of the Neandertals," Wilson observed blithely, twisting another molecular knife into the guts of any theory giving ancestral status to Neandertal Man.

If Di Rienzo's more recent dates for Eve were right, another fascinating phenomenon demanded explanation. Human mitochondrial DNA appeared to be evolving at a faster rate than that of other species, confounding all expectations. "I'm surprised that the human clock seems to be ticking faster," Wilson said. "But if it is, then humans offer a unique possibility to understand what determines the rate of molecular evolution in the first place. But, of course, the anthropologists don't care about that."

Wilson had a hypothesis about why the human molecular clock may be accelerating, which he proceeded to elaborate for me, accelerating himself as he probed its intricacies. His voice was delicate, sometimes fading almost to a whisper, but the force and rhythm of the ideas pulled me in, and I listened, fascinated, the rest of my questions forgotten. What drove up the speed of evolution in humans and all large-brained creatures, he said, was the brain itself, pushing innovations rapidly through a population, favoring not only those with the minds to innovate but those who can "catch on" to the innovations as well. Ultimately it all came back to molecular mutations. In a brain there must be a *mutation for the ability to catch on*, to detect fellow creatures who have discovered something and to imitate it if it is good. Among those in the population who catch on, some do so because they have more neurons to begin with, increasing average brain size in the next generation, and thus creating a positive feedback loop generating ever-faster evolution.

"A finding like that convinces me that molecular biology has introduced a new way of understanding the evolutionary process," Wilson said. "But it is not the past we are trying to understand, it is the future. The rate and direction of change."

"And what do you see in the future?"

"I am not alarmed by it. The future is going to be grand. But it will be a hell of a lot different than now. The life force may well have moved elsewhere. It may not, for instance, involve human bodies anymore."

Wilson gazed out across the Berkeley campus toward San Francisco Bay. After a moment he looked up with a start, as if he was surprised to find me still sitting there. "Well," he said, and simply walked away, without a handshake or a good-bye, gliding out the door and off down the hall, seeming to thin out and disappear, even before he turned the corner.

A year and a half later, Allan Wilson died of acute leukemia.

*   *   *

A couple of weeks after my second visit to Berkeley, Wolpoff telephoned to encourage me again to attend the American Association for the Advancement of Science annual meeting in New Orleans. "We'll tell you all you need to convince you that the Eve hypothesis is dead," he said.

As it happened, I had already bought my plane ticket. It was February 1990, and the modern human origins debate was still the hottest topic in science. The AAAS annual conference had scheduled time for two full sessions dealing with it: Milford's mass gathering of multiregional continuity advocates, and a seminar organized by Erik Trinkaus, the Neandertal expert at the University of New Mexico. Alan Thorne was flying up from Australia to join Milford's seminar, and Chris Stringer was coming from London to join Trinkaus's. Linda Vigilant would be in New Orleans too, with Mark Stoneking accompanying her. I could not afford to stay home.

The two seminars were scheduled on the same day, with Stringer slotted to speak first in Trinkaus's morning session, and Wolpoff's group presenting after lunch. Milford had also organized an early morning press conference, and the room was already packed when I arrived. At a table by the entrance were neat piles of press releases and a stack of eight by ten glossies of Milford, courtesy of the University of Michigan information office, looking solid and authoritative before a rack of skulls. The continuity advocates were ready at the table in the front of the room: Geoff Pope, Fred Smith, Alan Thorne, David Frayer, and a geneticist named James Spuhler, with Milford himself in the middle. Chris Stringer was in the audience. I asked him if he had any surprises in store.

"Not much that I didn't tell you already," he replied, "though I imagine it will be new to a lot of reporters in this room. Of course, they'll have to choose between getting the scoop pre-digested here or rushing over to the other side of the hotel to hear my more technical talk. Nice planning, Milford."

Trinkaus's seminar was more evenly weighted, including both a staunch Out of Africanist like Stringer and a geneticist in the continuity mold named Ken Weiss, as well as Trinkaus's own hybrid viewpoint. He had little use for Eve but did not think much of Neandertal's level of "humanness" either. "Chris and Milford *both* accuse me of betraying their causes," he said with a grin.

I was anxious to hear the talk by Geoffrey Long, Trinkaus's colleague at the University of New Mexico who was about to go public with a new study of the beta-globin gene—the same bit of nuclear DNA that had led Oxford's Jim Wainscoat to point to Africa as the human homeland. Even the mitochondrial people were admitting that the *real* answers to modern origins were cached away in the cell nucleus, where most of our genes reside. This was Long's territory. Maybe what he had to say could break the deadlock.

Long's analysis of over eight hundred different individuals—Senegalese, Nigerians, British, Italians, Thais, Indians, Polynesians, and others—revealed sixteen common versions of the gene distributed among the different geographic populations. The most common versions appeared far more often in Africa than anywhere else. In theory, these types should also be the oldest, so his work was pointing to an African ancestor. Even more provocatively, there was a remarkably sharp contrast between the Africans and non-Africans in the frequency with which certain haplotypes appeared.

"Fully half the Africans carry a gene type that I can't find anywhere in a sample of two hundred Europeans," he told me after the talk. "That's saying something really meaningful."

Meaningful—but not conclusive. Long suspected that if the two populations had been separated for half a million years or more, as the continuity theory holds, gene flow would have erased the profound contrast between them. The fact that it is still there today argues for a more recent, common African ancestor. But he refused to put a date on that ancestor. In fact, he did not believe anyone could. The information needed to establish the time of divergence by looking at the beta-globin gene is hopelessly muddled by changes in population size and migration patterns over the centuries. The same would be true for the Wilson group's attempts to define Eve's age with mitochondrial genes.

"Intuitively, I would think the common ancestor was recent," he said. "But I have no real basis for that. All I can really say is that it looks like there was a migration, but I can't tell you when."

Wolpoff's partisan gathering commenced after lunch. Fred Smith poked at the supposedly secure fossil evidence for the early appearance of moderns in Africa, and left it looking not so secure. Geoff Pope and Alan Thorne paraded the regional fossils of China and Australasia. Dave Frayer pleaded the Neandertal case for ancestry in Europe. Wolpoff and Spuhler went after Eve directly, ticking off the reasons why molecular clocks don't tock. In the audience, Mark Stoneking twisted in irritation.

"I'm really surprised, Milford, by all the inaccuracies in your presentation," he said, when Wolpoff had concluded his talk. "It would take up the whole question-and-answer period just to spell out the inaccuracies—."

"If you have a question, Mark, then ask it," Wolpoff interrupted from the lectern.

Later, Wolpoff bridled when I suggested that he might have cut off his critic unfairly: "As the seminar organizer, I was given explicit instructions not to allow people to make speeches during the question-and-answer period. Mark wanted to make a speech, and it wasn't the appropriate place for it."

By the time the seminar ended at five, my brain was so full of regional features and arguments against Eve that I could barely think. The glazed eyes of others in the audience revealed the same state.

"A friend of mine compared Wolpoff's seminar to a sales pitch for a time-share condo," Linda Vigilant said afterward. "They brought in one guy after another to hammer at you until you wanted to sign on the dotted line out of sheer exhaustion."

"It was *meant* to be a sales pitch," Wolpoff told me. "We planned the whole thing, rehearsed it, worked over the exact phrasing. We felt that we had to do this, because we were the victims of the complexity of our own ideas."

The most heartfelt pitch of the day was given by Dave Frayer, another Wolpoffian protégé, now at the University of Kansas. Frayer has made a career of championing a direct ancestral connection between European Neandertals and modern Europeans. Frayer argued that paleoanthropology's knee-jerk preference for "sensational catastrophism over mundane gradualism" had led most of his colleagues to cozy up to the new genetic evidence for replacement without so much as a second glance at the fossils themselves. "Unborn historians of paleoanthropology," he admonished his audience, "will one day marvel at the willingness of some human paleontologists in the late 1980s to eliminate the rich, premodern human fossil record from any kind of contribution to the course of human evolution."

Frayer's own reading of the record reveals a number of overlooked traits that clearly and specifically link the Neandertals to the Cro-Magnons. One such trait is the shape of the opening of the nerve canal in the lower jaw, a spot where dentists often give a pain-blocking injection. In many Neandertals, the upper portion of the opening is covered by a broad bony ridge, a curious feature also carried by a significant number of Cro-Magnons. But none of the alleged "ancestors of us all" fossils from Africa have it, and it is extremely rare in modern people outside Europe. By far the simplest explanation is that the Cro-Magnons inherited the trait directly from their Neandertal forebears. Frayer knows, of course, that it takes more than a single, rather obscure feature like this, or even an assortment of features, to convince people that Neandertals deserve a place in our ancestry. But he isn't about to stop trying. "If there is a Neandertal heaven, maybe they are up there smiling down on me right now," he concluded. "Nobody else is."

Frayer took questions after his talk. Again Mark Stoneking stood up in the audience. If these morphological features under discussion are genetically determined, he asked, could Frayer tell *which* genes determine *which* traits?

"Of course I can't do that, whoever you are," Frayer snapped back. "Nobody can."

I was surprised by the professed lack of recognition, because surely the two must have bumped into each other at some point through the months of battling over Eve. When the afternoon seminar was over, I asked Frayer whether he was being disingenuous.

"I knew who he was all right, if that's what you mean," he answered. The put-down was a reaction to what he felt was a deliberate insult on Stoneking's part—a thinly disguised jab at the futility of drawing any conclusions at all from mere morphology. "I hope I made him look like a fool."

Frayer was due to meet his colleagues for a drink and invited me to come along. The anthropologists had collected in a bar full of potted palms beside a synthetic brook in the hotel lobby. All of the continuity people were there, including Milford with his young wife and colleague, Rachel Caspari. I was surprised to find Mark Stoneking and Linda Vigilant in the group as well, squeezed together along a bench between Fred Smith and Alan Thorne. Chris Stringer was there too, fighting for space with a wayward palm frond. Milford was handing out drinks and cradling his new baby. The presence of the infant helped soften the scene. A basketball game was playing on an overhead TV, and we all watched it. I was lulled into the sense that some kind of rapprochement was under way, that no matter how bitterly the academic wars raged, at the end of the day they were, after all, only academic. At heart, these people were all fellow seekers. The only real outsider here was myself. But then I looked again at Mark and Linda. They weren't saying anything, just sitting there pressed side by side, as if seeking mutual comfort, holding their beers stiffly in their laps. They looked like a couple of missionaries in some old movie, invited to dine with the natives but not sure if they themselves were the planned main course. Stringer, too, looked as if he'd rather be drinking alone. When they had the opportunity, the two geneticists made their excuses and left. There was a movement toward going out to dinner, but I had had enough.

For a while, Wolpoff's persistent efforts to claim a hearing for his views paid off, at least in the media, where I sensed a slow shift toward his position. Then, when Linda Vigilant's expanded study using PCR hit the pages of *Science* in the summer of 1991, the Eve hypothesis gained new momentum. But there was a sad note in the triumph; the paper was dedicated to the memory of Allan Wilson. Six months later, the Eve hypothesis faced another identity crisis, over an issue not even Wolpoff had predicted. I followed these shifts, read the reports, even dropped into a few more meetings. But there was a limit, finally, to how much I could learn by watching the watchers, and that limit was reached in New Orleans. I was restless to hear another voice, to get away from the bones *and* the genes, and approach the mystery through a wholly different entrance. Neither fossils nor DNA can tell us what our ancestors were actually *doing* in those silent centuries. Whatever it was that coaxed us over that final threshold to humanity had to be something embedded in the lives of the Neandertals and the anatomical moderns who shared the same time, in another world. Ultimately this story, like any good story, is about how people behave.

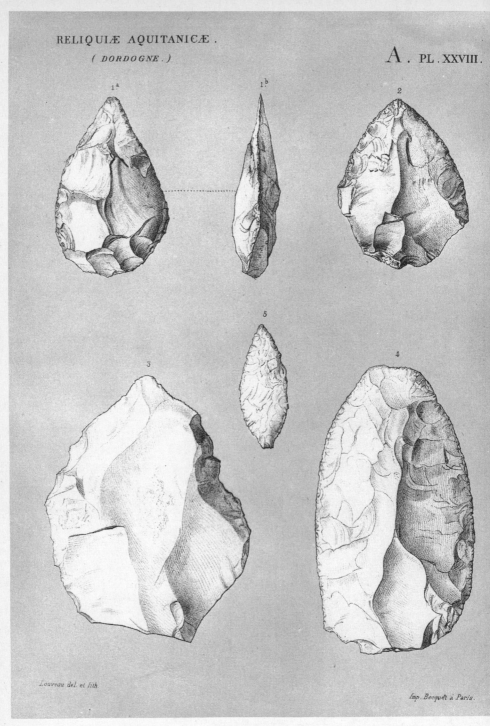

1ª   1ᵇ   2

5

3   4

Louveau del. et lith

Imp. Becquet à Paris.

An 1875 engraving of stone tools from the rock shelters above Le Moustier in southwestern France, the village that gave its name to the "Mousterian" culture of the European Neandertals   FROM LARTET AND CHRISTY 1875

Chapter Six

WELCOME TO THE STONE AGE

The first rule of anthropology is that if everybody believes what you've said, you've probably got it wrong.   —OWEN LOVEJOY

In one corner of my desk I maintain a little shrine to my ancestors. Striding confidently toward my Rolodex is a family of australopithecines. Made of molded plastic, they walk on bases painted rainy-season green. The female cuddles a chubby baby in her arms. Walking ahead and slightly to the left, a muscular male holds a crude stone tool in his outstretched hand. Next to my early hominids, I keep a couple of real stone tools. One is a shapely teardrop, about two and a half inches long by two inches across at its widest, composed of black obsidian. Whoever made this tool clearly knew what he, or she, was up to. It is almost perfectly symmetrical, the edges tapering to an even point. The other tool might be described as a sharp rock. Larger by half than the other artifact, it is uglier by two thirds—nothing but a gray flint flake of inscrutable proportions, not very different from something you might pick up in your driveway. The flake is comfortable to hold and bears a cutting edge sharp enough to slice paper or, as my daughter found out, to cut an inquisitive finger.

These objects came into my possession on a summer afternoon in Cambridge, Massachusetts, in a tranquil courtyard behind Harvard's Peabody Museum. I was escorted there by John Shea, a graduate student and expert stone knapper, who had been making stone tools for most of his life. (He is now an assistant professor of archaeology at the State University of New York at Stony Brook.) Shea started knapping at seven "because it was something neat to do with my hands," knapped his way through

Boy Scout survival training, and only later discovered that the stuff he had been banging out in his backyard for so long had meaning in a larger intellectual context; that there were academic traditions, schools of thought, swings of fashion, and heavy self-important debates emanating from these blank and innocent forms. Bearded, blue-eyed, with a woodsman's thick body, he was easy to imagine in a Boy Scout uniform, loaded down with merit badges. His style of archaeology, if not wholesome, was earnest and direct.

"I almost got kicked out of Harvard for throwing spears at dead deer," he confessed, referring to a series of tests he had conducted to analyze the damage to stone projectile points when they impact hide, muscle, and bone. "They were upset that the experiment might raise protests from animal-rights groups." Shea's habit of piss-tanning hides on the roof of the anthropology building was also noted with concern by the administration. While still undergraduates, he and another student conducted an experiment to see what kind of wear patterns resulted when an ancient tool was discarded and subsequently kicked around by passing animals. To telescope thousands of years of sporadic trampling into a few hours, they filled a wooden box with dirt, threw in some replicated flints, and stomped on them for a while, sending dirt flying through the graduate offices and waking up the preschoolers in the faculty day-care center downstairs.

I had come to Shea to learn how Neandertals actually made their tools. The technique most closely identified with the Middle Paleolithic throughout Europe and the Middle East is called "Levallois," after the Parisian suburb that started spewing up ancient tools during the construction of the Paris Metro system. To many archaeologists, Levallois represents a significant technical "leap" over the hand-ax technology that preceded it for a million years. But some leaps are greater than others.

"Middle Paleolithic tools were still pretty simple," Shea told me. "Basically, they were just trying to get more cutting edge from a core."

He had brought his knapping kit along in a drawstring bag. Picking a spot in the shade of a maple tree, he laid out a tarp and began to unpack the tools of this most ancient of trades: well-worn hammerstones for striking flakes off a core of flint, soft billets of boxwood and antler, all-purpose wooden shafts waiting for points or blades to define their use, grindstones for abrading edges, pressure flakers made of bone for hard-squeezing flakes off a tool's edge, metal files("OK, sometimes I cheat a little"), and leather protectors for his knees and hands. Last out of the bag was a tin of Band-Aids.

All that was missing was the raw material. He wandered into a patch of myrtle and picked up a dusty glob of obsidian. This glassy volcanic rock is not a natural part of the glacial lag of eastern Massachusetts; John had cached it in the myrtle himself. All around us, in fact, the courtyard was peppered with hidden hunks of raw flint and obsidian. Where other stone knappers might heap their raw material in an old orange crate in the corner of a lab, Shea had strewn his about the Harvard grounds. Every time he began a knapping session he played out a miniaturized version of the first requirement of the ancient knapper: Go out and get some rock.

"Flint knapping is more than a research tool," he told me as he squatted down on the tarp. "It's a way of getting inside the mind of the guy who did this because his life depended on it."

He began to work the piece of obsidian, at the same time telling me blow by blow what he was doing. I had seen plenty of Levallois step-by-step drawings, so I knew, abstractly, what he was after. The hallmark of the technique is control. Rather than just whittling a lump of flint down to a usable shape, the Levallois knapper preworks the core into a form that will deliver a flake of predetermined shape and thickness. Typically this is done by first striking off a series of small waste flakes, or *débitage*, around the periphery of the core. The desired result is a roughly tortoise-shaped thing with a rounded bottom and a flattened top. The next step is to prepare a "striking platform" at an acute angle to the tool-to-be, which will receive the final blow releasing the flake from the core. After the flake is struck, the core can be remodified with a few well-placed taps, preparing the way for another flake to be removed. In the hands of a good knapper, a core can deliver several strong, thin flakes, greatly increasing the amount of cutting edge available from a given hunk of raw stone. Each flake can be further modified, or "retouched," to the desired final product.

While he talked, John's piece of obsidian was shrinking under the sharp raps of his hammerstone. He picked up a "soft" hammer made of antler, gave a series of quick little taps at a chosen spot, turned the core, and tapped again. He explained what he was doing, studied his shrinking lump, then worried a spot with a grindstone. I got the impression that his knapping and his explaining were interfering with each other. He wanted to make his verbal points clear and his stone edges sharp, but voice and hands were at odds, competing for his concentration. He cut a finger, slapped on a Band-Aid, and got back to work. But it was not going well.

"The trick with knapping," he said, "is always to have two solutions to a problem, in case the first one fails." He labored it some more, but the rock wasn't cooperating. Irritably he pitched what was left of the tool-in-progress into the myrtle. Then his Scout-calm smile returned. "You know, I swear at some sites you can find mistakes—broken arrowheads, flawed cores, and so forth, just an angry toss away from a pile of *débitage*."

Once more Shea foraged for raw material and came back with another piece of obsidian. This time the prep of the stone went better. Sitting cross-legged with the core held on his thigh, he tapped, ground, and tapped again, switching implements gracefully when necessary. Finally the core looked the way he wanted it: a rough disk with a steep shoulder all around it. He held the core cupped in his left hand, with the striking platform ready to receive the blow.

"This is when prehistoric man got ulcers," he said. "After this, there are no second chances." He gave it one sharp *thwack*. When he unfolded his left palm, the desired flake dropped off underneath the core. A few light strokes along the edge, and it became the tear-shaped point that would later grace my desktop. Rolling up a trouser leg, he shaved off a swatch of leg hair in one stroke.

"Here," he said, passing the tool over. "Go get a mammoth."

The undersurface of the flake was glassy-smooth. Up close, I saw the textbook-perfect "bulb of percussion" at the butt, a swelling created by the physics of impact when the core was struck. I tested an edge with my finger.

"Be careful," John warned. "The edge of a fresh flake of obsidian can be as thin as a single molecule." A couple of weeks before, he said, Irven DeVore of the Harvard anthropology department had been scheduled to have surgery to remove a couple of minor melanomas on his face. He asked Shea to make a scalpel for the surgeon. John knapped some small blades out of obsidian, melted pine rosin from a tree, and dipped the blade and a wooden shaft into it. Then he jammed the blade and shaft together and lashed them up with sinew. DeVore came out just fine. The surgeon told Shea later that his obsidian scalpel was better than a metal one.

No one, not even Shea, would argue that Neandertals were producing surgical-quality scalpels back in the Middle Paleolithic. He is convinced, however, that they *were* hafting stone points.

"It's part of the received wisdom that Neandertals couldn't haft tools," he told me. "If you ask why not, the answer is 'They weren't smart enough.' If you ask how we know they weren't smart enough, the answer is 'Because they couldn't haft stone tools.' "

Simply hafting a tool, however, is not the same as making a spear that can be thrown accurately from a distance. Not everyone agrees with Shea that Mousterian points were used as projectiles. Many archaeologists believe they would have been too heavy.

"If you put one of the things that people call Mousterian points on the end of a stick and try to kill an animal, I venture to say the animal will kill you first," says Randy White of New York University. Another objection focuses on the butt end of the tool. In the course of hunting, spearpoints inevitably go dull or snap off. What is valued and kept by a hunter is not the point but the shaft, much the way you keep your safety razor around but periodically throw away the blade. Like razor blades, spearpoints need standardized butt ends that fit into a shaft without wiggling around. And by the look of them, Middle Paleolithic points did not. The Neandertals who made them left the bulb of percussion intact, which would have created a very poor contact between a point and an unmodified shaft tip.

Shea discovered, however, that cutting a beveled shelf out in the tip of a spearshaft and laying the opposite, concave side of the point against it created a very cozy fit. So perhaps the Neandertals were modifying the *shafts* to fit the unwieldy shape of their Levallois points. "Slop in some pine pitch, lash it tight with leather or sinew, and you have a very effective projectile. Especially if you happen to be a Neandertal with the upper body strength to make an Olympic javelin champion look puny," he explained.

Demonstrating that the tools *could* have been hafted does not, of course, mean that they were. To support his notion, Shea had to find out what would happen to replicated points that were hafted and thrown into prey-flesh: hence the research heaving spears into dead horses and goats. Shea also used knapped tools of various sorts to butcher meat, carve bone, hack and scrape at plant material, work wood, dig earth, and perform dozens of other activities to which the Neandertals might have subjected their own tools. This collection of experimentally worked artifacts was then examined under a microscope for traces of microwear—the patterns of tiny pits, scars, fractures, and polish that, in theory at least, betray the functions that the tool performed.

With this catalog of *known* functions and their micro-effects to refer to, Shea turned his microscope onto the tool collection at the key Neandertal site of Kebara in Israel. Out of over seven thousand tools examined for microwear, 7 percent bore traces of some kind of use, woodworking being the most prevalent. Projectile points accounted for less than half of 1 percent of all tools examined. But fully half of the Levallois points Shea checked for damage showed the telltale pattern of impact fractures and haft wear. Or at least Shea thinks so. Other archaeologists I have talked to point out that, even when carefully interpreted, microwear is fraught with problems. Five minutes of tanning hide, says one, can make a tool look a lot like one used for a minute working bone.

All the time Shea was telling me about step fractures and impact damage, he was busy prepping another core, a gray, pearly flint, discovered under the back stoop of the Zoology Department. Either the flint was easier to work, or else Shea had warmed to the task; but the core virtually melted into shape, a shower of *débitage* falling on his leather apron. His hands seemed to have gone off on their own, working in a happy digital excitement while the rest of him glided along in a lower gear. Watching the delight in John's hands, an urge began to tug at my own fingers. Something in my posture must have given it away.

"I'll let you finish this one," he said, handing me the core and a well-worn hammerstone. The real work was already done. All I had to do was whack it at the right angle on the striking platform. But that proved more difficult than it seemed. First I couldn't find exactly where the striking platform was, so he took the core back and drew a little red X on it with a felt marker. Then my grip on the hammerstone was wrong.

"Hold it like you're throwing a baseball," Shea told me. I wrapped two fingers around the smooth stone and snapped it down on the marked core, using more wrist. It took a few swipes, but then, *thock*, I felt the flake release into my palm. I lifted up the core, and there it was—not a pretty thing, but a good, usable tool even so, cool and comfortable against a curled finger, something that could slice through fresh meat and leave its mark on the bone beneath. I was surprised by a sudden, unaccountable shiver of pride.

"Congratulations." John smiled. "Welcome to the Stone Age."

If only it were that easy. The problem with understanding Neandertal behavior, why the complexion of their lives remains just beyond our reach, is that we have nothing to compare them to but ourselves. Which, if you think about it, is a dangerously limited reference point, opening a wasp's nest of opinions on how much they approached or fell short of the modern human way of doing things. Left unknowable are all the ways in which they were off behaving in another realm altogether. Curiously, this conundrum is less of a problem when looking at hominids who are further removed from *Homo sapiens*, the australopithecines and the early *Homo* species who endured their millennia on the other side of the gartel. As sources of analogy for their behavior—however inexact—we have not only ourselves but the behavior of great apes, various monkeys, and social predators like wolves and lions. Like the way your eyes supply depth to what you see by coordinating images from two points of reference, this dual perspective on early hominid behavior—the frankly biological, the supremely cultural—gives these distant ancestors an illusion of three-dimensionality. Not so the Neandertals. They press against us with their distended features and their heavy presence, too close for our eyes to bring them into focus.

To see Neandertals, you must look at them obliquely. The traditional way of doing this is to peer closely at their tools. A week after visiting John Shea, I was walking along a high, wooded cliff, five thousand miles farther into the Stone Age. The path was slippery and steep. I grabbed gnarled arms of live oak and boxwood to keep from sliding downslope. Somewhere below me the tiny Céou River flowed toward its confluence with the Dordogne, but the Céou, its valley, and the rest of the modern political entity called France were deeply buried in cotton-thick morning fog. The fog sealed off all trace of the post-Paleo world save one: the pale ramparts of the castle of Castelnaud, which rose sun-gilded out of the mist across the valley, like an apparition misplaced from a comic-book fantasy.

This is a good place, a classic place, to look for Neandertals. Long ago, the plethora of limestone caves along the Dordogne and Vézère rivers and their tributaries acted like magnets on the landscape, luring Neandertals as they had lured their ancestors and would entice the Cro-Magnons who followed them. Within thirty miles were literally hundreds of Middle Paleolithic sites, including some whose very names excite the hearts of paleoarchaeologists. *Le Moustier*, just north of the Vézère, whence comes the label Mousterian for the Middle Paleolithic industries stretching across Eurasia, and the burial site of a Neandertal youth.* *La Ferrassie* a

*The skeleton was excavated in 1909 by the infamous German archaeo-hustler Otto Hauser, who

few miles west, common grave of two adult Neandertals and six children, rich in artifacts. Upstream again, *Régourdou*, another famous burial site and alleged sanctum of bear-cult Neandertals. Straddling the Dordogne to the south, *Pech de l'Azé* and *Combe Grenal*, with their deep-piled layers of time. Farther east, *La Chapelle-aux-Saints*, where Marcellin Boule started all the trouble for Neandertal Man. If you add to this hallowed litany a few choice names from the even more numerous Upper Paleolithic sites in the same triangle—*Cro-Magnon, Abri Pataud, La Madeleine*, and a hidden wonder called *Lascaux*—it is easy to succumb to the common illusion that modern human evolution is something that happened in the Dordogne.

Ahead of me on the path were my hosts, Jean-Philippe Rigaud, Director of Prehistoric Antiquities for the Department of the Aquitaine (analogous to Director of Coal in Newcastle), and Jan Simek of the University of Tennessee. Rigaud is local raw material, brought up in Bergerac just a few miles to the southwest. He has been frequenting these cliffsides since he was a boy visiting his grandparents' farm, later becoming the student, assistant, and longtime colleague of the great French archaeologist François Bordes. Simek, by contrast, is exotic to this area: born in Great Neck, Long Island, to Czech parents, raised in the mountains of California, and now the proud owner of twelve acres in the Tennessee woods. Both men were wearing blue work shirts and jeans, but where Rigaud's seemed like the working attire of a gentleman scholar, Simek exuded back-country ruggedness, with a face knapped from stone and polished with use.

Walking jauntily beside Rigaud was a third man, white-haired, shorter, spectacled, and wearing a short-sleeve dress shirt, Bermuda shorts, black socks, and sneakers: the summer uniform of the indoor scientist venturing outside. Henry Schwarcz of McMaster University in Ontario is a veteran innovator in uranium-series and electron spin resonance dating, perhaps the most important figure in the dating revolution now taking place in modern human origins. He is busy and ubiquitous, constantly working on improvements in methodology while eagerly, even hungrily, dating sites in Europe, Africa, the Middle East,—wherever his precious, arcane services are demanded. I have heard it rumored that Schwarcz will date your site whether you want it dated or not.

The limestone cliffs we were traversing, known as the Massif du Conte,

---

frequently arranged to have it reburied and discovered all over again whenever some important personage visited the site, before selling it off for profit to a museum in Berlin. In a sad twist of irony, the skeleton below the neck was blown to pieces by an Allied bomb during World War II.

run for a cave-pocked mile above the east bank of the Céou. Many of the twenty-odd recesses along the massif have yielded traces of prehistory, but only two proved worthy of extensive excavations. Above us, obscured by a lip of backfill, was the opening of Grotte 15, better known as Grotte Vaufrey, excavated by Rigaud from 1969 to 1982, and now dormant. Farther down the path lay Grotte 16, Rigaud's active dig. Taken together, the two sites cover a time span equal to most of the human occupation of the continent. The lowest levels of Vaufrey bear tools from the Acheulean period, perhaps as old as 400,000 years. Above these lies one of the oldest Mousterian levels ever found, dated now—by Schwarcz—at 250,000 years, more than doubling the conventional notion of Mousterian time. The deposits at Grotte 16, just a few steps farther along the ridge, overlap at the bottom with the youngest ones at Vaufrey, about 60,000 years before the present, then progress on through the various industries of the Upper Paleolithic, into Neolithic times. Topping it off is a *remanié* of ahistorical detritus spewed up by water action, insect and rodent-burrowing action, and human puttering-around action. In this archaeologically useless foam one can find everything from Mousterian flakes to broken wine bottles and candy wrappers.

In a few minutes we reached the dig, a mitten-shaped cavern occupied by an international assortment of archaeologists and students, who looked up from their crouched attitudes, squinting into the sun to check out the late arrivals. Rigaud took us back into the dark thumb of the mitten, where two longtime colleagues, Roy Larick and Todd Koetje, were removing blocks of travertine from a rich Magdalenian layer—the last part of the Upper Paleolithic period. What was needed was a uranium-series date from Schwarcz on the travertine, to give an upper limit to the age of the Magdalenian level. Larick held up a block he had found that happened to contain a pocket of carbon, fortuitously targeting it for radiocarbon dating too. Schwarcz, faced with the prospect of two independently derived dates for the same sample, was in raptures.

"Don't bother to wrap it," he exclaimed, "I'll eat it here."

While Rigaud and Schwarcz discussed the geology in the back of the cave, Simek took me back to the main section where most of the students were digging. He wanted to show me where a hearth had been found in a level almost at the bottom of the sequence, down in the Mousterian period—Neandertal time. (It was later dated at 58,000 years ago.)

"Aren't hearths rare in the Mousterian?" I asked him. "Some people have told me there aren't any."

"That's what some people say. I find them all over the place."

Another archaeologist once told me there are two ways to look at Neandertals: You love them or you hate them. Jan Simek can clearly be counted among the lovers. Like any faithful admirer, he is quick to defend their accomplishments, and impatient with those who belittle them. He will concede that Neandertals were replaced in Europe by people with a more modern anatomy, but not without making their contribution to the gene pool first, and certainly not because they were stupid. To Simek, the idea that Neandertals were overrun in Europe by a *biologically* superior people is anti-evolutionary, politically suspicious, and just plain wrong. In his view, there is little evidence for a "great leap forward" at the beginning of the Upper Paleolithic. The arrival of the Aurignacian people—the first Upper Paleolithic folk—into Western Europe was just one movement among many, each of which brought its own culture: the Gravettians, the Solutreans, the Magdalenians, and so on into the agricultural revolution of the Neolithic. Whether one group of people had larger browridges or more muscle mass, Simek believes, is totally irrelevant to the movements of cultures across the landscape.

"These days, most people would be loath to equate biology and culture," he told me.

"But until a few years ago, wasn't that exactly how everybody was explaining the suddenness of the change from the Middle Paleolithic to the Upper Paleolithic?"

"Sure. And not so long ago, they hung black men for going out with white women too."

Out in the field, Simek's faith in Neandertal humanness translates into a sharp eye for continuities in behavior across the transition, even where others declare sharp breaks. Hence the excitement over the hearth at Grotte 16, where others might see merely a clot of charcoal, burned soil, and ash. But the distinction, though crucial, is at heart semantic. It all depends on what you call a hearth. As Simek explained, an unstructured pocket of charcoal and ash like this one is more properly called a "combustion area." Nobody doubts that Neandertals used fire, but there is a loaded argument over how *well* they used it, compared to the Cro-Magnons who came after them. Some archaeologists see signs of true hearths—pits lined with stones, modified with tunnels to improve draft, and so on—at Aurignacian sites like Abri Pataud, right at the beginning of the Upper Paleolithic. Others—and in this category you find a high percentage of Neandertal lovers like Simek—don't believe that such organized site structures came in until very late; for instance, at the site of Pincevent to the north, only some 12,000 years ago. If so, damning distinctions in social organization between Neandertal and modern humans based on site structure tend to blur. In archaeology, the facts rarely speak for themselves. What you see depends on what you are looking for.

"I'm a child of the sixties," Simek proclaimed. "I don't look for differences. *They* do. I look for similarities. Similarities go right over their heads."

I didn't have to think hard to imagine who "they" were, having interviewed many of them already, towered in Berkeley or pacing their basement lairs at The Natural History Museum in London. And having met *them*, I have to wonder whether the opposite of a Neandertal lover can be called a Neandertal hater or, more to the point, whether there really is entrenched political content in the way we choose to look at the past. Like Simek, I am a child of the sixties. But must my belief in human equality extend to my extinct brothers and sisters, if extinction was indeed the Neandertal fate? Is somebody a closet racist if he believes Neandertals were less than fully human? Or, turning the question upside down, are the real "racists" those who cannot imagine a world inhabited by two kinds of human, who cannot allow the vanished version the latitude to enrich the past as a wholly separate, completed human experiment?

Simek descended into the cave to help some students piece-plot a find, and I headed back down the trail to take a look at Vaufrey, alone. I knew that science was the only way I can get to where I want to go, but I needed a break from it. I climbed a near-vertical path, skirting the hardened backfill that hung from the mouth of the cave like a dry tongue. The cave entrance bristled with chain link and barbed wire thrown up against the inevitable pothunters. Behind the fence was a ravaged dusty hole, enlivened only by the squeaks of quick-dipping swallows. The place seemed unnaturally gutted, like a cadaver left behind after its organs have been removed. The act of archaeology had transcribed "Vaufrey" from the name of a cave into a body of data, printed up in a thick monograph on satiny, expensive paper. This abandoned cavity, a little rank and streaked with bird droppings, was no longer Vaufrey, but what was left after Vaufrey had been systematically extracted.

Still, to this spot the ancient ones had come, over and over through the Stone Age, pressing with their padded feet on the long-cold hearths of those who had come before. The past, immune to our spins upon it, *happened*. I sat down on a rock in front of the cave and looked out through the trees and over the valley. The castle was out of sight, and with the fence and the sad, violated cave behind me, there was nothing in my perspective that would not have been shared by a Neandertal squatting on the same spot 100,000 years ago: twisted trunks, swallows, the blanket of distilling fog below, the bright blue sky above.

"The fact remains there is a truth out there," Simek had told me earlier. "The only way to get at it is to get your hands dirty in the stuff."

Four miles to the east lies Combe Grenal, tucked in a dry valley about three hundred yards from the Dordogne. In 1953, François Bordes began what was intended to be a quick season's excavation there. Thirteen years later, he could look up from a massive hole in the ground some forty feet deep, an excavation that had uncovered sixty-five distinct archaeological levels spanning 100,000 years of human occupation. In all, Bordes meticulously mapped the location of some nineteen thousand artifacts from the site. Coordinating the massive amount of information from Combe Grenal with what he knew from previously excavated Mousterian sites, he essentially reinvented the way people look at Stone Age tools.

B.B.—Before Bordes—the prehistory of Man was looked on as a linear series of stages, each exemplified by a particular tool type, or *fossile directeur*. Bordes did away with this simplifying emphasis on "star" tool forms (*the* hand ax, *the* flake, *the* blade, and so on), substituting in its place a much more realistic appreciation of archaeological sites as *assemblages* of tools, which were the result of explicit human behaviors. Looked on in this dynamic way, the differences between tool assemblages suddenly became a potent source of information about human behavior and evolution. Bordes reorganized the Middle Paleolithic record in Europe into more than sixty tool types: sidescrapers, end scrapers, borers, blades, crude notched tools called denticulates, points, and so on, classifying the tools not only by their form but by the kind of technology, like Levallois, that produced them. He discovered that the assemblages from the various sites and stratigraphic layers, far from being mere jumbles of flakes and tools, fell into provocatively recurring patterns. The entire Mousterian, covering a period of perhaps 100,000 years, could be reduced to four basic "tool kits," distinguished from one another by the relative percentages of particular tool types represented.

The "Mousterian of Acheulean Tradition," for example, is characterized by a relatively large percentage of hand axes, in some ways reminiscent of the earlier Acheulean bifaces. Such hand axes are rare or absent in "Typical Mousterian" assemblages, which are instead distinguished by a high number of sidescrapers and carefully worked points. In the "La Quina-Ferrassie" tool kit, sidescrapers are even more dominant, making up as much as 80 percent of some assemblages, and the "Denticulate" Mousterian, named after the roughly sawlike edges found on some flakes, is recognized mainly by the absence of many tool types found in the other kits: There are no hand axes or backed knives, few points and scrapers, and all of them are mediocre in quality.

Had Bordes been content simply to recognize these patterns, he would still have made a great contribution. But he pushed on to propose what the patterns meant. For him, the four tool kits represented *distinct Neandertal tribes*, all existing at the same time in the same region of southern France, yet each bearing separate origins, histories, and traditions. Within an area about the size of New Jersey, these four ethnic groups came and went for tens of thousands of years, sometimes living in close proximity, sometimes replacing each other at a given site, yet never mingling their cultures. For example, throughout the twenty thousand years of the Würm I, the cool, early phase of the last Ice Age, the people of Combe Grenal used Typical Mousterian tools, while those of Pech de l'Azé, just six miles on the other side of the Dordogne, remained Denticulate Mousterians. At the same time, another group of Typical Mousterians was living at Le Moustier, though it was three times the distance from Combe Grenal as Pech de l'Azé.

Bordes believed that what united the two Typical Mousterian groups, in spite of distance, was the same powerful force that kept the Combe Grenal and Pech de l'Azé neighbors apart: tribal identity. People of the Denticulate Mousterian made tools their way not because such tools were inherently better adapted to the time or place they were living in but because the customs of their people dictated that tools should be made that way. The same goes for the other tribes. While Bordes himself confined his interpretations to southern France, other investigators applied his methods and meanings to Neandertals in other parts of Europe and the Near East.

Glimpsed through this window, Neandertals appear endowed with at least the basic outline of a three-dimensional *people*, possessed with a sense of group style, of self in relation to a larger social community. The obscure monolith of their lives differentiates into human populations, each with characteristic quirks of personality. At the same time, it is dif-

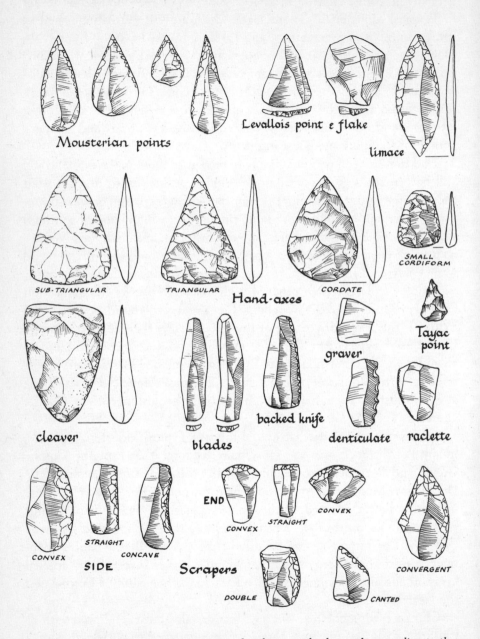

Mousterian points

Levallois point & flake

limace

SUB-TRIANGULAR

TRIANGULAR

Hand-axes

CORDATE

SMALL CORDIFORM

cleaver

blades

backed knife

graver

Tayac point

denticulate

raclette

CONVEX

STRAIGHT

CONCAVE

SIDE

END

CONVEX

STRAIGHT

CONVEX

Scrapers

DOUBLE

CANTED

CONVERGENT

A selection of Mousterian points, scrapers, hand axes and other tools, according to the classification of François Bordes COPYRIGHT © 1982 BY J. WYMER, REPRINTED BY PERMISSION OF ST. MARTIN'S PRESS ficult to embrace Bordes's interpretation without also believing that Neandertals must have been living very differently from any known modern hunter-gatherer group. How can two tribes exist side by side for hundreds, even thousands, of years and never get to know each other, never interact enough to influence each other's cultures? Martin Wobst, an archaeologist at the University of Massachusetts, has shown that hunter-gatherer groups continually try to *maximize* contact with their neighbors. But Bordes's Neandertals seem to have passed their days in almost unrelieved isolation. This becomes more believable if you keep in mind that the relative number and small size of Neandertal sites suggests extremely low human population densities. In the whole Aquitaine area, for example, there may have been only a few hundred people living at any one time.

"A man may well have lived all his life without more than a rare meeting with anyone from another tribe," Bordes wrote, "and it is very possible that these contacts, when they did take place, were not always peaceful and fruitful."

Bordes's *cultural* explanation of Middle Paleolithic variability was immensely influential, changing the methodologies of archaeologists working not just in the Mousterian but in all time periods. It helped that Bordes was a man of inexhaustible energy and scholarship, who took enormous delight in intellectual combat, waving his arms and shouting, overwhelming his adversary with the power of his ideas and his vast knowledge of facts.

"If at the end of a discussion with Bordes you felt that you had made even a single point, you felt good," Jean-Philippe Rigaud told me later at the Grotte 16 camp.

Who could be a worthier intellectual adversary for such a man than Lewis Binford? In the middle 1960s, long before he set out to destroy the image of man the mighty hunter in the Lower Pleistocene, Binford and his then-wife, Sally, threw down a challenge to Bordes's cultural view of the Mousterian. The Binfords praised Bordes's recognition of patterns, but they proposed quite a different interpretation of their meaning: Instead of representing traces of tribes, the various tool kits simply reflected different activities. It was not the *people* who varied, but the *functions* they happened to have been performing on the landscape, functions that could be tied in with changes in climate and season. Assemblage variability thus offered a reflection of human adaptation to an environment, rather than a cultural identity fixed through time. An assemblage rich in denticulates might be the spot where someone stripped bark, whittled a tool, or engaged in some other kind of woodworking, while a conglomeration of sidescrapers could be evidence of food preparation or hide scraping.

Not surprisingly, Bordes bridled at this American incursion into his turf. The Bordes-Binford Debate that ensued is one of the loudest and most colorful in the history of archaeology. Sally Binford characterized the first meeting of the two men as "like two male dogs circling and sniffing, trying to find out if the other was really an enemy." Lew Binford, never one to underplay a scene, recalled the encounter like this:

> We must have argued for more than an hour, our voices getting higher and higher, standing up, sitting down, pacing back and forth, leaning over charts, big clouds of smoke issuing from his pipe, equally big clouds pouring from my cigarettes. . . . All of a sudden Bordes jumped up and came around face to face. I stood up almost automatically. He put his hand on my shoulder, looked me directly in the eyes, and said, "Binford, you are a heavyweight; so am I."

The heavyweights continued to raise smoke in academic journals for years, as the two men met during field seasons to argue themselves into a froth of mutual exasperation. In the process, they cross-fertilized each other's ideas even as they condemned them.

"Bordes used facts to control Binford's imagination," Rigaud remembered, "but Binford used his imagination to enrich Bordes's facts." In the end, Bordes invited Binford to collaborate on a two-volume work on Combe Grenal, one volume authored by each. Before the project could get started, Bordes suddenly died. For Binford, it was "one of the saddest events of my life."

And who was right? Most archaeologists on both sides of the Atlantic would agree that for the moment the answer lies somewhere in the triangle defined by the points "both," "neither," and "impossible to say." Students of Bordes in France have shown that there do indeed appear to be cultural traditions persisting through time in geographically defined regions, though the enduring traditions are expressed more in the way tools are made than in the resulting shapes of the implements. But the tribal hypothesis still cannot explain what would keep groups culturally distinct for thousands of years, with nothing between them but, say, an easily fordable stretch of the Dordogne River. Binford's "activities" hypothesis sounds more plausible, but microwear analysis, which can link a particular type of tool to a particular function, has given no support to his theory. On the contrary, microwear studies like John Shea's show that no matter what the tool form in the Mousterian, when you find microwear it is usually the kind associated with woodworking.

A quarter of a century later, the Bordes-Binford Debate continues to generate heat. Perhaps the issue on which it turns is a particularly expressive example of a larger conflict. For Bordes, working out of the French paleontological tradition, the differences between Mousterian tool kits became reflections of discrete historical entities. They could not mingle and blend any more than separate species can freely crossbreed to produce a mush of hybrid organisms. But it is part of the whole "melting-pot" American ethos that cultural lines can be smudged, crossed, and even erased. For Binford, who grew up hunting muskrats in the swamps of coastal Virginia, *who you are* can never be as important as *what you do*. What is meaningful—and therefore what one should look for in the archaeological record—is adaptive behavior, the way individuals act to survive under the conditions in which they live.

While French and American heavyweights were playing out these grand themes, a British archaeologist named Paul Mellars was politely proposing another explanation. According to Mellars, the secret hand behind the variability in the Mousterian was neither style nor function, but *time*. If one eliminated from the analysis the "ragbag" assemblages that were essentially defined by negatives—the absence of backed scrapers, the absence of hand axes, and so forth—the pattern was simplified into three distinct, successive industries, or technological cultures: the Ferrassie, the La Quina, and the Mousterian of Acheulean Tradition.

"I was always a bit peeved that the debate was looked on as a 'battle of giants,' " Mellars told me when I visited him in Cambridge, England. "It was all based on a false premise: that the variations Bordes and Binford were talking about were synchronous. But I could see that there was a chronological sequence. If you've got fifteen sites, all sharing the same pattern of stratigraphy, you don't have to be a bloody genius to figure out what's going on."

Recently, dating by thermoluminescence—a powerful new way of determining the age of artifacts—has firmly supported Mellars's chronological argument. But to confirm that time, and time alone, is responsible for the variation, the same sequence of industries must appear at every site, and so far this is not clearly the case. Mellars's theory also continues to suffer from a sort of killjoy plainness. If tool industries changed simply as a function of chronology, the whole question becomes a sort of non-issue, saying nothing about human behavior except the obvious, that it changes through time. It is an argument with little explanatory power behind it.

"To explain the whole thing with chronology was, well, boring," Mellars admits. "So my ideas never got a fair hearing. But the new dating work has borne them out."

If Mellars's theory throws water on the Bordes-Binford conflagration, that of a younger, brasher archaeologist threatens to deep-six the whole business. "It's time the matter was dropped," says Harold Dibble of the University of Pennsylvania. "Neither Bordes nor Binford ever had any data. They were both locked into a fatal assumption: that these tool types were made voluntarily, that the people who made them *did them on purpose*. If that's true, then it has to be that the variation is either a matter of function or style. I mean, there isn't anything else. But what if the initial assumption is wrong?"

Where Bordes and Binford saw the lithics as scrapers and points and so on, Dibble insists that we look at them for what they are: anonymous trash. If ancient populations behaved like modern ones, most of the artifacts in an archaeological assemblage are not the fresh, keen-edged tools that bit into flesh or whittled a wooden spear, but the discarded end products of a history of use. A tool might be used initially to chop up a tuber, then resharpened along another edge to scrape a hide, and finally, after that edge went dull, retouched again to form a point that could be used to dig marrow out of a bone. All through its life cycle, the same piece of stone assumes different forms, any one of which might be the last—and hence the one that an archaeologist would pick up a hundred millennia later, confident that what he had found had some intrinsic formal meaning. Ironically, it might be the only part of the original stone that *couldn't* be used as a tool.*

*Tony Marks of the University of Texas once told me how a student of his had planned a study of raw-material use in the Paleolithic of Ethiopia: "The first day in the field, the student was examining an outcrop of obsidian when out of nowhere this guy comes up, knocks off a core, and walks off. So the student follows the guy back to his village. It turns out he's using the stone to tan hide. They make a scraper and fix it into a big block of wood, for weight, and get to work. But every five minutes, the thing needs sharpening. Eventually what was once about the size of a pancake gets worn down to a tiny butt end. When it is too small for the block to hold, they chuck it out and go back for a new one. It's that tiny scrap that an archaeologist in the future would see."

With this in mind, Dibble decided to look at the classic Bordean typology. More than a third of Bordes's sixty-odd flake-tool types were various kinds of *racloirs,* or scrapers. These could be divided up into four
main classes: simple single-edge scrapers retouched along one side; double-edged scrapers retouched on both sides; convergent scrapers where the
sharpening had extended down along the two sides until they met, forming a point; and transverse scrapers, where the sharpening was not along
the side but on the edge opposite the striking platform. Examining tools
from sites in both France and the Levant, Dibble discovered that all the
scrapers could be fitted into one of two "reduction sequences." Single
scrapers were resharpened to become double-edged, and retouched further
until the edges converged. Or single scrapers were continually retouched
along one edge until the artifact was "transformed" into a transverse
scraper. In either case, the tools Bordes interpreted as real, discrete entities
were, according to Dibble, simply episodes in a continuum. Grouping the
artifacts into tribal tool kits makes no more sense than conducting a cen-
sus in a region and calling all four-year-olds one tribe, all teenagers another, all mature adults a third, and so on. Reading different activities
into the tool assemblages, à la Binford, is equally futile.

Needless to say, Dibble's theory has not endeared him to his colleagues
on either side of the Atlantic. "When I first presented my reduction sequence argument at a conference in Liège in 1986," he told me, "a French
archaeologist sprang out of his seat and shouted, 'Who is this grotesque
example of American archaeology who has come over here to insult François Bordes?' That still hurts."

Binford responds less emotionally, but with equal warmth. "If I wanted
to play a practical joke on archaeology," he told me, "I could not have
done better than to invent Harold Dibble."

Dibble's message is neither a joke nor an insult, but a caution. Most
of the material evidence for human behavior—not just for the Mousterian, but for the bulk of our time on earth—amounts to pieces of worked
stone. Understandably, archaeologists try to suck every drop of meaning
possible out of these ancient implements. Perhaps in their thirst for data,
they go too far. Consider the extremely important question of whether
Neandertals communicated in a fully human language. A common argument in support of their lingual abilities is the articulated variety of their
tools. If the "canted sidescrapers," "alternate burinating becs," "Mousterian tranchets," and other ware in the Bordean kitchen represent the
arbitrary imposition of form onto rock, it follows that their makers' minds

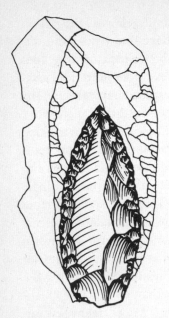
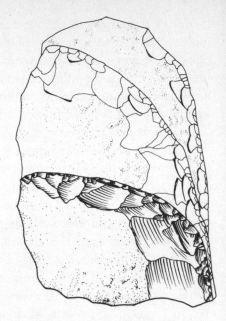

Tools within tools within tools: According to archaeologist Harold Dibble, Middle Paleolithic stone tools passed through a continuous series of forms as they were knapped, used, and knapped again to sharpen the cutting edge. Thus the objects found by archaeologists millennia later are not distinct tool types but merely the random end products of a history of use. COURTESY HAROLD DIBBLE

could operate and communicate in the arbitrary, abstract code of linguistic symbols. It is hard to imagine how else a young Neandertal could learn the difference between a double bi-convex sidescraper and a double bi-concave one, unless an older Neandertal could *tell him.*

This argument dibbles away, however, if both artifacts are merely the nameless byproducts of one last attempt to squeeze a little more cutting edge out of a rock. In that case, Neandertal linguistic needs can be streamlined. To borrow from Gertrude Stein, a tool is a tool is a tool.

"I'm not saying Neandertals didn't talk," protests Dibble. "I'm only saying that stone tools don't show that they did. The typology reflects *our* language, not theirs."

I met Harold Dibble in Bordeaux, when he was giving a seminar at the university. Red-faced, with a fair beard, he took up space with a kind of jolly obstinacy. He lectured in fluent French but stuck resolutely to his native phonetics; when a word was the same in both languages, he delivered it in pure, hard American: defiant consonants bobbing in the sea of vowels. The audience was loaded with what might be called neo-Bordeans, including archaeologist Jacques Pelegrin, perhaps the most skilled and scholarly flint knapper in the world, learning literally at Bordes's knee, beginning when he was only sixteen. But Dibble defused any open rancor with his easy humor.

"All right, we have time for some questions," he said after the talk, "but only from people who agree with me. And Jacques can't talk at all."

Pelegrin did have his chance to argue, however, and Dibble to respond and others to add their voices, and as the room began to devolve into pockets of localized dispute, I sat in the back and despaired of ever catching a glimpse of a Neandertal life through such a maze of approaches and ensnarled traditions. Later, I brought my despair to Simek, back at Grotte 16.

"Well, maybe stone-tool shapes aren't the thing to look at," he said with a shrug, showing a commendable matter-of-factness for a man who spends much of his time looking at them.

There is indeed more to a tool than its shape. Most archaeologists today think of a finished artifact as merely part of a *system*, a process that began when someone picked up a piece of dusty flint intending to modify it into something he or she could use. The system includes raw-material procurement, the manufacture of the tool through various stages, its transport across the ancient landscape, its primary use, resharpening, reuse, and so on until its final tossing.* Strictly speaking, the "history" of the tool continues, as it is buried, jumbled up, moved around by water, erosion, or other natural agents, and then, if it is a lucky tool, brought to light by a squatting grad student, brushed off, mapped, bagged, labeled, pondered, classified, photographed, ink-rendered, computer-imaged, and otherwise made much of.

All along the use-history of the tool, there are points where humans have left clues to the nature of their lives. Consider the procurement of raw material. The best minerals to make tools, as any hominid knows, are the fine-grained, silica-rich minerals like flint and chert, which produce

---

*Sometimes a tool can even be brought back into use after a vacation away from human hands for a few millennia. When flint is split open by a toolmaker, the exposed surface discolors over time, developing a "patina" like the green skin of a copper roof. Occasionally tools are found bearing a "double patina," showing that a discarded tool simply lay around for many years before being picked up and retouched again and used. Phil Chase, a colleague of Harold Dibble's at the University of Pennsylvania, told me of a Neolithic ax in the Levant that had been crafted out of a Mousterian flake struck from its core perhaps 50,000 years earlier.

clean sharp edges when struck. If you were a Neandertal living in the Dordogne, finding such stuff would be no problem; you would have to go out of your way to avoid it. The same is true in other parts of France. Walking along some row houses above a train station near Amiens, I saw flint doorstops and flint garden walls, great knobby nodules of flint lining vegetable beds, flint piled up on porches for no apparent reason at all. My four-year-old regularly returned from a walk in the Fontainebleau Forest with a couple of raw hunks and bashed them up on the sidewalk, perhaps responding to some primitive imperative set free by the smooth promise of siliceous stone.

So getting flint is not an issue; what matters is the quality. Low-quality flint—smaller nodules, or rock lacking the interior consistency that will produce a long clean break when struck—can be found by walking along riverbeds. But the better-quality stuff occurs in regional patches: the burnt-rose–tinted Bergerac flint, for instance, or the famous chocolate flint of the Holy Cross Mountains of Poland. In the Acheulean, occupation sites are usually parked near to the source of flint used in the tools they contain, be it high or low quality. Beginning in the Middle Paleolithic, on the other hand, flint starts to travel across the landscape.

"Middle Paleolithic people did not just go out and grab some flint, make a tool, and move on," says Jean-Michel Geneste, a colleague of Rigaud's in Bordeaux. "The production of a tool begins to require not only time, but distance."

Geneste's careful studies of raw-material movement in the Dordogne show that superior-quality rock like Bergerac flint moves much farther, often turning up fifty or more miles away in Mousterian sites. These "exotic" flints are not found in the form of cores, but as Levallois flakes and finished scrapers, points, and bifaces—by and large tools requiring more investment of labor. The best-quality material is thus being reserved for tools that seem to be more carefully conserved. In contrast, the discarded cores found in Mousterian occupation sites come from less than half a mile away. This local, low-quality material serves well enough for the amorphous edges and denticulates that can easily be banged out—the Mousterian equivalent of plastic forks and other disposable ware.

For Geneste, the shift to a lithic system involving transport represents a cognitive leap, a "new state of knowledge." Just as the Levallois technique testifies to the toolmaker's ability to imagine the future tool before he or she begins to shape the core, flints found many miles from their source reveal that people were imagining future needs enough to carry tools around, even when there was not yet a specific purpose in mind for them. Though Neandertals moved flint around more than their Acheulean ancestors, however, they were not in the same league as the people who came after them. The Aurignacian period, which ushers in the Upper Paleolithic Age in Europe, shows no great jump in the *distance* flint is moved through the system. But there is an abrupt change in the *amount* of long-distance movement. Whereas flints from farther than twenty miles away are very scarce in Mousterian sites, making up only 1 percent of a given assemblage, in the Aurignacian, these exotics balloon to fill up as much as 20 percent of an assemblage. Not a single Mousterian site comes close to yielding that much exotic flint. Nonutilitarian objects—the odd curious fossil or shell that somebody picked up and capriciously kept around—move even less often and a shorter distance from their source.

The difference is critical. If raw material was moving around the landscape, then so, of course, were people. And if people were moving more, they were much more likely to be encountering *other* groups of people. It doesn't really matter whether one group met with another in the course of obtaining flint, or picked up some pretty flint when they had really set out looking for social contact, or perhaps encountered both flint and fellow humans while pursuing some other short-term goal—following the blood trail of a wounded aurochs, perhaps, or checking out what might be attracting a circling trio of buzzards. Whatever the reason for more long-distance movement, as long as population densities were the same, it would mean more social contact.

In contrast, the dominance of local flints in Mousterian sites suggests that individual Neandertal groups were socially isolated, each hunkered down within a given range. Of course, there was some form of contact and movement of individuals between groups; Neandertals did, after all, succeed in producing other Neandertals for hundreds of thousands of years. But there is nothing in what has been left behind to suggest a structured social network within a region, a web of alliances and dispersed relationships like those characterizing all modern human societies. Living in the Perigord 50,000 years ago, you might *know* about the people over the hill or farther downstream, but that knowledge was, apparently, not all that essential to your survival. Modern human groups, in contrast, would wither up and perish in such stark social isolation.

"Judging by the raw material, either the Mousterians were more territorial, or they were moving around themselves but leaving their tools behind," says Lewis Binford. "Either way, they sure aren't behaving like us."

*　*　*

People who love Neandertals do not love Lew Binford. At any point in the lithic system where you choose to crack open a window on Neandertal lives, Binford will be there, waiting to show you how little they resemble us. Most archaeologists, Binford contends, begin by *assuming* that Neandertals were "watered down versions of ourselves," and then frame their hypotheses accordingly—a path of reasoning that can only circle back on itself and end up "proving" what was assumed in the first place: the essential humanity of Neandertals. It is only by actively dismantling the initial assumption that one can even hope to get at the true nature of Middle Paleolithic behavior.

I encountered Binford early in my travels, back in his office at the University of New Mexico, a sort of smoke-filled Command Central of Dismantlement. (Binford has since moved to Southern Methodist University in Dallas.) Handsome, gray-bearded, his powerful frame slightly paunched, he sat behind a deskful of documents like an aging field general pondering which front to attack next. Binford does not actually hate Neandertals, of course, but the characterizations derived from his methodology of skepticism are about as unforgiving as one can find in the science. What he thinks Neandertals and other archaic *sapiens* lacked most acutely is a distinctly human quality he has named "planning depth." Modern humans constantly perform actions whose benefit will be realized only in the future. Fishing societies, for instance, will move themselves to a camp by a seasonal salmon run well before the salmon get there themselves. By the time the fish arrive, the fishermen will have ready the specialized gear needed to capture them, and when the salmon have been scooped up in abundance, they will be dried into quantities that can help feed the group for months to come. All these actions bear witness to the human ability to plan well into the future for needs that have yet to arise.

On the other hand, Middle Paleolithic hominids—Binford pointedly refuses to call them human—seem to have subsisted in the now. Some tools may have been carried around a bit, as Geneste's studies demonstrate, but not to an extent that suggests such "curation" was a systematic part of their lifeways. In the archaeological record such tools may show signs of being recycled, à la Dibble, but that doesn't mean they were *necessarily* being carted around for reuse. In Binford's view, it may mean only that "poorly equipped groups" of hominids arrived at repeatedly used sites, and only when a need arose did they pick up and resharpen whatever junk the last, equally badly outfitted group had left behind. Because they "curate" their tools intensively—hanging on to them for future use—modern hunter-gatherer groups wind up leaving behind in a given site far fewer tools than faunal bones from their meals. In the Mousterian, this pattern is strikingly reversed: During 75,000 years of occupancy at the Combe Grenal cave, for example, the Neandertals left behind 17,389 tools and only 6,932 bone fragments.

"What seems to be suggested," Binford says, "is that tools did not stay in the system very long. That's a clue to the lack of planning depth."

Nowhere is this shortcoming more evident than in the Neandertal diet. Binford specifically uses salmon as an example of a rich food source exploited by modern people endowed with the ability to plan for its seasonal exploitation. Salmon were plentiful in the springtime in the rivers of southwestern France throughout the Paleolithic Age, and their bones are found in Middle Paleolithic cave deposits. But they weren't brought there by hominids. They were put there by bears.

"There's plenty of salmon in the bear caves," Binford says. "But in all of Combe Grenal, there's just one salmon vertebra. And the Dordogne River is sitting right there!"

The implication is that while the bears took advantage of the salmon, the Neandertals of Combe Grenal did not, either because they lacked the tactical means to catch and preserve the fish, or because it never occurred to them to move their bodies to this annual nutritional bonanza. Either way, they seem oddly shallow in their planning. They similarly failed to exploit the annual reindeer migrations, an even more abundant, if less predictable, source of protein. Even from the beginning of the Upper Paleolithic, at such sites as Abri Pataud and Roc de Combe, modern human hunters specialized in reindeer. At some later sites deer made up as much as 99 percent of the fauna. In western France, Upper Paleolithic sites also tend to cluster in a few locations—such as certain stretches of the Vézère Valley—that would have served as migration routes for reindeer. Mousterian sites, in contrast, show no such concentration on reindeer or any other single species, at least not in Binford's eyes. Mousterian sites are furthermore scattered randomly through a region, rather than lined up along migration routes. If their inhabitants were hunters, they did not seem to be responding much to movements of the hunted. Instead, they were taking whatever happened to wander inside their territories.

Binford asserts that Neandertals were opportunistic feeders who killed prey only on an "encounter" basis. While he is the most vocal proponent of this view, he is not alone. "These were people who were sort of sitting in an area, eating whatever is available there, and having really no concept of what is behind the hill," says Olga Soffer of the University of Illinois at Champlain. "The feeling I get from them is that every day is the first day of their lives."

The "catch-as-catch can" view of their daily itineraries gains further support from the bones of the Neandertals themselves. Working with Chris Ruff of Johns Hopkins University, New Mexico's Erik Trinkaus has discovered a possibly telling difference between the thighbones of Neandertal and those of modern humans. Generally speaking, bones grow thicker in response to habitual stresses put on them. A professional tennis player develops not only bigger muscles in his playing arm but much thicker bone structure there as well. All humans begin life with femurs that are roughly doughnut-shaped when looked at in cross-section. Hunter-gatherers and other individuals who customarily walk long distances develop pronounced "pilasters"—a thickened ridge along the trailing edge of the bone—which gives their femoral cross-sections a teardrop shape. Compared to contemporary humans, both Cro-Magnon and Neandertal thighbones are remarkably robust. But whereas the Cro-Magnons' legs reveal the expected hunter-gatherer teardrop in cross-section, the Neandertals' are much rounder. The bone is evenly and thickly distributed all the way around the circumference of the shaft. Given that bone grows in response to loading, Trinkaus surmises that the rounded shafts reveal that Neandertals were much more accustomed to moving continuously and in all directions, side to side, up terrain and down, in an irregular pattern quite unlike the straight-on gait of modern hunter-gatherers.

The implication is that modern humans would be more likely to begin their day with a destination in mind. In contrast, a Neandertal waking up from an overnight in Combe Grenal would set out with little plan in his head except eating, wherever and whenever Neandertal and Neandertal-food happened to meet each other. Not surprisingly, this image of aimless foragers is not universally shared.

"One has difficulty seeing them randomly tripping over the landscape looking for dead animals," says Tony Marks of Southern Methodist University. "This is not a scenario which suggests a quarter of a million years of survival."

Perhaps not—but species do not need to pass any tests for foraging excellence; they need only forage well enough to survive. If their anatomy is any indication, Neandertals may have made up for their lack of planning depth by sheer hard work.

"Neandertals and earlier hominids were behaving in such a way as to stress their bones to the breaking point," Binford says. "Our species never does that. It's particularly striking in the juveniles—they look like Olympic athletes."

The most curious witness to this legacy of hard work is not Neandertal

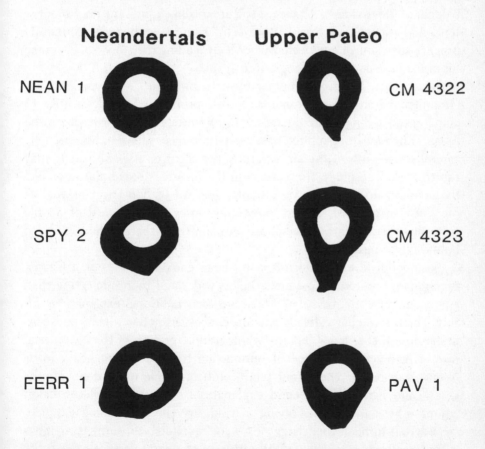

The shape of dead bones speaks of the behavior of the individual in life. Cross-sections of Neandertal thigh bones on the left, compared with those of Cro-Magnons, on the right. See text. COURTESY ERIK TRINKAUS

musculature, however, but the Neandertal nose. I have already mentioned the distinctiveness of this hyper-appendage, the crowning glory of the Neandertal face. To Erik Trinkaus, what is important is the nose's functional importance.

There is a story by Nikolay Gogol, called "The Nose," in which a self-important minor official wakes up one morning to find that his nose has disappeared. Unbeknownst to him, the nose turns up in a loaf of bread, and later in the story is seen riding about St. Petersburg in a well-appointed carriage. By the time the official catches up to his nose, it is dressed in the uniform of a higher-ranking official, so it is only with trepidation that the poor man approaches his organ and politely reminds it that, after all, it rightfully belongs in the middle of his face. "You are mistaken, sir," the nose snootily replies. "I am not your nose. I am myself." Eventually the nose resumes its proper place, and all ends well. But nobody reading the story can ever take his or her nose completely for granted again.

Noses are not simply fleshy air funnels; they are highly sensitive, environmentally crafted instruments. Nose morphology is very sensitive to climate, controlling the temperature and humidity of the air reaching the lungs. When repairing injured noses, plastic surgeons have to be extremely careful to reconstruct the air dynamics just right or else the person may run the risk of respiratory infections in the future. A broad, flat nose, like those found in tropical environments, gets rid of body heat as well as moisture. In contrast, angled, projecting noses, more typical of people living in drier climates, precipitate out moisture that is then ready to humidify the next breath.

Neandertal noses, while not quite large enough to fill out a Czarist functionary's uniform, were nevertheless wider and more projecting than any noses currently in vogue. There are two traditional explanations for this. The first we have already encountered—it is not so much the Neandertal nose that is large, but the whole middle portion of the Neandertal face, a biomechanical piece of remodeling to support the heavy loads Neandertals put on their front teeth, probably while using their mouths as a "third hand." The second explanation, offered by Carleton Coon among others, is climatic. Living in cold, dry climates, Coon reasoned, Neandertals evolved their large noses as "radiators" to warm their inhalations, which is especially important given the nasal passage's proximity to the highly temperature-sensitive human brain.

Trinkaus and his student Robert Franciscus believe that the Neandertal nose does function as a radiator—but not the kind of radiator Coon meant. People living in very cold, dry conditions, such as those of the last glacial period in Europe, should have narrow, projecting noses, the better to conserve heat and moisture. (Modern Eskimo noses are quite narrow.) But from side to side, the Neandertal nose is far wider than climate alone would predict—wider, indeed, than even tropically suited humans today. Such a nose would not *heat* incoming air, like a room radiator, it would *dissipate* heat, like a car radiator. The mystery is why a creature living under some of the coldest conditions any hominid has seen would want to get rid of body heat. The answer, for Trinkaus, lies in the steady physical activity Neandertals were putting out, just to get by.

"Anyone who has worked in the Arctic knows that a paradoxical danger there is from overheating," Trinkaus explains. "If you overheat, you sweat. If that sweat freezes, you are really in trouble."

This car-radiator theory may reveal, which Coon's hypothesis could not, why Neandertal noses are equally broad in both the harsh cold of Europe and the more temperate regions of the Middle East—in fact, wherever you find Neandertals. The explanation for the forward projection of the appendage is more self-evident. Noses that are angled down and long, front to back, contain more baffles, twists and turns, which create "turbulence" with each breath. The rush of air over mucous membranes precipitates moisture into the nasal passage, humidifying the air reaching the lungs. All in all, a constantly active hominid in a cold, dry climate could not ask for a better respiratory appliance. It fit better on a hominid who was burning more calories to achieve the same result—survival— than on the gracile moderns who arrived later.

"Either Neandertals were doing things very different from us," Trinkaus told me, "or they were doing things like us, but not doing them as well."

Back across the hall in Command Central at the University of New Mexico, Lew Binford dismantled for me the evidence for Neandertal hunting prowess. Binford, whose analysis of early hominids at Olduvai Gorge turned them from mighty hunters into marginal scavengers, has only a slightly higher view of their Middle Paleolithic descendants. "Neandertal sites are heavily biased toward scavenged meat," Binford told me. "Not until late in the Mousterian can you make a good argument for killing."

At Vaufrey, in the very layer where Jean-Philippe Rigaud found an early eruption of Mousterian technology, Binford sees precious little evidence to suggest a subsistence based on hunted meat. Even though Vaufrey Level VIII contains impressive fauna like red deer, horse, and aurochs, there are no cut marks on the bones that suggest hominids had removed them from an animal they had themselves killed. On the contrary, the selection of body parts in the site and gnaw marks on the bones indicate to Binford that the real hunters of Vaufrey were wild dogs. The hominids had only scavenged the bones from preravaged carcasses. He found evidence of equally ignoble human involvement with the fauna in the English sites of Hoxne and Swanscombe.

All three of these sites are, however, at least 250,000 years old. One can only guess at the nature of the inhabitants there, but they were probably either *Homo erectus* or some kind of pre-Neandertal. What about the Neandertals themselves? Among the animal bones of Combe Grenal dating from 85,000 and 45,000 years ago—Neandertal prime time—Binford did find new patterns of meat procurement. Medium-sized animals like reindeer and red deer had been brought into the cave by hominids, and the pattern of anatomical parts represented—as well as the absence of gnaw marks on the bones—suggested that the animals had indeed been hunted for their meat. But the large animals, like horses and wild cattle, continued to be scavenged. And the main staple of the Neandertals at Combe Grenal, according to Binford, was not flesh at all. Judging from the traces of pollen left on flake tools at the site, it was aquatic plants plucked from the canyon stream. Cattails, to be exact.

"At present the inevitable conclusion seems to be that regular, moderate-to-large-mammal hunting appears simultaneously with the foreshadowing changes occurring just prior to the appearance of fully modern man. . . . " Binford wrote in 1985. "Systematic hunting of moderate to large animals appears to be a part of our modern condition, not its cause."

What is an inevitable conclusion to Lew Binford, however, is an insupportable exaggeration to others. Most archaeologists grant Neandertals and other Middle Paleolithic people more hunting prowess, and considerably deeper planning abilities. If John Shea is right and Neandertals were throwing spears, it is hard to imagine what they were throwing them at, if not game. In some places, such as the famous Neandertal site of La Quina, many archaeologists see the piled-up bones of bovids, horses, and reindeer beneath a steep cliff as the aftermath of a "cliff drive," a cooperative and therefore planned hunting technique often used by Paleo-Indians of the New World.

Binford's most aggressive critic on the question of Neandertal hunting is Philip Chase, a colleague of Harold Dibble's at the University of Penn-sylvania. Chase points out that while there may be no reindeer-specialized sites in the Middle Paleolithic, you can't find any such sites in the early Upper Paleolithic either, at least beyond the Périgord. The true "Reindeer Age" begins some twenty thousand years later. It seems unfair, therefore, to dismiss Neandertals for failing to possess skills that modern humans would not develop themselves for twenty thousand years.

More to the point, excavations scattered across Europe and western Asia have turned up Middle Paleolithic sites dominated by other prey, often animals much rougher and tougher to hunt than reindeer. Excavations at Mauran in the French Pyrenees have yielded no less than 108 bison, over 90 percent of the total faunal bones found—not exactly the easy pickings of cattails, rabbits, and scavenged bigger beasts that Binford imagines in the typical Neandertal larder. At Vogelherd, Germany, the specialization is horse; at Teshik-Tash in Uzbekistan, Siberian mountain goat. And so on for Middle Paleolithic sites at Starosel'e in Crimea, Volgograd in southern Russia, and Ehringsdorf in Germany, among others.

"It is possible that some of these sites represent the activities of groups heavily dependent upon a single species of game," says Chase. "It is more likely that most of them represent the activities of peoples who, while exploiting a range of resources, tended to go to a specific location in order to hunt a specific species, probably at a particular time of year."

That sounds like deep planning to me. But Chase may be falling into the trap of allowing the assumption of their humanity to color his interpretation of the evidence. If a particular species of animal was abundant in a given place, other archaeologists believe, people would subsist on that animal not because they had consciously planned strategies to do so but simply because the animal was there to be eaten. If people returned year after year, the bones of the common species would pile up. In Lew Binford's view, such sites may have nothing to do with humans at all.

"Any paleontological site can show single species concentration, but that doesn't mean the animals were hunted," he says. "Just four years ago, seventy red deer jumped off a cliff during a thunderstorm. They were spooked. The lead animals got pushed off by the followers. This sort of thing happens all the time. At Wretton in England, hundreds of bison bones were found together; maybe those animals drowned crossing a river. Mass death is a part of nature."

True enough. But if Neandertal lovers must prove rather than assume that Neandertals were hunters, Chase thinks skeptics like Binford have an equal responsibility to prove they were not. According to Chase, the bones of Combe Grenal do not agree with Binford's own criteria for identifying the meals of scavengers. It stands to reason that the predator who initially kills an animal will consume the meatiest parts, leaving behind for scavengers—hominid or otherwise—the less nutritious pieces like heads, feet, and lower leg bones. These "marginal parts" should show signs that the scavengers went to some effort to extract what little bits of marrow and meat were available. But Chase maintains that at Combe Grenal the carcasses of large animals like horse and bison are often represented by meaty limbs. And rather than the rough hacking marks one might expect from a scavenger trying to scratch a bit of marrow out of a dried limb bone, the bones frequently bear cut marks like those that would be produced by slicing through fresh flesh. Some of the animals, moreover, were in their prime when they were killed, suggesting that the hominids were not simply scavenging the carcasses of individuals who had died naturally from sickness or old age.

"I'm not saying they weren't scavenging sometimes, or that they didn't demonstrate opportunistic behavior patterns," says Chase. "But Lew endows Neandertals with less planning depth than you can see in chimps or wolves. I think they were opportunistic when they needed to be, when it paid off. Not because they were dumb."

"I never once have said they were dumb," Binford protests. "All I've said is that they were not like us."

Smart, then, but not human? What, then, were the Neandertals doing—or not doing—that brought them to this oxymoronic impasse? Binford's answer is also back at Combe Grenal—or more accurately, in the reams of raw data from the cave that reside in his study. To begin with, he points out that quibbling over how much meat Neandertals hunted and how much they scavenged obscures the point that there is not enough meat in the site to feed anybody in the first place.

"All told, there are about a hundred and twenty-five reindeer in the whole site," he told me, "along with ninety red deer, seventy-odd horses, and sixteen—that's right, *sixteen*—bovids. Now, Combe Grenal goes from one hundred thirty thousand to fifty-five thousand years ago. What does that tell you?"

I worked out the math—75,000 years of occupation, divided by about 300 medium to large animals—and I came out with one meat meal every 250 years. Granted, the carcasses obviously aren't distributed evenly through the strata, some bones may not have been preserved, and so forth, but even so, these smart not-humans were not exactly barbecue-dependent. At least not in the cave. "Contrary to what everybody assumes, we are not looking at diet in these early Mousterian sites," Binford explained. "Everybody is imagining this stuff is the product of family meals. But it just isn't happening that way."

It didn't happen that way, I was shocked to learn, because there were no Neandertal families to begin with: no pair-bonded couples, no Mr. and Mrs. Combe and their dear friends the Grenals, no bands of male hunters bringing food home from the field to their waiting wives and children. The horizontal layers of Combe Grenal bespeak a quite different social organization, over and over again with a numbing redundancy through those 75,000 years. In each level, there seem to be two separate sorts of activity areas. In the middle of the cave, one finds what Binford calls "the nest." Clustered within a ten-by-ten foot area of soft, ashy material are concentrations of denticulates, amorphous notched tools, and utilized flakes. Along with these throwaway tools are broken splinters of animal limb bones that would be marrow-rich, together with teeth, jaws, and head parts. But assemblages of a different kind are scattered around the periphery of the cave and tucked in among boulders: "Tiny little bull's-eye" groupings that contain scrapers and other high-investment tools as well as animal jaw fragments. Whereas the identifiable fauna in the "nests" tend to be old or juvenile red deer, pig, and other stream-margin mammals, those in the outer pockets of activity are bovids and other grassland grazers found in the uplands farther from the cave.

Binford concedes that this pattern of differentiation in the fauna is "fuzzy." But there is nothing fuzzy about the raw materials used. The "nest" tool kits are made from limestone from the cave, or from cherts found a short distance away. In the locations with scrapers, the raw material comes from upland plateaus, usually a couple of miles away, sometimes as much as thirty. Put all the variables together—raw material, time-investment in the tools, animal species and body parts, the ashy evidence of fire in the nests—and you come out with two distinct strategies of land use. It would appear that the cave is being shared by one group of people who are highly mobile, moving over substantial territory, and another group of people who are "sitting" on the resources available in the immediate area, eating plant foods and scavenged animal bits.

To Binford, these two groups of people look a lot like men and women. "Generally speaking, among primates, females sit on resources while males run around locating resources," he told me. "Among modern hunter-gatherer groups, the females are always foraging closer to home. They exploit low-risk resources like plants that they can count on as reliable sources of nutrition for their offspring. The men go after high-risk food, like hunted meat, and they don't always find it."

The dividing difference between Neandertal males and their modern counterparts is that moderns bring what they catch back to be shared with the women and children. The Neandertals—at least the ones who chose to occupy Combe Grenal for seventy-five millennia—were not so generous. They might well have been *catching* game, or scavenging successfully—but they weren't transporting it back to share with the females and the young. Most of the time, the males were dining—and living— somewhere else. The odd head parts and marrow bones that they *did* bring into the cave may simply have been tidbits requiring more processing to squeeze out their food value: limb joints that could be heated to release their succulent grease, bones to be dried out before marrow could be extracted, skulls to be broken open so the rich brain-food could be fingered out. Perhaps some of these scraps were shared with the nest-dwellers, but not to the extent that they could depend on them. Neandertal women and children, Binford believes, were pretty much making it on their own. Erase from your mind all those four-color magazine illustrations of happy Neandertal families: The Male Provider was simply not in the picture. All Neandertal fathers were absentees.

Binford presented this astonishing idea to me in the deceptively off-hand, elliptical way in which he is accustomed to debating colleagues: strings of facts and arching insights tossed like lassos above the listener's head. If his hypothesis is right, Neandertals could not possibly have been "like us." Every modern human society is organized around some version of family. The organization may be loose or tightly knit, pair-bonded, serially mated, polygamous, polyandrous, whatever. Males needn't be of the "Honey, I'm home!" variety. They might be bonded more tightly to their brothers than to their wives, or they might spend years as warriors isolated from female contact, or in some extreme patriarchal societies they might even be sanctioned to execute their spouses if they prove displeasing. But in every known society, males play some role in provisioning the family and raising the young. It is the human solution to a general mammalian dilemma: how to get juveniles safely through their most vulnerable time, from the moment they are weaned to the moment they can forage competently on their own. These Neandertals had no such assistance. Or so Binford believes.

"Males were simply not part of the solution," he said. "They were part of the problem."

Binford readily admits that his interpretation of Combe Grenal is far from secure, and Combe Grenal is only one site in one part of the vast Neandertal range. Olga Soffer, however, points out that juveniles do make up a significantly greater portion of remains in Neandertal caves than in cave sites of early modern humans, suggesting that they were indeed more vulnerable to the perils of disease, starvation, and, perhaps, menacing male intruders. This does not necessarily mean that they were not being provisioned or defended by resident males. But placed in the context of other clues to Neandertal socioeconomics—the small site sizes, the lack of organization, the exceedingly low population densities, the extreme muscularity of the women as well as the men, the lack of long-distance movement and equivocal evidence for food sharing—Soffer sees no reason to cling to the idea that Neandertals lived in familiar, male-protected family arrangements. Rather than opt for food sharing, Neandertals may have solved the juvenile-vulnerability problem by settling in places where more diverse resources would be compacted within the restricted day range of females with young.

Outside of Soffer's office and Dismantlement Command Central itself, however, I found precious little support for this radical revision of Neandertal life-style. "Lew might be able to document different use of space," Trinkaus suggested, "but claiming that the differential use represents males and females is just inference. I don't know how you could ever tell that."

Others were even less kind to the hypothesis, if they had heard of it. When I asked Jan Simek for comment, he simply looked at the ground and shook his head. I asked Richard Klein of Stanford University to respond to Soffer's arguments, and he laughed. "Well, there's room for every idea in archaeology," he said. Illinois's Geoffrey Pope warned me: "You have to be careful when you listen to Olga. She's been terminally binfordized." And when I later brought up the matter with Jirí Svoboda, a Czech archaeologist, he stared at me as if I were myself slightly deranged.

"You must understand," he said, "for forty or fifty years we have been taught in this country that Neandertals were our ancestors. This was not presented as a theory, but as truth. Now we are learning that this is not necessarily the only way to look at things. But what you are saying, this is impossible to believe."

Normally, when qualified people tell me that a hypothesis is unbeliev-able, laughable, or unworthy of comment, I let it go. But something about this one held on to me. Lew Binford is arguably the most influential archaeologist of his generation, certainly one of its most galvanic theoriz-ers; even his offhand ideas have value, like napkins doodled on by Picasso. More important, I heard in the quick-draw dismissals an edge of defen-siveness, a kind of inner bristling, as if the hypothesis crossed a line that everyone, even Binford, should know was not supposed to be crossed. I sensed the objection was not so much that he was wrong, but that he *couldn't be right*, that the idea of resident females and males who lived separate wandering lives ("visiting firemen," Binford called them) cut across the grain of what was real or imaginable. Today, there are, in fact, large, intelligent, social beings who live in matriarchal societies, their males lurking on the fringes only during mating season. But these crea-tures have great floppy ears and trunks. One can accept that earlier hom-inids might have lived this way, divorced as they are from us by sanitizing immensities of time. But *Neandertals*? So recent, so familiar, so bound to us that their blood may still flow warmly in our veins? Impossible—it simply isn't human. That shock of nonrecognition threatens our precon-ceptions, the intuitive commonsense beliefs and easy inferences, like china rattling in an earthquake.

Which is, of course, precisely Binford's intention.

Far from Combe Grenal, in a back room of the Natural History Mu-seum in Moscow, two Neandertals stand on either side of a dusty dis-play case. The life-size plaster reconstructions are the work of Mikhail Gerasimov, the famous artist-anatomist who was the model for the fo-rensic genius in the thriller *Gorky Park*. One of these Neandertals is no-ble-featured, the other brutish in aspect; while the first gazes into the future, the second, according to one archaeologist, "looks like he couldn't find the doorway."

I had hoped to come to France and leave with a coherent portrait of the way Neandertals lived their lives. Instead, I found a split personality, a schizophrenic Neandertal, alternatively endowed with foresight or in-capable of planning; a cooperative big-game hunter who could not hunt; a full-fledged human with the family life of an elephant. Like one of those little trinkets with an image that shifts back and forth as you turn it to catch a different light, the same Neandertals either wave a human greeting from the past, or grunt back in sullen animal distrust. Discovering which one is real would require more distant travel.

One thing I learned in the Dordogne had the unequivocal ring of a clue. I left Grotte 16 and headed for Paris in the late evening. As I drove through silent villages, each nested alone under the stars, it occurred to me that in a matter of minutes I was bisecting landscapes that in Neandertal times were never crossed in a lifetime, or twenty lifetimes. Joined as we are today by limitless mobility and vast electronic webs of communication, we take connection and exchange for granted—locally, regionally, even globally. It is difficult to conceive of what life must have been like, lived in the isolation suggested by the movement of tools and stone through the Mousterian. Isolation breeds conservatism: Why change, in a thousand or in fifty thousand years, when innovation and influence do not flow streamlike from group to group, but spring up and die in still pools? And so the Neandertals carried on the same traditions, hunted or scavenged much the same animals, in the same ways, through millennia.

Later humans do not act like that. Right from the start of the Upper Paleolithic, things began to pop and jump like water splashed into a frying pan—suddenly there was art, and body ornament, and bone tools, and a hundred new ways of doing the same simple chore, all of it shifting and spreading through trade, migration, or conquest. But it is a dangerous error to measure the "humanness" of Neandertals by comparing their culture to that of later peoples. By that standard, the traveler on the road to Paris a hundred years ago would be less human than I because he was riding a horse instead of driving a Renault 5. The only valid comparison is between the Neandertal culture and whatever modern humans were doing *at the same time*. If there was a stirring in Eve's homeland of something new, some evident leap to a higher cultural plateau, then perhaps the Neandertals really were an inferior breed born to obsolescence, their browridges protruding clownishly like the fins of fifties cars. On the other hand, if they were doing things in much the same ways as their neighbors to the south and east—or doing things differently, but just as well—then they remain an enigma to be reckoned with.

The answer to that question could not be found in France, or anywhere else in Europe. Seven hours after leaving Grotte 16 I was back at De Gaulle airport, boarding a plane for Tel Aviv.

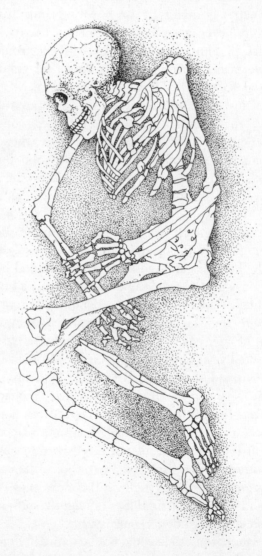

A skeleton from Qafzeh, Israel, one of the earliest known modern Homo sapiens
DAVID DILWORTH, REDRAWN COURTESY BERNARD VANDERMEERSCH

Chapter Seven

MYSTERY ON MOUNT CARMEL

Behold, Esau my brother is a hairy man, and I am a smooth man.
—GENESIS 27:11

In Israel, on the far side of the Neandertal range, a wooded rise of lime-stone issues abruptly out of the Mediterranean below Haifa, ascending in an undulation of hills to the south and east. This is the Mount Carmel of Solomon's Song of Songs, where Elijah brought down the false priests of Ba'al, and the prophetess Deborah put the Canaanites to rout. In sub-sequent centuries, armies, tribes, and whole cultures tramped through its rocky passes and over its fertile flanks, bringing Hittites, Persians, Jews, Romans, Mongols, Muslims, Crusaders, Turks, the modern meddling of Europeans, one people slaughtered or swallowed or subjected by the next, but somehow springing up again and gaining strength enough to slaughter or swallow in its turn.

My interest here is in more ancient confrontations. Mount Carmel lies in the Levant, a tiny hinge of habitability between the sea and the desert, linking the two great landmasses of Africa and Eurasia. A million years ago, a massive radiation of large mammals moved through the Levant from Africa toward the temperate latitudes to the north. Among these mammals were some ancestral humans. Time passed. The humans evolved, diversified. The ones in Europe came to look very different from their now-distant relatives who had remained in Africa. Then, still long before history began to scar the Levant with its sieges and slaughters, the ancient humans from Europe and those from Africa wandered into this link between their homelands, leaving their bones on Mount Carmel.

What happened when they met? How did two separate kinds of human being respond to the sight of each other, after a hundred thousand years of isolation? When they looked into each other's eyes, what message passed between them? Was one's countenance a threat to the other, a horror, a bad joke? Were they instantly enemies, struggling until one destroyed the other? Or did they sit at one fire, swapping information, expertise, and mates? How different *were* they? How much the same?

Reaching the Stone Age in Israel is easy; I simply rented a car in Tel Aviv and drove a couple of hours up the coastal road. My destination was the cave of Kebara, an active excavation hunched above a banana plantation on the western, sea-weathered slope of the mountain. The hard part was escaping the present. When I got lost and arrived at some kind of military checkpoint, a soldier gestured politely with his Uzi, inviting me to turn around. A few minutes later I found the plantation and parked on its upper edge. Standing at the bottom of a parched, sun-baked path, I followed with my eyes a man descending from above, with the rhythmic, rapid progress of someone on skis. As we approached each other, what appeared to be ski poles turned out to be crutches. The man was hopping down the slope on his left leg, his right leg only a stump.

We met on a level piece of ground about halfway between the plantation and the cave. The man told me he was a paleobotanist, helping to identify ancient seeds and pollens in the Kebara excavation. There was a pistol strapped to his waist. It suddenly occurred to me that of the three people I had encountered so far in this country, the only unarmed one was the woman who rented me the car. The paleobotanist explained that he doubled as an inspector of forests, a job that often took him to remote areas on his own. "I haven't needed the pistol yet," he said, "but the time that I don't have it will be the time that I need it the most." We wished each other good luck, and he skipped on down the hill, deftly working the crutches as much for balance as for support. His right leg, I learned later, was blown off in the 1973 war.

I trudged on up the slope and soon reached the cave mouth. Inside, there was no Israel and no Palestine—only a cool, sheltered emptiness, greatly enlarged by decades of archaeological probing. A wire grid hung down from the vaulted ceiling, bestowing a spatial logic on the layers of deep time beneath. Scattered through the excavation's three dimensions were a dozen or so scientists and students; an equal number were working at tables along the rim. The atmosphere was one of hushed, almost monkish concentration, like the reading room of a great library. A couple of people looked up as I came in, but most were absorbed in the square meter of promise directly under their eyes.

Kebara is a mature, dignified enterprise, with ladders and walkways running through a descending series of exposures cut cleanly into the reddish-gold earth. The excavation began ten years ago, picking up on the previous work of Moshe Stekelis of Hebrew University in the 1950s and early 1960s. Stekelis exposed a sequence of Paleolithic deposits and, before his sudden death, discovered the skeleton of an infant Neandertal. A greater treasure emerged in 1983. After Stekelis's time, the sharp vertical profiles of the excavation crumbled under the feet of a generation of kibbutz children and assorted other slow ravages. A graduate student named Lynne Schepartz was assigned the mundane task of cleaning up the deteriorated exposures by cutting them a little deeper. One afternoon she noticed what appeared to be a human toe bone peeking out of a fused clod of sediments. The next morning her whisk broom exposed a pearly array of human teeth: the lower jaw of an adult Neandertal skeleton. Stekelis's team had missed it by two inches.

Lynne Schepartz is no longer a graduate student, but she was still spending her summers at Kebara. I found her and asked how it felt to uncover the fossil.

"Unprintable," she said. "I was jumping up and down and screaming."

She had reason to react unprintably. Her discovery turned out to be not just any Neandertal but the most complete skeleton ever found: the first complete Neandertal spinal column, the first complete Neandertal rib cage, the first complete pelvis of any early hominid known. She showed me a plaster cast of the fossil—affectionately known as "Moshe" to everyone who has passed through Kebara since—lying on an adjacent table. The bones were arranged exactly as they had been found. Moshe was resting on his back, his right arm folded over his chest, his left hand on his stomach, in a classical attitude of burial. The only missing parts were the right leg, the extremity of the left, and except for the lower jaw, the skull. Nobody knows exactly how Moshe lost his head, but it probably happened soon after his use for it had expired. Perhaps erosion exposed it to passing hyenas. Or perhaps the people who had buried him came back a couple of years later, and when they left, they took along something of Moshe to remember him by. Nobody knows.

Schepartz led me down ladders to Moshe's burial site, a deep rectangular pit near the center of the excavation. On this July morning, the Neandertal's grave was occupied by a modern human named Ofer Bar-

Yosef, who peered back up at me from behind thick glasses, magnifying my sense that I had disturbed the happy toil of a caverniculous hobbit. I had met Bar-Yosef before—I was here, in fact, on his invitation—but that was back in his bright, spacious office at Harvard's Peabody Museum, where he was not allowed to wear thick smudges of dirt on his face and an expression of such childlike contentment. Another archaeologist confided to me that Bar-Yosef was "one of the two or three real geniuses in his field." I had no reason to doubt that, but right now I was looking at a man who basically loved to dig in dirt. He seemed evolved to the task, nimble and gnomically compact, the better to fit into cramped enclosures.

Later, over a coffee in a damply cool recess in the back of the cave, Bar-Yosef told me that he had directed his first archaeological excavation at the age of eleven, rounding up a crew of friends in his Jerusalem neighborhood to help him unearth a Byzantine water system. He had not stopped digging since. Kebara was the latest of three major excavations under his direction.

"My daughter has been coming to this site since she was a fetus," he told me. "She used to have a playpen set up, right over there."

Throughout his career, Bar-Yosef has dug for answers to two personal obsessions: the origins of Neolithic agricultural societies, and—the point where our obsessions converge—the twisting conundrum of modern human origins.

The story in the Levant never really made much sense. In the old days, back when everybody *knew* that modern humans first appeared in Western Europe, where the really modern folks still live, you could spot a hominid by the kind of tools he left behind. Bulky Neandertals made bulky, Middle Paleolithic flakes, while svelte Cro-Magnons made slim, Upper Paleolithic "blades." Narrowness is, in fact, the very definition of a blade, which in paleoarchaeology means nothing more than a stone tool twice as long as it is wide. In Europe, a new, efficient way of producing blades from a flint core appeared as part of the "cultural explosion" at the beginning of the Upper Paleolithic, coinciding with the appearance of the Cro-Magnon people. Here in the Levant, however, it has long been conceded that there were no fancy new tool technologies to mark the arrival of anatomically modern humans, not to mention painted caves, beaded necklaces, or other evidence of exploding Cro-Magnon *couture*. On the contrary, the tool kits of the earliest modern humans are the same humdrum Mousterian products that had been churned out by the local Neandertals for millennia.

In this part of the world at least, how modern a hominid *looked* in its body said nothing about how modern it *behaved*.

In 1982, Arthur Jelinek of the University of Arizona made an inspired attempt to massage some sense into this nagging paradox. He had been digging for years at the famous Tabun cave, just a couple of bus stops up the coastal road from Kebara. With over eighty vertical feet of deposits, Tabun had been regarded as a sort of yardstick of human occupation for the rest of the Levant, spanning over 100,000 years. In the Mousterian part of the sequence, roughly 75,000 to 40,000 years ago by his reckoning, Jelinek noted that there was indeed a change in the tools through time— though a more subtle one than the classically abrupt shift in Europe. While no blade-based industries suddenly burst upon the scene, the Mousterian flake tools themselves seemed to get flatter, relative to their width, through time, without departing from the basic Mousterian tradition. Jelinek believed that this trend represented "a fundamental underlying pattern of increasing manual dexterity and control."

The pattern made even more sense when Jelinek matched the actual human fossils in the Levant against the tools found alongside them. The fattest Mousterian flakes came from a layer near the bottom of the Tabun sequence where Dorothy Garrod had found the skeleton of a Neandertal woman. If flake thickness was a true measure of time, the Tabun woman was the oldest Mousterian hominid in the area. The next oldest proved to be the Neandertal infant that Stekelis had found at Kebara, beside tools that matched those in a higher level on the Tabun yardstick of change. The robust modern human skeletons of Skhul cave, literally around the corner from Tabun, yielded flake tools that were flatter still. But the flattest of all belonged to the not-so-robust moderns of Qafzeh cave, a few miles to the east. Thus an "acceleration" in the trend toward flatter flakes appeared at precisely the anatomical border between the Neandertals and moderns. Although the physically modern Skhul-Qafzeh people might not have crossed the line into full-fledged, blade-based humanness, they appeared, as Jelinek put it, "on the threshold of breaking away.

"Our current evidence from Tabun suggests an orderly and continuous progress of industries in the southern Levant," he wrote, "paralleled by a morphological progression from Neanderthal to modern man."

Jelinek's scenario had a familiar echo to it. As in Europe later on, tools got thinner along with the bodies of the people who made them. Only in this case, the reduction in tool thickness was front to back rather than side to side. The scheme did not explain away all the anomalies in the Near East, but it was enough to make most of his colleagues sigh in relief.

But there was trouble brewing. Also in the early 1980s, Ofer Bar-Yosef, physical anthropologist Bernard Vandermeersch of the University of Bordeaux, and paleontologist Eitan Tchernov of Hebrew University quietly announced a more disturbing chronology for human evolution in the Levant. The clock they used to chart human progress was based not on tool shapes but on evolutionary changes in "microfauna," the little rodents that had been killed and consumed by cave-dwelling owls, their bones regurgitated, buried, fossilized, and rediscovered in the sieving screens of archaeologists. Through time, the rodent species in a given region change along with shifts in climate. Making the logical assumption that owls eat only the rodents who exist in their immediate area, Bar-Yosef and his colleagues reasoned that the tiny rodent bones found in the caves, and identified as to species, were a ledger of ancient climate in the Levant. For example, two species of African rat found in the hominid levels of Qafzeh cave indicated a period of warm climate. At Tabun, these two species were present only in the very lowest levels, while the bones of gray hamsters were found only in higher levels at Tabun and Kebara. Gray hamsters are a Eurasian species, suggesting that the climate had turned colder, driving the African rats out of the Near East and allowing more cold-adapted species to gain a foothold.

When the three men plotted the comings and goings of the rodents against known changes in climate through time, a bizarre view of human evolution in the Levant emerged. The modern human skeletons from Qafzeh cave ended up at the bottom of the Mousterian sequence. They appeared to be roughly 100,000 years old, rather than the sensible 40,000 that Jelinek believed. Against all expectation, the thin-boned Near Eastern moderns were considerably *older* than the barrel-chested Neandertals who were supposed to be their own ancestors. As Eitan Tchernov put it, "A fully modern Man roamed around the Levant before any Neanderthal came for a visit." If so, he was an audacious trespasser—not only intruding into Neanderthal Man's allotted time frame, but using his tools as well.

This theory suffered the fate of any idea that makes a difficult problem even harder to explain. It was ignored. Conventional wisdom much preferred Jelinek's familiar match of thinning men with thinning tools. Conventional wisdom wanted to have its modern humans arriving in the

Levant *after* the Neandertals, just as they did in Europe, where the conventional wisdom had grown up in the first place. In short, conventional wisdom would have liked Bar-Yosef and his friends to take their little rat bones and go away. Instead, they came here to Kebara. "The only way we could persuade people that our suggestion was serious was to put together another project," Bar-Yosef told me. Partially excavated, but with a deep lode of additional deposits left untouched, Kebara was the perfect choice. They gathered together an impressive team of ten scientists and went to work.

The first task was to figure out the complicated stratigraphy inside the cave. In archaeologists' dreams, the deposits in a site are lined up horizontally, like the layers of a cake, with the oldest on the bottom and the youngest on top. Kebara is more like an archaeologist's nightmare. "These Israeli caves don't look like rock shelters in the Dordogne," Bar-Yosef complained, waving a hand at the exposures below us, as if admonishing a favorite child. "Please, give me a rock shelter any day."

Like the other Mount Carmel sites, Kebara is a natural "karstic" formation, meaning that it was created by water's dissolving action upon limestone. Bar-Yosef pointed up. The cave walls, water-slick and nobbled with lime, tapered sharply and rose in a natural chimney, open to the sky some sixty feet above. Chimneys like these, he told me, are common features of limestone caves. Less visible are the "swallow holes," a kind of natural drainage sump that forms underneath. As time goes by, debris builds up inside the cave—stones, animal bones, bat guano, fossilized hyena turds, silt and sand blown in—plugging up the swallow hole. Originally, all this stuff might have been deposited in a relatively organized way. But every once in a great while the swallow hole comes unplugged.

"It's as if you pulled the plug in your child's bath," Bar-Yosef explained. "Everything is drawn down toward the center, all at once." As in a tub, the drain's sucking action is strongest right over the swallow hole, weakest on the fringes of the cave. Thus, if an unsuspecting geologist should stroll in a few thousand years after one of these unpluggings, he might assume that all the debris in a single horizontal level had been deposited at the same time. But he would be wrong. In the past when the swallow hole came unplugged, it would have pulled hardest on the debris above it. Younger material that had originally been deposited at a higher level would have been sucked down, coming to rest on a level with debris from the cave's fringe that had been deposited thousands of years before. To complicate matters further, any given cave may have several swallow holes, as well as water channels running between them, picking up material originally deposited in one place and dropping it in another.

The two experts who interpreted this nightmare at Kebara were geologists Paul Goldberg and Henri LaVille. (I could see Goldberg's heavily bearded face poking up from a trench below me; LaVille was squatting on a rise near the cave mouth, directing the work of some students.) They divided the stratigraphy into twelve units. The three most recent contained an Upper Paleolithic tool technology, made within the last 30,000 years. The next three revealed a mixture of Upper Paleolithic and Mousterian tools, which proved to be not evidence of two intertwined cultures but simply the result of the natural blending processes described above. The most enlightening units were farther down: six successive layers filled with Mousterian tools and hearths. A few of the tools were found *in* the hearths, tinged rose and crackled on the surface from the heat of the fire. Similar burned flints had showed up at Qafzeh too.

There was nothing particularly mysterious about the origin of these burned flints. Somebody had discarded a tool. Later, people being as careless in the Middle Paleolithic as they are now, somebody else came along and idly kicked it into the embers of a hearth. One would have to think hard to imagine a more meaningless gesture. Yet upon such trivial events—a flick of a toe, the serendipitous acquaintance of flint with fire—the whole issue of *time* in the Levant finally came to rest. And upon that issue, ultimately, hangs the mystery of modern human origins.

This is a tale of two caves. In 1965, a young Bernard Vandermeersch began digging at Qafzeh, a few miles inland from Kebara on a hill in lower Galilee. People marvel at the luck it takes to find a single hominid fossil; on the terrace outside the cave mouth, Vandermeersch found a veritable Middle Paleolithic cemetery. Finding the bones proved to be the easy part. Long after the people had been buried, the swallow hole of Qafzeh had become permanently flooded, forming a drainage channel running out of the cave mouth. Sediments that had been soft enough for burying deceased relatives became soaked, and later hardened into the rocklike conglomerate called breccia. The removal of the discovered dead thus became *une galère*, the French word for a problem so drudging that it conjures up images of time spent rowing a galley ship.

"It was all very difficult," Vandermeersch told me, in the cool shadows

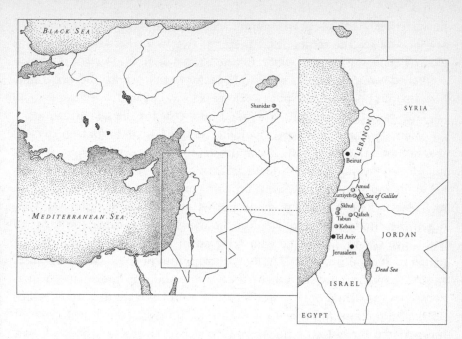

A map of the Levant, showing key Middle Paleolithic fossil sites   DAVID DILWORTH

of Kebara cave. "We worked there until 1980, and in all that time I excavated maybe thirty-five square meters. In fifteen years, a space smaller than my office! We had no choice but to go at it with hammers and chisels. My colleagues used to call it Sing Sing." The greatest find also presented the greatest problems. The treasure was a marvelously complete skeleton of a young woman, known as Qafzeh 9, found buried together with an infant huddled at her feet. "I could follow the bones to find the limits of the skeletons, but to excavate them in the field would have been impossible," he said. "This could only be done back in the lab, with the help of a high-speed dental drill."

The biggest dilemma was how to get a block of breccia weighing nearly a ton down the side of a steep ravine. As it happened, Vandermeersch had a connection to somebody high up in the Israeli Army. His name was Moshe Dayan. So when the block was ready, a large military helicopter dropped from the sky above the cave mouth.

"The pilot descended slowly, with a net hanging down for us to put the block in. But before he was low enough, the rotors began to kick up all the dust, so much that the pilot couldn't see where he was going. I looked up, and I could see the rotors just a meter or two from the wall and coming closer. People say I turned white. But he backed away, and we got the block securely down. Afterward, I went to the pilot to thank him, and I told him how anxious I had been. 'Not as anxious as I was,' he said."

This was in the summer of 1967, a busy year for the Israeli Army.

A few days after talking to Vandermeersch, I went to Qafzeh myself. It is not a tourist attraction. The approach to the cave is through a booming, dust-billowing quarry. Mechanical monsters with tires dwarfing my car lumbered around Jurassically, browsing on rock. I parked the little Fiat in a place where I hoped it would not be spotted and ingested, and we began the climb. My guide was Tal Simmons, a postgraduate student at the University of Tennessee who grew up in Israel. Three students had joined us for the fun of it. The hot, choking fun of the quarry dust soon gave way to the pungent fun of the wadi leading up the ravine, which doubled as a natural open sewer for the town of Nazareth, a mile farther up the hill. We were lucky. Our excursion did not coincide with one of the periodic flushings of the town's collective cloaca, so the wadi was odorous but dry. A film crew the previous year, Tal told us, had to pass their equipment hand to hand across a clotted running stream. Casually, she also reminded us to watch out for scorpions. We left the wadi and followed a narrow path along the edge of the ravine, rewarded at last by a view of the oval gape of Qafzeh in the opposite cliff face. A few minutes later we collapsed in the cool shadows inside.

I looked back down the slope, following the snaking trail of the wadi until it issued out into cotton fields in the Esdraelon plain below. Behind me was the cave interior, an unprepossessing space that has never yielded much of archaeological interest. At my feet was the cave terrace, which has yielded a treasure of immeasurable interest to me. In modern human origins, Qafzeh is the wild card; where it falls in time determines the course of the game. If Jelinek's chronology based on slimming tool forms was right, the twenty four human fossils found here could be "proto-Cro-Magnons," the evolutionary link between a Neandertal past and a Cro-Magnon future—and thence to the present moment. But if the rat bones were telling truer time, such linear logic was overthrown. Both dating methods were *relative*, merely inferring an age for the skeletons by where they fell in an overall chronological scheme. What was needed was a new way of measuring time, preferably an *absolute* dating technique that could label the Qafzeh hominids with an age in actual calendar years.

Such absolute dating methods have been available for decades. The most celebrated is the geochronological clock of radiocarbon dating, which measures time by the constant, steady "decay" of radioactive carbon atoms into nitrogen. Developed in the 1940s, radiocarbon dating is still one of the most accurate ways to pin an age on a site, so long as it is *younger* than around 40,000 years. In older materials, the amount of radioactive carbon still left undecayed is so small that even the slightest amount of contamination in the experimental process leads to highly inaccurate results. Another technique, relying on the decay of radioactive potassium instead of carbon, has been used since the late 1950s to date volcanic deposits *older* than half a million years. Radiopotassium was the method of choice for dating the famous East African early hominids like Lucy and *Homo habilis*, as well as the new "root hominid" announced in 1994. Everything that lived between the ranges of these two techniques, however—including the moderns at Qafzeh, the Neandertals at Kebara, and indeed, all the far-flung sites mapping the birth of *Homo sapiens* fell into a chronological black hole.

A way to illumine the hole had been lying around for more than three centuries, waiting to be discovered. One night in October 1663, the great English physicist Robert Boyle, for reasons that remain obscure, took a diamond to bed with him. Resting the diamond "close upon a warm part of my naked body," Boyle noticed that it began to glow. So taken was he with his responsive gem that the very next day he delivered a paper on the subject at the Royal Society in London. His surprise at the glow was all the greater, he told his assembled colleagues, because "as you know, my constitution is not of the hottest."

Boyle's body heat had triggered the atomic phenomenon now known as thermoluminescence, or "heat-light." If you were to pick up an ordinary rock and try to describe its essential rockness, phrases like "frenetically animated" would probably not leap to mind. But in fact, minerals are in a constant state of inner turmoil. Minute amounts of radioactive elements, within both the rock itself and the surrounding soil and atmosphere, are constantly bombarding its atoms, sheering electrons out of their normal orbits around the atomic nucleus. All this is normal electron behavior, and after gallivanting around for a hundredth of a second or two, most of the electrons dutifully return to their proper positions. A few, however, become trapped en route—physically captured within crystal impurities or electronic aberrations in the mineral structure itself. These tiny prisons hold their electrons until the mineral is heated, whereupon the traps spring open and the electrons return to their stable state. In the process, they release energy in the form of light—a photon for every homeward-bound electron.

Three hundred years after Boyle, Martin Aitken of Oxford University developed the methods to turn thermoluminescence (TL) into a geophysical stopwatch. The watch works because the radioactivity bombarding a mineral is fairly constant, so electrons become trapped in their crystalline prisons at a steady rate through time. If you crush the mineral you want to date and heat a few grains to a high enough temperature— about 900 degrees—all the electron traps will release their captive electrons at once, creating a brilliant puff of light. In a laboratory, a device called a photomultiplier can easily quantify the intensity of that burst of luminescence as a spike on a graph. The higher the spike, the more trapped electrons have accumulated in the sample, and thus the more time has elapsed since it was last exposed to heat. Once a material is heated and all the electrons have returned "home," the stopwatch is set back to zero, and the process starts over again.

In most cases, the amount of TL stored in the electron traps is simply a measurement of the time elapsed since the mineral was originally forged in its volcanic furnace, or since it lay on the surface, exposed to sunlight sufficient to spring open the traps. But sometimes a mineral has been reheated by human hands. Aitken and his Oxford colleagues first used the technique to date ancient pottery, since firing clay to make ceramics supplies more than enough heat to blow open the electron traps and reset the stopwatch to the moment the pottery was made. Neither Neandertals nor early moderns knew how to make pottery. But they *did* know how to make tools out of flint, a mineral that happens to be very lively, in a thermoluminescent sense. In the Middle Paleolithic, flint tools also happened to lie around in the path of careless feet, opening up an exquisite opportunity for geochronologists of the future.

In the early 1980s, Hélène Valladas, an archaeologist at the Center for Low-Level Radioactivity of the French Atomic Energy Commission, coaxed thermoluminescence out of flints from Kebara and Qafzeh caves. By 1987, she and her physicist father, George Valladas, had squeezed an age of 60,000 years out of burned tools found beside Moshe at Kebara. That number pleased everybody, since it agreed with *both* time schemes arrived at through relative dating methods. The shocker came the following year, when Valladas and her colleagues announced the results of her work at Qafzeh. According to the Valladas team, the skeletons were 92,000 years old, give or take a few thousand. Suddenly, the disturbing hypothesis of Bar-Yosef and his friends was not so easy to ignore. But that did not make it any less disturbing.

"If these TL dates are correct, what does this do to what else we know, to the stratigraphy, to fossil man, to the archaeology?" worried Tony Marks of SMU, echoing the sentiments of many of his colleagues. "It's all a mess."

TL is still considered experimental, and has fiendish complications to overcome. For instance, to convert the burst of light given off a heated flint into calendar years, one has to know both the sensitivity of that particular piece of flint to radiation and the dose of radioactive rays it has received each year since it was "zeroed" by fire long ago. The sensitivity of the sample can be determined by assaulting it with artificial radiation in the lab, and the annual dose of radiation received from within the sample itself can be calculated fairly easily by measuring how much uranium or other radioactive elements the sample contains. But determining the annual dose rate from the environment *around* the sample—the radioactivity in the surrounding soil, and cosmic rays from the atmosphere itself—is an iffier proposition. At some sites, fluctuations in this environmental dose rate through the millennia can turn the "absolute" date derived from TL into an absolute nightmare.

According to Valladas herself, "Those of us who are actually doing the dating don't like to use the word 'absolute.' We prefer just to call it physical dating."

Fortunately, most of the radiation dose for the Qafzeh tools came from within the flint itself, which eliminated many of the problems. Several other Neandertal and modern human sites have since been dated with TL, and the one at Qafzeh remains not only the most sensational but the surest. Key sites in the Levant have also been dated by a "sister" technique called electron spin resonance (ESR). Like thermoluminescence, ESR fashions a clock out of the steadily accumulating electrons caught in traps. But where TL measures that accumulation by the strength of the light given off when the traps open, ESR literally counts the captive electrons themselves while they still rest undisturbed in their crystal jail cells.

One must be endowed with a special kind of brain to understand fully how electron spin resonance works. I am not, but I can offer a layman's summary. All electrons "spin" in one of two opposite directions—let's call them up and down. Metaphors are a must, because the nature of this electron "spinning" is quantum mechanical, and "words cannot describe it, only huge mathematical equations can," according to one expert I talked to. The spin of each electron creates a tiny magnetic force pointing in one direction, something like a compass needle. Under normal circumstances, the electrons are paired so that their opposing spins and magnetic forces cancel each other out. But trapped electrons are unpaired. By manipulating an external magnetic field placed around a sample to be dated, the captive electrons can be made to flip around and spin in the other direction. When they flip, each electron absorbs a finite amount from a microwave field also applied to the sample. This energy loss, called resonance, can be measured with a detector. It is a direct count of the number of electrons caught in the traps, and thus a measure of time. ESR has been used to date everything from stalagmites to stale potato chips, but it works particularly well on tooth enamel, with an effective range of from a few hundred to two million years.

Using electron spin resonance to date large mammal teeth found near the Qafzeh skeletons, Henry Schwarcz, the ubiquitous site dater at McMaster University, and his colleague Rainer Grün at Cambridge came up with a date even older than Valladas's thermoluminescent surprise. The skeletons were at least 100,000, and perhaps 115,000 years old. Schwarcz and Grün then took their ESR techniques to Skhul cave. The moderns there had been given an age of around 40,000 years by relative dating methods. Schwarcz and Grün coolly tagged on another 60,000.

"People said that TL dates at Qafzeh had too many uncertainties," Bernard Vandermeersch told me at Kebara. "So we gave them ESR. They doubted that too. So we gave them Skhul. By now, it is very difficult to dispute that the first modern humans in the Levant were here by one hundred thousand years ago."

Clearly, if modern humans were inhabiting the Levant 40,000 years before the Neandertals, they could hardly have evolved from them. This news was cheerfully lapped up by replacement advocates. Chris Stringer, in the same issue of *Nature* where Hélène Valladas published her first Qafzeh date, boldly announced, "Evolutionary models centered on a direct ancestor-descendant relationship between Neanderthals and modern *H. sapiens* must surely now be discarded." *Surety* is a very rare commodity in this most insecure of sciences. But *if* the dates are indeed correct—

and in spite of Vandermeersch's confidence, that *if* remains—it is hard to see what else one can do with the venerated belief in Neandertal ancestry but chuck it, once and for all.

Case closed? On the contrary, the dates only give another twist to the mystery on Mount Carmel. Presuming that the moderns did not just come for a visit 100,000 years ago and then politely withdraw from the Near East, they must have been around when the Neandertals arrived 40,000 years later—if the Neandertals as well weren't there to begin with. Either way, two distinct *kinds* of human, sharing *one* culture, were apparently squeezed together in an area not much larger than the state of New Jersey. Rather than resolving the paradox, the new dating techniques only teased out its riddles: If two different kinds of human were behaving the same way in the same place at the same time, how can we call them different? If they looked so different, then how can they have been doing the same things? If modern humans do not descend from the Neandertals but replace them instead, why did it take them so long to get the job done? Can one call so slow a process "replacement"?

What would help answer these questions would be some new clue to the lives of these two ancients, something that would separate them *behaviorally* as crisply as someone like Stringer believes they can be distinguished in their bones. This is precisely what those studious people at Kebara were searching for.

One thing was sure. They were not likely to find it by looking at stone-tool forms. With Jelinek's theory of "increasing manual dexterity" blown apart by the new dates, there is nothing left that clearly says, *"These* tools here were made by Neandertals, and these others, by moderns." There are subtle distinctions, perhaps. A higher percentage of points at Kebara. A penchant there for knocking flakes off a core in one direction only. A preference at Qafzeh for working the stone in a circular fashion instead, toward the center of the core. But these are quirks, embellishments, local traditions. The thrust through it all is pure, unalleviated Mousterian.

Furthermore, there is little suggesting that the neighboring humans were *using* their tools any differently either. One of the dividing lines between Neandertals and moderns used to be considered the technological innovation of hafting: attaching a stone tool to a shaft and greatly amplifying its striking force, be it a hammer or a projectile. Such mechanically generated power could have relaxed the need for all that Neandertal muscle bulk, allowing the genes for modern body types to spread across the human landscape. Mousterian tools, so the argument goes, did not have the standardized butt ends that would have fitted a haft.

I found this theory convincing right up to the moment at Harvard when John Shea fixed a razor-sharp Mousterian point onto a spear shaft with pine pitch and waved the thing in my face. After that, I became a believer in the haftability of Neandertal tools. But while some other researchers have found traces of hafting on Mousterian tools, others question whether these were the true projectiles that Cro-Magnon Man clearly had at his disposal later on, or merely "hand-held thrusting spears." Bordeaux's Jean-Michel Geneste actually performed a series of experiments launching hafted Mousterian points into dead animal targets with dismal results. "Maybe Shea can show that the points were hafted," Geneste told me, "but they cannot be thrown. They are too heavy."

When Geneste told me this, I took stock of his physique. Though he is somewhat bulkier than the average French male, I have to wonder whether he is a good judge of what a Neandertal in his prime might have considered "too heavy." More to the point, John Shea's microwear studies, described in the previous chapter, make a strong case that the stone points at Kebara and those at Qafzeh revealed the same kinds of damage marks, including about an equal number of impact fractures from points that had been hafted and heaved at some resistant target—presumably prey.

There is more to Kebara than what can be gleaned from stone tools. The day after my trip to Qafzeh, Harvard graduate student Dan Lieberman, digging down in Moshe's grave, uncovered a fireplace. The sediments of Kebara are thick with these ancient hearths, so dense in places they turn the walls of the excavation into marble cake, with swirls of black and gray ash lacing the brick-red sediments. Seeds found among the ashes have been identified; apparently the people of Kebara were eating a lot of wild peas. Meanwhile, geologist Paul Goldberg hardened samples of the ash in polyester and sliced off very thin sections to be examined microscopically. The work revealed traces of grass, wood, and bone in the ash, and the residue left from burning bat guano.

Neandertals here, like those in Europe, apparently did not know how to coax more heat from a fireplace by lining it with stones or digging ventilation channels. The most impressive feature of the Kebaran hearths is their sheer abundance. From the very bottom of the cave, around 75,000 years ago, and especially around Moshe's level at 60,000 years ago, the number and thickness of the ashy swirls imply that the cave was intensely occupied by humans. "Mount Carmel is a good place to live today, and back then it may even have been better," Bar-Yosef told me, poking at the ash with a flat trowel. "A nice climate, plenty of game and plants to eat. So if you want, maybe what you are seeing here is a kind of permanency, a sedentism. Or what Lew Binford calls a 'magnet site,' because he has to have a new name for everything."

Other traces of the cave's former "magnetism" were nearby. About ten feet away from Moshe's grave and a few thousand years higher up in time, Tal Simmons squatted among a dense clutter of broken animal bones and flint. Simmons had been sitting in more or less the same spot for the last four years, slowly digging down in time. In 1986, she uncovered a clutter of bone and tool debris. The clutter became a pocket, the pocket, a layer, and finally the layer, an archaeological lode almost a foot deep. There were other dense concentrations of animal bones along the northern wall of the cave. But those bore gnaw marks, and were mixed in with a profusion of coprolites—fossilized hyena turds—and even a few baby hyena skeletons. This is a classic profile of a hyena den. Simmons's jumble of debris, in contrast, showed a distinctly human touch. "This is the only place in the cave where the bones bear a lot of cut marks," she told me. "My feeling is it's some kind of refuse dump."

Faunal expert John Speth of the University of Michigan had been brought in to help re-create the meaty part of the Kebaran diet, identifying the human garbage represented in the "refuse dump" and elsewhere in the cave. So far, a fairly typical Neandertal pattern was emerging. "It's pretty much of a mixed bag," said Speth. "You find all kinds of medium and small animals—deer, horses, tortoise, birds, gazelle."

Judging by this menu, the Kebaran people were hunting or scavenging whatever food was walking or flying around in the immediate vicinity. Taken in concert with the density of hearths, you end up with a portrait of a highly territorial people, sitting tight in an area of relative plenty rather than migrating seasonally in pursuit of some particular kind of game. As in the Dordogne, I got the sense of a people hunkered down into the landscape, isolated but knowing well their immediate surroundings and taking from them what they need.

If the traces left behind at Qafzeh showed a different pattern—hearths more like the modified contrivances of the European Upper Paleolithic, or a diet revealing a specialized focus on a particular kind of game—then we might have found the clue to a behavioral difference we are looking for. But they don't, at least not conclusively. The hearths at Qafzeh are no fancier in design than the ones at Kebara. In the faunal record, there is a hint that the people there were concentrating on a single animal species—red deer—rather than the smorgasbord of Kebara. But Qafzeh is inland from the coast, where red deer might simply have been the most available thing to eat. Besides, this preference is not confined to the modern human burial level at Qafzeh, but runs right through the Mousterian sequence. Which of our two human types was doing the deer-eating all that time is anybody's guess.

After four days at Kebara, I had found no cultural grounds for distinguishing moderns and Neandertals in the Near East—no innovative edge on the one hand, no lingering mental dullness on the other. But something else in Moshe's cave might still yield a clue—Moshe himself.

When hunter-gatherers bring home a kill, the carcass is shared among the group. The hunter might keep the head and marrow bones and give his in-laws a hindquarter, the forequarters to someone else, and so on. Much the same fate awaited Moshe when he was found. The four physical anthropologists on the Kebara project each took a piece of him home to analyze and describe, according to their special interest. Vandermeersch got the shoulder and arm. Ann Marie Tillier, a colleague of Vandermeersch's at Bordeaux, was awarded the lower jaw and teeth, Baruch Arensburg of Tel Aviv University, the vertebral column and ribs. That left only Yoel Rak, also of Tel Aviv University. Rak had spent this summer at the Institute of Human Origins in Berkeley, where I talked to him before coming to Kebara.

"I was working on faces when the skeleton was found," he told me, "and since it had no face, I was left with what nobody else wanted—the pelvis. And that's how I got interested in pelves."

Before Moshe was discovered, the few tidbits of Neandertal pelvis already known were struggling to support a crushing weight of implication. The front of the pelvic opening in humans is composed of two joined pubic bones, one on each side. In modern humans, they are short and stubby, while all the known Neandertal pubic bones were long and slender—fully 20 percent longer, side to side, than those of the modern humans who followed. According to New Mexico's Erik Trinkaus, the reduction in pubic bone length decreased the distance between the two hip joints, streamlining the human stride. But this improvement in walking efficiency did not take place in a biomechanical vacuum. The pubic bones also form the front part of the female birth canal. If they were shortened, the birth canal would have lost as much as 25 percent of its total diameter. No human infant could have squeezed its big-brained skull through a channel that small. Yet the fact that we are still on the planet confirms that successful births were taking place.

Trinkaus pondered this riddle and came up with an elegant theory: Modern human babies were simply born sooner, before their heads grew too large to pass through the birth canal. Our species' nine-month pregnancy thus represents a reduction from the primitive Neandertal norm, which Trinkaus estimated at eleven to twelve months. The hypothesis was bursting with implications. First, an infant born three months "early"— relative to the previous standard—requires far more maternal support to survive, which would seriously restrict the movement of its mother. Where females might once have been out hunting with the males, they must now have been spending more time caring for their helpless newborns back at the home camp, setting up an enormous change in hominid social organization and division of labor. Second, a decrease in gestation length exposes the newborn to all sorts of environmental stimuli, at an age when its Neandertal counterpart was still wallowing in the languid sensory world of the womb. Human brain growth is very quick just after birth, so the modern baby's three-month-early matriculation into the world would have conferred, as Trinkaus put it, a "headstart on neurological development." Finally, everything else being equal, shorter pregnancies mean shorter intervals between births, more children per mother, and possibly faster population growth—which alone might account for why one kind of human could replace another on the landscape.

And all this, gleaned from the 20 percent of extra pubic bone in the few bits of Neandertal pelvis known.

"Erik took a few clues and built a really interesting hypothesis," Yoel Rak said to me. "As far as I'm concerned, this was science at its best." Then he shrugged. "But as soon as I saw the Kebara pelvis, I knew he was wrong."

Like other known Neandertal pubic bones, Moshe's was long and slender. But with a complete pelvis in hand, Rak could see that the length of the pubic bone did not affect the size of the pelvic opening, which was no larger than that of the average modern human male. By inference, the birth canals of Neandertal females were no more capacious than those of women today, effectively killing the notion that the difference between them and us had to do with the amount of time spent in the womb.

Instead, the longer pubic bone merely changed the location of the inlet, which in Moshe was farther forward than it is in modern human pelves.

Yoel Rak thinks that although the Neandertal's forward placement of the pelvic inlet carries no reproductive repercussions, it does suggest a significant difference in the way Neandertals walked. The Neandertals' center of gravity would have been directly over the hip joints, rather than somewhat behind as in humans today. This would have brought the full weight of the Neandertal pounding against its hip joint with every stride. Rak considers that the pelvic inlet's backward migration in moderns was a locomotor step forward, with the muscles in our thighs and backside acting as "shock absorbers" against the stress of long-distance walking. "From a biomechanical point of view, modern humans are greatly adapted for walking," he told me. "There is very little doubt that Neandertals were less efficient."

So Moshe does have something to say about how Neandertals might have been doing things differently from their modern human contemporaries. The locomotor inefficiency of their pelves fits neatly into the pattern of a people who were adapted to a less mobile life-style than modern human hunter-gatherers, including their own contemporaries in the Middle East. Several pelves recovered from Qafzeh and Skhul bear the modern stamp. The femurs also display pronounced "pilasters"—the thickened ridge along the trailing edge of the bone which Erik Trinkaus chalks up to a life spent walking a great deal. But if Neandertals were more restricted in their movements, it does not mean they were less human. Bedouins and Bushmen are more active on the landscape than most Bantus or Bostonians, and their skeletons show it. But this does not mean the Bushmen are more highly evolved, or that Bostonians fail to meet some essential standard of humanness. The same goes for Moshe and his kin, whatever the shape of their femurs and hips.*

*Recently, Dan Lieberman and John Shea underscored this contrast in mobility between Neandertals and modern humans in the Levant, using a different approach. Mammal teeth contain a bonelike tissue called cementum, which is deposited incrementally around the roots of teeth through the animal's lifetime. In some species, these cementum layers are opaque if they are deposited during a wet season, translucent if they formed during the dry season. The outermost layer of a fossilized mammal tooth can thus reveal at what time of year the animal died. Lieberman and Shea examined the outer layer of cementum of prey species' teeth found in Kebara and Qafzeh and discovered a fascinating contrast. At Kebara, about half the prey animals had been killed during the wet season, and half during the dry. At Qafzeh, however, *all* the outer layers of cementum were translucent, indicating that the cave had been used by hunters only during the dry season. Apparently, the modern humans at Qafzeh were using the cave during

Another of Moshe's bones *does* seem to show on which side of the human-animal divide Neandertals fall. Among his share of the Kebara treasure, Baruch Arensburg of Hebrew University received a tiny, horseshoe-shaped structure of the throat called the hyoid. This one little bone says more about the Neandertal's place on the evolutionary scale of behavior than all the rest of Moshe put together—perhaps as much as any single bone ever found.

At issue is the one human trait that most scientists agree draws a clear line between us and the rest of creation: complex spoken language. If language is indeed the *sine qua non* of humanness, then no paleoanthropological quest could be more important than finding out when language first appeared. Unfortunately, the organs of verbal communication don't turn to stone as bones do. Paleoanthropologists are left trying to infer what soft throat tissues like the larynx, the vocal cords, and most of all, the brain itself might have looked like in our ancestors. There are only indirect clues, like the delicate, fossilized impressions made by the brain on the inside of the skull and the shapes of the jaw and the skull base. The hyoid bone, anchoring muscles of tongue, larynx, and jaw, should be particularly revealing. But hyoids are fragile things, and none had ever turned up in the fossil record of humankind. Until Moshe.

For years, American anatomists Jeffrey Laitman of Mount Sinai School of Medicine in New York City, Philip Lieberman of Brown University, and Edmund Crelin at Yale had been forcefully insisting that Neandertals lacked the requisite throat anatomy for fully human speech. Their arguments were based primarily on computer models of the shape of the Neandertal vocal tract, reconstructed from classic fossils like La Chapelle-aux-Saints and La Ferrassie. The key to these computer models is the extreme "flexing" of the base of the skull in adult modern humans—the way our skulls angle down sharply in back of the mouth. Other mammals, and also human infants, have a much flatter skull base, which forces the larynx up high in the throat. This makes it possible to swallow and breathe at the same time but does not give sufficient room above the larynx to

---

only one season, while Kebaran Neandertals were staying in the area and hunting all through the year—a conclusion reinforced by a large proportion of broken projectile points at Kebara. This contrast in land use, Shea says, might explain why modern human skeletons were conditioned to walking around more, as they circulated seasonally from one area to another, while the thick Neandertal bones were better built for foraging through a limited area, season after season. It says nothing, however, about whether one or the other human kind was more apt to survive in the long run.

make rapid human speech possible. According to Laitman and his colleagues, the skulls of most early hominids show the flat, mammalian pattern that implies they did not have fully modern human speech capabilities. Surprisingly, some of the flattest skull bases belong to those closest to us in time: the Neandertals. None of these anatomists claims that Neandertals couldn't talk, but they believe that their "essentially non-human airways" could not produce the rapid-fire speech of modern humans, and that this was their downfall. "I propose that the extinction of Neanderthal hominids was due to the competition of modern human beings who were better adapted for speech and language," Lieberman concluded.

Laitman and Lieberman's work lent anatomical support to what replacement theorists believed *had* to be true: that moderns must have had a major, critical advantage over the Neandertals in order to replace them, and language was the most often cited candidate. At the same time, a host of other experts were stubbornly convinced by the sheer size of Neandertal brains and traces of brain structure that Neandertals would have been perfectly capable of speech. What was needed to break the deadlock was a new bit of evidence. "If you have the jaw, and you know the shape of the hyoid bone," Arensburg told me, "then you know the hyoid's position in the throat. And if you know the hyoid position, then you know where the larynx is positioned, too. And in every major respect, this Kebara man had a hyoid built just like ours today."

Jeffrey Laitman protests that Moshe's vocal tract cannot be reconstructed from just his hyoid and jaw, and the rest is forever missing. In any case, for him and his colleagues, hyoids alone do not settle the argument. For Arensburg, there is more than enough of Moshe to decide the question of language. "The idea that Middle Paleolithic people were mute 'village idiots' is ridiculous. This person was intentionally buried by his fellows. You cannot tell a gorilla to go bury his grandfather. You cannot learn abstract things through imitation. If people do not have language, they do not have traditions. And if they have traditions, then they must have had language."

At the end of my last day at Kebara, I crouched at the rim of the excavation to survey its three dimensions. Near the center, Tal Simmons was on her knees among the ancient refuse, dusting off the thousandth bone chip with the same care she had bestowed on the first. In the back of the cave, Paul Goldberg burned bits of grass in an aluminum pie pan, continuing his probe into the ash of the Kebaran hearths. John Speth, sitting on a raised plank, bent over some charts. Deep in Moshe's grave, Dan Lieberman sweated, sighed, laid down his shovel. "I've never been so dirty," he muttered, and began to dig again. A younger student, frustrated by the difficulty of plotting an artifact in his workspace, threw down his measuring stick in disgust. Ofer Bar-Yosef climbed down to help him. All over the site, layers of soil that took generations to collect were skimmed off, dumped in buckets, and taken outside to be screened.

And in all this, nothing had so far conferred the slightest stigma of disadvantage on the Neandertals. At the same time, nothing dug up at Qafzeh or Skhul marked the people there as modern in anything but their physical body shape. Two anatomical types in one place, acting equally human. Perhaps for tens of thousands of years.

I took the paradox with me to mull over outside, in a still summer afternoon in Israel, where the horizon manifested the present moment in the silhouette of an oil tanker, far out to sea. There was a quick and enticing resolution to the problem of two contemporary humans to consider: They were not two, but one. If the names "Neandertal" and "modern human" are meaningful distinctions, if they have as much reality, say, as the oil tanker pasted onto the horizon, then they cannot be blended, any more than one could blend the sea and the sky. But what if the names are mere edges after all, edges that might have had firm content in the antiquity of France and Spain but not here, not in *this* past; edges whose contents spill over and leak into each other so profusely that no true edges can be said to exist at all?

In that case, there would be no more mystery. The Levantine paradox would be a trick knot; pull the rope gently from both ends, and it unravels on its own. Think of one end of the rope as cultural. Every species on earth has its own ecological niche, its unique set of adaptations that persists through time. The "principle of competitive exclusion" states that two species cannot squeeze into the same ecological niche. The slightly better-adapted one will drive the other out. Traditionally, the human niche has been defined by culture, so it would be impossible for two kinds of human to coexist, using the same stone tools to compete for the same kinds of plant and animal resources. One of the human types would drive the other into extinction, or never allow it to gain a foothold.

"Competitive exclusion would preclude the coexistence of two different kinds of hominids in a small area over a forty- or fifty-thousand-year period unless they had different adaptations," says Geoffrey Clark of Arizona State University, a passionate advocate for a Levantine "Oneness" in the Middle Paleolithic. "But as far as we can tell, the adaptations were identical at Kebara and Qafzeh."

Clark adds to the list of common adaptations the use of symbols—or lack of it. Perhaps Neandertals lacked complex social symbols like beads, artwork, and elaborate burial ritual. But so, he believes, did their skinny modern contemporaries down the road at Qafzeh. If *neither* was littering the landscape with signs of some new mental capacity, by what right do we favor the skinny one with a brilliant future, and doom the other to dull extinction?

This leads to the morphological end of the rope. If the two human types cannot be distinguished on the basis of what they left behind in the archaeological record, then the only valid way of telling a Neandertal from a modern human is to declare that one looks "Neandertalish," and the other doesn't. If you were to take all the relevant fossils from the Near East and line them up, could you really separate them into two mutually exclusive groups, with no overlap between? A replacement advocate might think so, but a believer in continuity like Geoffrey Clark insists that you could not. He thinks the lineup might better be characterized as one widely variable population, running the gamut from the most Neandertal to the most modern. The early excavators at Tabun and Skhul saw the fossils there as an intermediate grade between archaic and modern *Homo sapiens*. Perhaps they were right. "The skeletal material is anything but clearly 'Neandertal' and clearly 'modern,' " Clark maintains. "Whatever those terms mean in the first place, which I don't think is much."

"It is difficult to understand," writes the arch-continuity champion Milford Wolpoff, "how the presumably earlier Qafzeh folk of the Levant were able to coexist with the possibly intrusive Neanderthals, given that they lived at the same time in the same places, manufactured the same tool industries, utilized the same technology, and adapted with the same subsistence patterns *unless they became the same people*."

But for all its appeal, the Oneness solution to the Levantine paradox is fundamentally flawed. Nobody disputes that the tool kits of the two human types are virtually identical. But it does not follow logically that the toolmakers must be identical as well. Middle Paleolithic tool kits are associated in our minds with Neandertals because they are the best-known human occupants of the Middle Paleolithic. But if people with modern anatomy turn out to have been living back then too, why *wouldn't* they be using the same culture as the Neandertals? There is nothing inherently inadequate about a Mousterian tool relative to the blade tools of Cro-Magnon moderns later in Europe. A sharp Levallois flake will cut, scrape, poke, or slice just about anything you want. The big advantage of blade technology is economy; a given hunk of flint will deliver ten or twenty times the amount of cutting edge to a blade maker as the same hunk in the hands of a Levallois knapper. This could be a critical advantage if you happened to live in an area where flint was scarce, or where your trips back to the flint source were restricted because there was too little water or game on the landscape. But there are outcrops of high-quality flint scattered all over Mount Carmel, and adequate fresh water too. In short, Neandertals and moderns in the Levant did not have to be the same people to be making the same tools, if neither one needed to make anything else.

"If you ask me, forget about the stone tools," Ofer Bar-Yosef told me. "They can tell you nothing, zero. At most, they say something about how they were preparing food. But is what you do in the kitchen all of your life? Of course not. Being positive people, we are not willing to admit that some of the missing evidence might be the crucial evidence we need to solve this problem."

Whatever the tools suggest, the skeletons of the Neandertals and moderns *look* different. Erik Trinkaus's stress on the functional aspect of ancient anatomy has made clear that the skeletal differences between Neandertals and moderns reflect two distinct patterns of behavior, however alike their archaeological leavings may be. He admits that the contrasts he sees—some extra bone mass here, a ridge missing there—do not add up to a complete account of what Neandertals and moderns were doing differently, at least not yet. But the pattern is too consistent to dismiss as arbitrary. Furthermore, the two physical patterns do not follow one from the other in time, nor do they meet in a fleeting moment before one triumphs and the other fades. They just keep on going, side by side but never mingling.

In his behavioral approach to bones, Trinkaus purposely disregards the features that might best discriminate Neandertals and moderns from each other genetically. This allows him to float blissfully above the continuity versus replacement debate. By definition, these traits are poor indicators of the effects of life-style on bone, since their shape, size, and so forth are decided by heredity, not by use. But there is one profoundly important aspect of human life where behavior and heredity converge: the act that allows human lineages to continue on in the first place.

Humans love to mate. They mate all the time, by night and by day, through all the phases of the female's reproductive cycle. Given the opportunity, humans throughout the world will mate with any kind of other human. While they naturally tend to choose mates from within their own immediate region, the choice of partner is theoretically limited only by the number of sexually mature individuals of the opposite sex on earth. The barriers between regions, races, and cultures, so cruelly evident in other respects, melt away when sex is at stake. Cortez began the systematic annihilation of the Aztec people—but that did not stop him from taking an Aztec princess for his wife. Blacks have been treated with contempt by whites in America since they were forced into slavery, but some 20 percent of the genes in a typical African American are "white."

Consider James Cook's voyages in the Pacific in the eighteenth century. "Cook's men would come to some distant land, and lining the shore are all these very bizarre-looking human beings with spears, long jaws, browridges," archaeologist Clive Gamble told me. "God, how odd it must have seemed to them. But that didn't stop the Cook crew from making a lot of little Cooklets. One thing you can count on with humans—whether they can interbreed or not, the first thing they do when they meet is try to find out."

Project this universal human behavior back into the Middle Paleolithic. Assuming that humans of whatever kind were as free in their choice of mating partner as humans are today, when Neandertals and modern humans came into contact in the Levant, they would have interbred, no matter how "strange" they might initially have seemed to each other. If their cohabitation stretched over tens of thousands of years, the fossils should show a convergence through time on a single morphological pattern, or at least some swapping of traits back and forth.

But the evidence just isn't there, not if the TL and ESR dates are correct. Instead, the Neandertals stay staunchly themselves. In fact, according to Trinkaus, they become increasingly *less* like the moderns, in spite of prolonged contact with them. According to some recent ESR dates, the least "Neandertalish" among them is also the oldest: the relatively delicately built woman of Tabun. The full Neandertal pattern is carved deeper at Kebara and Amud caves, around 60,000 years ago, and blossoms even more in the still younger, though poorly dated, late Neandertals of Shanidar Cave in Iraq. The moderns, meanwhile, arrive very early at Qafzeh and Skhul and never lose their modern aspect. When they show up again in the fossil record 35,000 years ago—two little-known skulls from higher levels at Qafzeh, and the boy "Egbert" from Ksar' Akil—they are essentially contemporary and indistinguishable from the Cro-Magnon moderns of Europe. Of course, even with the dating wizardry of TL and ESR, the ledger of morphological change through time in the Near East is still faintly written. And new fossils may be revealed at any moment, conclusively demonstrating the emergence of a "Neandermod" lineage. Judging from the evidence in hand, however, the most likely conclusion is that Neandertals and modern humans were not interbreeding in the Levant.

This goes to the core of the paradox. Humans mate with all other humans. These two creatures were equally human, judging by every standard we use to measure that word in the past. Their brains were the same size. Their cultures had reached the same level of sophistication. They both may have spoken—perhaps even to each other. Their lives, though different, show no advantage of one over the other. And they apparently lived together in the same place for a very long time. But they did not mate with each other, at least not so as to produce fertile offspring. But all humans mate with all other humans. . . .

Of course, to mate, you first have to meet. Some researchers have contended that the coexistence on the slopes of Mount Carmel for 50,000 years is merely an illusion created by the poor resolution offered by the archaeological record. If moderns and Neandertals were *physically* isolated from each other, then there is nothing mysterious about their failure to interbreed. The most obvious form of isolation is geographic. Several years ago John Shea suggested that the greater bulk of Neandertals, and their relatively inefficient locomotion, might have kept them confined to the forested lowland coastal areas of the Levant, where game and plant resources were abundant year-round. At the same time, the anatomically modern people could have been light-stepping their way through the drier, savanna-steppe highlands of the Negev and Jordan, following migrating game.

"The earliest moderns," Shea wrote, "may have distinguished themselves by colonizing regions of relatively lower productivity than did their archaic predecessors and contemporaries."

When Neandertals are found in harsh environments, the explanation is invariably that they were pushed there by superior moderns. Yet in this case, when moderns are seen in the same undesirable locations, they are viewed as distinguished colonists! The argument is more seriously undermined by the fact that no fossil remains of these intrepid modern pioneers have been found in the Negev and Jordan. Modern fossils are found in caves like Qafzeh, just an easy day's walk from Kebara no matter how your pelvis and thighbones are put together, not to mention Skhul, within shouting distance of the Tabun Neandertal site. In short, geographic isolation might explain away the Levantine paradox if there were more geography in the Levant to go around.

But imagine an isolation in time as well. As testified to by the shifting complexion of the rodent fossil record, the climate of the Near East fluctuated throughout the Middle Paleolithic, now warm and dry, now cold and wet. Perhaps modern humans migrated up into the region from Africa during the warm periods, when the climate was better suited to their lighter, taller, warm-adapted physiques. Neandertals, on the other hand, might have arrived in the Levant only when advancing glaciers cooled their European range more than even their cold-adapted physiques could stand. In other words, perhaps the two did not so much cohabit as "time-share" the same pocket of landscape between their separate continental ranges.

The fossil evidence offers encouraging support for this hypothesis. The Qafzeh-Skhul people appear in the Levant at just about the middle of the last warm interglacial period before the one we are presently enjoying. Moshe, on the other hand, is 60,000 years old, putting him and his kind in the Levant at one of the coldest periods of the last Ice Age. The fauna echo the refrain. Along with the African-built humans at Qafzeh, you find hippo, spotted hyena, ostrich, warthog, zebra, and other warm-adapted large mammals, not to mention the African rats who started the dating controversy in the first place. The Neandertals arrive in turn, with their own coterie of European fellow-travelers: brown bear, woolly rhino, wolf, and red deer. So perhaps the two hominids moved in and out with the climate, never really getting to know each other. Or so John Shea believes.

"That Neandertal and early modern fossils never occur in the same cave deposits," he writes, "suggests they did not occupy overlapping ranges for prolonged periods of time and may not have had the opportunity to interbreed on a significant scale."

While the solution is intriguing, there are problems with it. Hominids are remarkably adaptable creatures. Even *Homo erectus*, who lacked the

MYSTERY ON MOUNT CARMEL

large brain, hafted spearpoints, and other cultural accoutrements of its descendants, managed to thrive in a range of regions and under diverse climatic conditions. And while hominids adapt quickly, glaciers move very, very slowly, coming and going. Even if one or the other kind of human gained sole possession of the Levant during climatic extremes, what about all those millennia in the Near East that were neither the hottest nor the coldest? There must have been long stretches of time—perhaps enduring as long as the whole of recorded human history—when the Levant climate was perfectly well suited to *both* Neandertals and modern humans. What part do these in-between periods play in the time-sharing scenario? It just doesn't make sense that one human population should politely vacate Mount Carmel just before the other moved in.

Magnifying these doubts are the new ESR dates for the Neandertal woman of Tabun. Until this latest surprise, the "absolute" dates were keeping good time with the rodent paleontology, supporting the warm-weather modern, cold-weather Neandertal time-share idea. But according to these new dates, the Tabun woman and her people were in the Levant around 110,000 years ago, some 40,000 years before Neandertals were scheduled to occupy the place. In fact, she was found in deposits sugges-tively close in age to those of her fully modern neighbors at Skhul and Qafzeh. "The ESR age estimates for [the Tabun female] again raise the question of a general contemporaneity of Neanderthal and early modern humans in the Levant," wrote Rainer Grün and his colleagues when they reported the new dates.

If these humans were isolated in neither space nor in time, but were truly contemporaneous, then how on earth did they fail to mate? "When two human groups come in contact and do not interbreed, there is some-thing very strange going on," says Bar-Yosef.

There is only one solution to the mystery left. Neandertals and moderns did not interbreed in the Levant because they *could* not. They were reproductively incompatible, separate species, equally human per-haps, but biologically distinct. One of these species may have looked more like ourselves than the other did, but the resemblance is immaterial to any judgment of their cultural and behavioral sophistication. Neither was evolving into the other. Neither was ascending toward supremacy, nor fading into extinction. They may not even have been competing with each other, at least no more than they competed among their own kind. But in an area this small, they most certainly would have met, face-to-face, as a matter of course. They may have gestured greetings, sojourned together, perhaps even spoken. Two separate species, who both just happened to be human at the same time, in the same place.

At first glance, this seems the least likely resolution to the Levantine paradox. Humanity, by definition, is unique, the only presence allowed on *our* side of the great natural divide. Since the 1970s, anthropologists have grudgingly accepted the possibility that two or three different *hominid* species coexisted in Africa millions of years ago. But these distant beings weren't really human; they were hairy little apelike things running naked through their senseless millennia. They do not violate the apparent logic that the only human species that can exist are: 1) ourselves; 2) earlier hominids in the process of evolving into ourselves; or 3) failed "offshoots" from our direct lineage that went extinct, probably from competition with a superior species—ourselves.

Cohabitation in the Levant in the last Ice Age conjures up a chilling fourth possibility. It forces us to imagine two equally gifted, resourceful, communicative, curious, emotionally rich human entities compacted into a single tableau, weaving through one tapestry of landscape—yet so different from each other as to make the racial diversity of present-day human beings seem as nothing. We find such plots in novels and movies, not in serious prehistorical research. And even the authors of *The Clan of the Cave Bear* and *Quest for Fire* indulged their sundry human types with the one bridge needed to smudge the boundary between them: They mated. Take that sexual bridge away, and you end up with two fully sentient, fully sapient human species pressed into one place, as mindless of each other as two kinds of birds sharing the same feeder in your backyard.

If the Competitive Exclusion Principle holds true, such a coexistence depends upon some subtle, as-yet-undiscovered difference in the two species' adaptations. Who knows—perhaps the moderns *were* adapted to a drier inland ecological niche, at least during part of the year. Some biologists, however, do not think that competitive exclusion is quite the "principle" it has been made out to be. It might reduce to nothing more than a rather fancy bit of circular reasoning: Count the number of species in a given habitat to determine the number of niches the habitat contains, and lo and behold, there is one species for every niche. In any case, a species is not defined by its adaptations—how it gets its meals, fights off its predators and parasites, keeps warm or cool, finds space to build its nest or dig its den, and so on. If it were, a fox living in the forest would, by definition, be a separate species from one living in the meadow, who in turn would be distinct from ones living in the prairie, the swamp, and the city park. Bring all these foxes together during the females' heat, and in a few months you will have living, mewling proof of why species cannot be adequately defined by their ecological niches.

What *does* mark the edges of foxness, or of humanness? When paleo-anthropologists bicker over whether Neandertal anatomy is divergent enough to justify calling Neandertals a separate species from us, they are using a *morphological* definition of a species. This is a useful pretense for the paleoanthropologists, who have nothing but the shapes of bone to work with in the first place. But they admit that in the real, vibrantly unruly natural world, bone morphology is a pitifully poor indication of where one species leaves off and another begins. Ian Tattersall, an evolutionary biologist at the American Museum of Natural History, points out that if you stripped the skin and muscle off twenty New World monkey species, their skeletons would be virtually indistinguishable. Many other species look the same even with their skins still on. According to Tattersall, pine voles and meadow voles match feature to feature in every way except for the shape of a single cusp on one premolar tooth. At the other extreme *Homo sapiens* is a single species that expresses itself all over the planet in various sizes, shapes, colors, and bone structures—all of them emphatically *sapiens*, all of them potential mates for each other.

The most common definition of *biological* species, as opposed to the morphological make-believes paleontologists have to work with, is a succinct utterance of the esteemed evolutionary biologist Ernst Mayr: "Species are groups of actually or potentially interbreeding natural populations that are reproductively isolated from other such groups." The key phrase is *reproductively isolated*: a species is something that doesn't mate with anything but itself. The evolutionary barriers that prevent species from wantonly interbreeding and producing a sort of organismic soup on the landscape are called "isolating mechanisms." These can be any obstructions that prevent otherwise closely related species from mating to produce offspring who are themselves fertile. The obstructions may be anatomical. Two species of hyrax in East Africa share the same sleeping holes, make use of common latrines, and raise their young in communal "play groups." But they cannot interbreed, at least in part because of the radically different shapes of the males' penises. Isolating mechanisms need not be so conspicuous. Two closely related species might have different estrus cycles. Or the barrier might come into play after mating. The chromosomes are incompatible, or perhaps recombine into an offspring that is itself incapable of breeding, an infertile hybrid, like a mule.

Mayr's "biological concept of species" is much truer to the living, natural world than any notion based merely on differences in morphology. At the same time, it is easy to see why paleoanthropologists despair over trying to apply it to ancient hominids. The characteristics needed to recognize a biological species—the isolating mechanisms—are not the kind that usually turn up as fossils. How does an estrus cycle get preserved? What does an infertile hybrid, reduced to a few fragments of its skeleton, look like? How does a chromosomal difference turn into stone? Perhaps Neandertal and modern human males, hyraxlike, sported wildly divergent penises. Or perhaps the Neandertals had forty-eight chromosomes to our forty-six. There is no way of knowing. To my mind, the continuity of the morphological differences *through time* between the two humans of Mount Carmel hints that they were biologically different species. How much more satisfying it would be, however, if we could point to some feature, something as clean and hard as a bone, and say, "There—*that's* the thing that kept them biologically apart, no matter how much they were the same in other ways."

This is probably impossible, but there is another way of looking at species that might offer a tiny window of hope. The "biological species concept" is a curiously negative one—what makes a species itself is that it doesn't mate with anything else. Thus a species is defined not by any activity or property inherent to itself, but by the limits imposed around it by isolation mechanisms. This is a little like defining a pie filling by calling it "what's inside the crust." What if we could turn this definition inside out and look at the pie filling itself?

A few years ago, a South African biologist named Hugh Patterson uncovered what he believes is a subtle but debilitating flaw in the biological species concept. Most evolutionists—including Ernst Mayr—believe that a new species can originate only when a small population of individuals becomes geographically isolated from others of their kind, probably in a habitat unlike the one to which they are adapted. A classic example is a flock of birds who are blown off course and end up stranded on a new island. In most cases, the stranded population goes extinct. But if it manages to adapt and even flourish, its descendants may come to live in the same region with the descendants of its parent population. If the two populations are able to interbreed and produce fertile young, then no matter how different they may have become in other respects, they are still the same biological species. On the other hand, if isolation mechanisms have evolved to prevent interbreeding, then no matter how similar they look or act, they are a new species.

There is a great deal more to the process than is outlined here, but even this truncated summary is enough to hint at the hidden fault Hugh Patterson saw lurking in the biological species concept. If a population must be geographically isolated from its parent population in order to become a new species, how will it evolve reproductive isolating mechanisms against interbreeding with individuals of the parent population, since none are around to be isolated from? Let's say that a flock of birds—call them tree-nesting whippersnippers—is blown onto a desert island where no whippersnippers existed before. Ten thousand years later, the island rings with the song of whippersnippers. But since there are few trees, all of them make their nests on the ground. If some of these ground-nesting whippersnippers should follow an errant gust of wind back to the land of their tree-nesting kin, the difference in nesting sites might look like an "isolating mechanism" against their interbreeding. But reproductive isolation can't be the *cause* of their divergence, because there weren't any tree-nesting whippersnippers on the desert island to evolve reproductive barriers against. The real reason for the evolutionary change was not the presence of tree-nesters, but the scarcity of trees. In other words, the emergence of isolation mechanisms might be an *effect* of a population's divergence. But it is hard to see how it could be the defining *cause*.

With this problem in mind, Hugh Patterson turned the biological species concept inside out, proposing a view of a species based not on whom it *doesn't* mate with, but on whom it *does*. Species, according to Patterson, are groups of individuals in nature that share "a common system of fertilization mechanisms." With reproduction at its core, Patterson's concept is just as "biological" as Mayr's. But he turns the focus away from barriers preventing interbreeding and throws into relief the array of adaptations, spanning every level in the life of an organism, that together ensure the successful meeting of a sperm and an egg. Obviously, sex and conception are fertilization mechanisms, as is the genetic compatibility of the two parents' chromosomes and their recombination into an offspring that is itself fertile. But long before a sperm cell gets near a receptive egg, the two sexes must have ways of recognizing each other as potential mates. And therein, perhaps, lies a solution to the mystery of Mount Carmel.

Every mating in nature begins with a message. It may be chemically couched. Eggs of a brown alga *Ascophyllum nodosum*, for example, send out a chemical that attracts the sperm of A. *nodosum*, and no other. It may be a smell. As any dog owner knows, a bitch in heat lures males from all over the neighborhood. Note that the scent does not draw squirrels, tomcats, or teenage boys. The mating signal transmitted is received only by male members of the species *Canis familiaris*, who are well equipped by evolution with olfactory sensors to respond to it. Many bird species use vocal signals to attract and recognize the opposite sex, but only of their own. "A female of one species might *hear* the song of the male of another, or witness his courtship display," explains Judith Masters, a colleague of Patterson's at the University of the Witwatersrand in Johannesburg, "but she won't make any response. There's no need to talk about what *prevents* her from mating with that male. She just doesn't see what all the fuss is about."

Similarly, a taxonomist may be at a loss to discriminate between various firefly species based on the anatomy of the males. But on a summer night, a female firefly will have no such trouble. The males of her species will signal to her with a precise pattern of light flashes, distinguished from that of other species by their color, frequency, and duration. To her, the bioluminescent machismo of the other males is just so much noise. It is part of a wholly different "species mate recognition system." According to Masters, "a nice way of putting the difference [between Patterson's and Mayr's concepts of species] is that the recognition concept emphasizes the characteristics that the organisms *themselves* use to identify who is a member of the species and who isn't."

A species' mate-recognition system is extremely stable compared to adaptations to the local habitat. An individual sparrow born with a slightly too short beak may or may not be able to feed its young as well as another with an average-sized beak. But a sparrow who sings an unfamiliar song will not attract a mate, and thus is not going to have any young at all. He will be plucked from the gene pool of the next generation, leaving no evolutionary trace of his idiosyncratic serenade. The same goes, of course, for any sparrow hen who fails to respond to potential mates singing the "correct" tune. With this kind of price for deviance, everybody is a conservative. "The only time a species' mate-recognition system will change is when something really dramatic happens," Masters told me.

For the drama to unfold, a population must be geographically isolated from its parent species. If the population is small enough, and the habitat radically different from what it was previously adapted to, even the powerful evolutionary inertia of the mate-recognition system may be over-come. This sea change in reproduction may be accompanied by new ecological adaptations to the environment. Or it may not. Either way, the only shift that marks the birth of a new species is the one affecting the recognition of mates. Once the recognition threshold is crossed, there is no going back. Even if individuals from the new population and the old come to live in the same region again—let's say, in a well-trafficked cor-ridor of fertile land linking their two continental ranges—they will no longer view each other as potential mates. The same would apply, of course, to human species. So perhaps Neandertals and modern humans were able to coexist in the Levant for tens of thousands of years without interbreeding because they were not sending and receiving the same mat-ing signals. We have at least to consider the astonishing possibility that there were once two kinds of intelligent human beings, mutually uncom-prehending in the only area that could merge them into one.

One tantalizing advantage of looking at species this way is that there may be a trace of real, biological species differences lingering in the fossil record. The only traits that determine whether two organisms really rep-resent different species are the features associated with their mate-recognition systems. In most cases, mating signals do not show up as fossils. But some species, notably those who signal to potential mates mostly by sight, may have a bit of their mate-recognition system inscribed in their anatomy. Elisabeth Vrba of Yale has shown that antelopes and other bovids wear their species' identification tags on their heads in the shapes and sizes of their horns and antlers. ("The only difference between a screw-horn goat and a non-screw-horn goat is a screw horn," comments Ian Tattersall.) These hard cranial ornaments show up very conveniently in the fossil record, making identification of extinct species a much easier task.

Neither Neandertals nor early modern humans sported antlers, of course. Nevertheless, human mate-recognition systems are overwhelm-ingly visual. "Love comes in at the eye," wrote Yeats, and the upright, bipedal posture of hominids provides a lot of sexual signals for the eye to take in. The human penis, for example, is much larger than that of all other great apes. The breasts of the female are swollen out of all propor-tion to those of apes. But the locus of the human body that lures, cap-tures, and holds the eye most of all is the face. In one recent study, when American women were asked what part of a man's body they looked at first, 77 percent answered either the face or the eyes. The man's buttocks came in a distant second, with 12 percent. Among men, 64 percent were attracted first to the woman's facial region, 21 percent to her chest.*

Human faces are exquisitely expressive instruments. Behind our facial skin lies an intricate web of musculature, concentrated especially around the eyes and mouth, evolved purely for social communication—expressing interest, fear, suspicion, joy, contentment, doubt, surprise, and countless other emotions. Each emotion can be further modified by the raise of an eyebrow or the slight flick of a cheek muscle to express, say, measured surprise, wild surprise, disappointed surprise, feigned surprise, and so on. By one estimate, the twenty-two expressive muscles on each side of the face can be called on to produce ten thousand different facial actions or expressions.

Among this armory of social signals are stereotyped formal invitations to potential mates. The mating display we call flirtation plays the same on the face of a New Guinean tribeswoman as it does on the features of a *lycienne* in a Parisian café: a bashful lowering of the gaze to the side and down, followed by a furtive look at the other's face and the coy retreat of the eyes. A host of other sexual signals are communicated facially— the downward tilt of the chin, the glance over the shoulder, the slight parting of the mouth. The importance of the face as an attractant is underscored by the lengths to which humans in various cultures go to embellish what is already there. But the underlying message is communicated by the anatomy of the face itself—its shape and outline, the geometric arrangement of the features, their relative proportions and placement. " 'Tis not a lip, or eye, we beauty call, but the joint force and full result of all," wrote Alexander Pope. And it is that "joint force"— over generations—that keeps our species so forcefully joined.

This brings us back to the Levant: two human species in a tight space for a long time. The vortex of anatomy where Neandertals and early moderns differ most emphatically, where a clear line can be drawn between *them* and *us* by even the most rabid advocate of continuity is, of course, the face. The Neandertal's "classic" facial pattern—the mid-facial thrust

*Our species shares the use of the face as the focus of mate-recognition signals with many other primates. "It is a common Old World anthropoid ploy," says Judith Masters. "Cercopithecoid monkeys have a whole repertoire of eyelid flashes. Forest guenons have brightly painted faces with species-specific patterns, which they wave like flags in the forest gloom. Baboons yawn ostentatiously, and show their canines. Good old evolution tinkering away, providing new variations on a theme."

picked up and amplified by the great projecting nose, the puffed-up cheekbones, the long jaw with its chinless finish, the large, rounded eye sockets, the extra-thick browridges shading it all like twin awnings—is usually explained as a complex of modifications relating to a cold climate, or as a support to heavy chewing forces delivered to the front teeth. Either way it is assumed to be an environmental adaptation evolved to help its bearer survive in a particular habitat. But what if these adaptive functions of the face were not the reason they evolved in the first place? What if the peculiarities evolved instead as the underpinnings to a totally separate, thoroughly Neandertal mate-recognition system?

Although it is merely a speculation, the idea fits some of the facts and solves some of the problems. Certainly the geography of the European continent was such that the Neandertals' ancestors were cut off from other populations enough to overcome the inertia of the ancestral mate-recognition system, allowing something new to emerge. During glacial periods, contact through Asia was blocked by the polar glaciers and the vast uninhabitable tundra beneath them. In those cold times, mountain glaciers between the Black and Caspian seas all but completed a barrier to the south. "The Neandertals are a textbook case for how to get a separate species," John Shea told me, back in Cambridge. "Isolate them for one hundred thousand years, then melt the glaciers and let 'em loose."

Explaining Neandertal faces as part of a mate-recognition system rather than an adaptation to a harsh, glacial climate also accounts for the granitelike stability of the facial pattern through time. If the big noses and so on represented cold adaptations, why would they not fluctuate over millennia, reducing during the warm interglacials that the Neandertals experienced in Europe and western Asia? As the locus of mate recognition, regardless of what adaptive role the Neandertal face was playing, it would *have* to remain stable through time. Any significant deviation from the facial theme would limit that individual's chance of finding a mate.

If mate recognition lay behind a species-level difference between Neandertals and moderns, the Levantine paradox can finally be put to rest. We need search no longer for that hidden behavioral deficiency in Neandertals that made them either primitive versions of ourselves, or failed aberrations. Regarded as simply the product of a species-producing change in mate recognition, their cohabitation with moderns in the Levant no longer needs explanation. The Neandertals and moderns managed to coexist through those long millennia, doing the same humanlike things, but without interbreeding, simply because the issue never really came up.

The idea scarcely seems imaginable. But this, too, adds to its appeal; on this issue, the human imagination has become too cramped and contested a space. Continuity believers cannot credit the idea of two human types coexisting in sexual isolation. Replacement advocates cannot conceive of such a long period of coexistence without competition. They would rather see Neandertals and moderns pushing each other in and out of the Levant in an extended struggle finally won by our own ancestors. Of course, if the Neandertals were a biologically separate species, something must have eventually happened that caused their extinction. After all, we are still here, and they are not. Why they faded and we managed to survive is a separate story, with its own shocks and surprises. But what happened on Mount Carmel might be more remarkable still. It is something that people today are not prepared to comprehend, especially in places like the Levant, where botanists wear handguns and lose limbs in war. Two human species, with far less in common than any two races or creeds now on the planet, may have shared a small, fertile piece of land for 50,000 years, regarding each other the whole time with steady, untroubled, peaceful indifference.

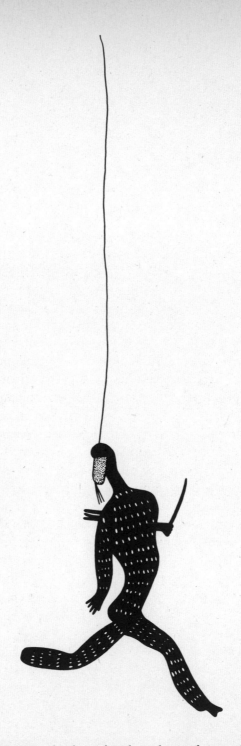

Bushman rock artists in South Africa often depict human beings with animal features, such as this "trance buck" shaman from Natal. The line running out the top of the figure's head may represent the shaman's spirit leaving on out-of-body travel. FROM LEWIS-WILLIAMS AND DOWSON 1989

Chapter Eight

A SENSE OF US

I left Kebara late one afternoon and booked on a plane from Tel Aviv to Johannesburg, with a short leg on to Cape Town. My goal was to find out more about the first physically modern people and how they got that way. With me I carried a caveat from Mount Carmel: Anatomy is not destiny. If two human types could coexist for 50,000 years, then the modern human form is no guarantee of superiority. The inhabitants of Qafzeh and Skhul may have looked like us, but without clear indications that they were behaving in a more modern way than their Neandertal contemporaries, what significance does their modern appearance really have? The very term "modern," attached to anatomy alone, loses all its sheen in the light of those 50,000 unexceptional years. Farther back in the hominid lineage, anatomy and behavior seem to evolve hand in hand. The tools of *Homo erectus*, for example, are far more elaborate than the crude cracked rocks left by its predecessor *Homo habilis*—and *erectus* had the bigger brain. But with the emergence of our own species, the body-behavior coupling breaks down. From here on in, calculating a fossil's evolutionary promise by the bulge of its brow or the thrust of its chin makes no more sense than judging a friend by the number of joints in his toes, or whether he can curl his tongue.

But even such a confirmed Neandertal lover as I had become has to face the fact that we are here today, and they are not. Judging by the kinds of stone tools associated with them, the last Neandertal population dwindled to nothing around 30,000 years ago. The number of human species on earth thereafter was reduced to one. If the survival of one kind of human and the demise of the other cannot be explained by anatomy, there must have been something else, a later development that infused the moderns with new vigor, renewed restlessness. Could this infusion, too, have arrived from Africa? Were the earliest African moderns actually *doing* modern things, before their Neandertal contemporaries? If so, did this behavior leave a trail behind, something to show that these stirrings were eventually exported out to the Middle East and beyond? Or did they simply flare up and die, unconnected to the cultural swell that would propel us into the present moment?

The first place to look for answers is Klasies River Mouth, alleged home of the first known anatomically modern people. Hung on the tip of the continent, with nothing between its open gape and Antarctica but the empty sea, it seemed shimmeringly remote, a cradle at the end of the world. At this time of year, unfortunately, the cave was just out of my reach. The field season was over, and with no one at the site to interview except sea otters and starlings, I had instead arranged a visit with Hilary Deacon of the University of Stellenbosch, the man who presides over the work at Klasies. In Deacon's hands, the site has been transformed from a hastily dug if productive hole in the ground to a richly detailed canvas of time, its importance extending far beyond the scatter of human bones gracing its deposits.

I waited for Deacon to pick me up at the airport in Cape Town, my eagerness to talk to him mixed with a certain measure of dread. Although I had never met him, I had formed an image of him from our correspondence: impatient and demanding, given to sudden outbursts of temper, usually directed against science writers and other worthless dilettantes. All this was evident in a letter I had received from him, responding to one of mine. It was my standard letter requesting an interview, but I had misplaced a couple of letters, spelling his place of work "Klaises River Mouth."

"From your spelling of the name Klasies I assume you have yet to read up about the sites," he wrote back. Galled, I had since crammed up on Klasies River Mouth and early African *Homo sapiens* in general. I paced alone in the airport pickup area, imagining myself riding back to Stellenbosch with Deacon quizzing me on South African Middle Stone Age site spelling the whole way. "If you want me to talk to you, spell 'Buffelskloof'! 'Heuningsneskrans Shelter'! 'Umhlatuzana'!"

From my reading, I knew that much has changed in African archaeology since the 1960s when Carleton Coon labeled the continent the "indifferent kindergarten" in the schooling of modern mankind. According to radiocarbon dates available at the time, the Middle Stone Age in

Africa, characterized by tools much like those of the Mousterian in Europe, was thought to have begun about 38,000 years ago, about the same time that Mousterian tools were fading out in Europe, giving the impression that people in Africa were getting around to discovering Levallois techniques and the rest of the Mousterian package just as their European counterparts were abandoning them for more sophisticated technologies. According to this dating scheme, the key, human-defining transition to the Upper Paleolithic—in Africa called the Late Stone Age—did not occur until as recently as 6000 B.C. By that time, settled populations in the Middle East and parts of Europe were already growing crops and forming complex political systems.

Then in 1972, archaeologists Peter Beaumont and John Vogel announced new radiocarbon dates taken at five South African sites. Rather than *beginning* 38,000 years ago, Beaumont and Vogel showed, the Middle Stone Age actually *ended* around that time, the Mousterian artifacts giving way to Late Stone Age tools in sync with the transition to Upper Paleolithic tools in Europe. The old-time scheme, it turned out, was based on contaminated samples and wrongly identified tool kits. Middle Stone Age tools showed up in archaeological strata far too old to date by radiocarbon methods—according to Beaumont and Vogel's reckoning, in excess of 100,000 years old.

In one stroke, the hominids in Coon's "indifferent kindergarten" skipped several grades to become the technological equals of any population on earth in their time. But there was an even sharper threat to the old Eurocentrism hidden in the new time scheme. Some of the redated MSA sites, notably Border Cave and Klasies River Mouth, harbored apparently modern human remains. Quite suddenly, these fossils were the oldest known anatomically modern human beings in the world. Just about everyone agrees that these African moderns were behaviorally the equal of their European contemporaries, the Neandertals. My question for Hilary Deacon, if he would talk to me, was whether they might have been a step or two ahead.

The dreaded Deacon arrived and quickly proved himself dreadfully polite, the very opposite of my fears. With an ingrained solicitude unknown beyond the English Commonwealth, he took my bags, insisting that I cancel my hotel reservation and stay with him and his wife, Janette, an expert on the Late Stone Age. On the way to Stellenbosch, he pointed out features of the terrain, the architecture of the Dutch colonial farmhouses, the squatters' settlements massed along the dunes against the sea.

Pale, with a sharply trimmed gray beard, he was so deferentially shy and soft-spoken that I had to strain to hear him over the car engine. Rather than test my knowledge of Klasies River Mouth, he apologized for not being able to take me personally to see the cave. (Given that it was in-active at the time and five hundred miles to the southeast, no apologies were needed.) He planned to devote the entire weekend to my visit and told me that Philip Rightmire of the State University of New York at Binghamton was also in town, collaborating on a new study of the Klasies River Mouth fossils. This was a stroke of luck. Rightmire is a leading authority on early *Homo sapiens* anatomy. Back in 1979, even before Chris Stringer popularized the Out of Africa scenario, Rightmire had developed his own version of the theory.

I met Janette Deacon and Philip Rightmire over lunch—the first of a string of personally warm, if thermostatically nippy meals in the Deacons' house. (It was winter, and along with the gene for politeness, it seemed the English in South Africa have inherited the British contempt for arti-ficial heat.) Afterward, we drove to the Archaeology Department at the university, locked and empty on a Saturday. In a cold, musty classroom on the first floor, full of dark displays and creaking floorboards, Deacon set up a projector and inserted a carousel of slides. A gorgeous cliffscape appeared on the screen: a gray-green mass of rock plunging in a sinuous curve down to a patch of surf-troubled beach. Formations of scarred rock thrust out of the sand and foam. A large cave opened like a wound on the cliff wall about a third of the way up, with another visible just below it. More caves, Deacon told me, lay out of sight around the bend. The whole complex was Klasies River Mouth.

This southern coastline is a dynamic, high-energy environment. The beaches and rocks teem with shellfish—brown, black, and white mussels, turban shells, oricrickle in the low tides, limpets and periwinkles. Cor-morants and oystercatchers nest in the cliffs. Dolphins play just offshore. The footprints of sea otter litter the beach in the morning, to be erased by the surf. Storms, sometimes lasting two or three days, cause violent transformations in the shore. Overnight, a rocky beach turns to sand, a sandy one to rock. The color of the sea itself constantly permutes, trans-lucent blues and greens erupting suddenly out of a leaden gray. Deacon described all this with wistful relish. He and a team of Stellenbosch ar-chaeology students had been working there since 1984, squeezing field seasons in between semesters, using whatever money could be found in an academic environment cut bare by years of economic sanctions against the apartheid South African government. There were no facilities of any kind. Firewood had to be collected to stay warm. The nearest store was twenty-five miles away. "We live totally isolated in a cocoon world, next to the sea and against the cliffs," Deacon said.

The caves at Klasies are a signature of the ocean's slow violence, a cluster of wave-eroded pockmarks in the cliff cut a million years ago. All trace of their earliest occupants has been erased. During warm, interglacial phases in the past, the sea rose high enough to flush out whatever debris, human or otherwise, had built up inside. Since the last such flushing, leopards and other carnivores had left behind the remnants of their meals, and predatory birds had disgorged a plethora of fish and rodent bones. The greatest contributors to this Pleistocene garbage heap, however, were people. As they came and went, they left behind the ashes of their hearths, burned plant-food wastes, animal bones battered open for their marrow, thick middens of shellfish collected from the shore. At the height of the present warm interglacial, about 6,000 years ago, the sea reached up and gobbled about two thirds of the collected sediments. Enough remained, however, to keep Hilary Deacon and his kind busy for years.

Klasies River Mouth was excavated previously in the late 1960s by Ronald Singer of the University of Chicago and British archaeologist John Wymer. Singer and Wymer were "hot for hominids," as Philip Rightmire put it, and they succeeded in finding what they came for. Deacon had added some hominid finds of his own, but his mission at Klasies is fundamentally different from that of his predecessors. In the six years he had been working the site, he excavated only a tiny fraction of the tons of earth Singer and Wymer dug out in fourteen months. Deacon's forte is detail. His micro-excavation techniques squeeze every possible drop of data out of each measured square. His team has plotted each hearth, bone, and chip of rock, often using sieving screens as fine as one tenth of an inch. Sediment samples have been removed to be examined under laboratory conditions, the sizes and weathering of individual sand grains carefully noted as a means to understand changes in sea level and climate. As a result of his meticulousness, the story of humanity in this spot has a context, a sense of change through paleoenvironmental time. And with Klasies as a reference point, other, more enigmatic sites in South Africa have a context too.

At the very bottom of the fossil-bearing deposits is a series of layers of light-brown sand. Deacon flashed a slide of this "LBS member" up on the screen. The sand, blown into the caves from the beach that lies just a few dozen feet away, is full of hearths, artifacts, animal bones, and deep pockets of discarded mollusc shells. These are much the same sorts of shellfish that can be picked up and eaten there today—various mussels, turban shells, and periwinkles. And eaten they most certainly were. Many of the shells still bear signs of being placed in a fire. "Klasies is the oldest seafood restaurant in the world," Deacon said.

The fact that the Klasies people were consuming mussels in such quantities is interesting in itself; for whatever reason, earlier humans avoided shellfish. But the seashells provide another essential element, unrelated to diet, in the Klasies story. Like all shelled animals, ocean molluscs use oxygen, which they obtain from the seawater around them, in the manufacture of their shells. Oxygen comes in different atomic forms, or isotopes. Oxygen in seawater is mostly the common isotope known as oxygen-16. But seawater also contains a tiny fraction of "heavy" oxygen-18, which contains two more neutrons in its nucleus. Chemically, these two forms of oxygen behave pretty much the same, but since oxygen-16 is physically lighter, it evaporates faster. So long as the global climate is relatively warm, this makes little difference—the evaporated oxygen is returned to the ocean in the form of water, and the proportion between the two isotopes remains constant. During glacial periods, however, much of the evaporated ocean water comes back as snow, accumulating on polar ice caps. Since a relatively higher amount of light oxygen-16 evaporated in the first place, it tends to mount up on the ice caps, leaving a significantly higher percentage of oxygen-18 behind in the ocean water.

All this chemistry is a matter of complete indifference to the shelled creatures of the ocean, who construct their housing by using whatever mix of oxygen isotopes happens to be in the seawater at the time. But time is what they illuminate. Among these shelled species are trillions of microscopic plankton. When they die, their shells drift to the bottom of the ocean—each perished plankton a tiny gauge recording the ratio of oxygen isotopes that existed in the ocean during its brief, magnificently insignificant lifetime. Thus there is another crucial yardstick of global change and earthly time, slowly mounting up beneath the ocean's bottom. Since the 1960s, Nicholas Shackleton of Cambridge University and others have been hauling that yardstick back up onto land in the form of deep-sea cores from various parts of the world's oceans. Through constant adjustment and comparisons, they have reconstructed the comings and goings of glacial periods over much of the last seventy million years.

When Hilary Deacon was looking for a way to date the LBS member at the bottom of the Klasies River Mouth deposit, he sent some of his mollusc shells to Shackleton. They proved to be "isotopically light," the comparatively high proportion of oxygen-16 revealing that they had been formed during a warm phase in the global climate. Using the record of deep-sea plankton as a guide, it was clear that the molluscs had lived during the last period between major ice ages, 120,000 years ago. This date, confirmed by a variety of other techniques, is etched into the bottom of the site with enviable surety. "We've got so many things going that there is hardly any question of this," said Deacon.

From that beginning point, Klasies tells a story of human occupation lasting another 60,000 years. The basic structure of the narrative is revealed in the excavation of cave 1A at the main site. Above the LBS member is a "break" in human occupation, followed by a long series of strata rich in people's leavings, embedded in fine-grained sand blown down from an ancient dune formation above the cliff. Oxygen isotope analysis and other dating methods suggest that these layers were laid down between 100,000 and 80,000 years ago. By that time, the earth's climate had cooled from the apogee of warmth of the interglacial, and the sea had receded a mile or two from the cave. It was still close enough, however, for the people living in the caves to include great quantities of shellfish, a rich source of minerals, in their diet. Judging by the ashes in the numerous hearths, they were also dining on geophytes—perennial plants that bear carbohydrate-rich buds beneath the soil—which grow in patches along the mountain ridges and coastal margins. The bashed-in bones of fur seals, penguins, and various terrestrial mammals reveal that the Klasies people supplemented their vegetable diet with hunted or scavenged meat. Among the bones is a remarkable number from the eland, a large, placid-tempered antelope. The preponderance of eland bones and the fact that animals of all ages are represented suggest that they might have died in "catastrophic" hunting incidents, perhaps driven over cliffs or into traps.

Littering these deposits are the same Middle Stone Age tools and debris that typify the LBS member below—Mousterian flakes, sidescrapers, and denticulates, which appear with numbing regularity throughout Africa at the time, as they do in Eurasia. But then something very strange happens. Above the SAS member, about 70,000 years ago by Deacon's reckoning, a new kind of tool industry appears. While the amorphous MSA flakes still show up, they are joined by much smaller, thinner blade tools, often refashioned into geometric shapes like crescents and triangles. Many appear to have been made by a more complex technique called "punched blade," in which one end of a pointed antler or other rodlike object is placed against a prepared core of stone, with the other end receiving the blows from the hammerstone. The back ends of the tools have sometimes been purposely blunted, almost certainly to prepare them to fit into a haft of some kind. Many are spearpoints. But perhaps the most startling change from the MSA tools is the raw material from which they are made. Rather than relying solely on local quartzite, people were importing high-quality, fine-grained silcretes from dozens of miles away.

This remarkable tool industry is known as the Howiesons Poort, from a site east of Klasies where such tools were first found. All of its features eventually appear in stone tools in Europe too. You just have to wait a long time.

"If I didn't know where and when it was coming from, I would unhesitatingly classify the Howiesons Poort as Upper Paleolithic," archaeologist Paul Mellars told me in Cambridge. "If it is seventy thousand years old, as Deacon says, it would be thirty or forty thousand years earlier than anything like it in Europe. And lo and behold, you've got anatomically modern humans first appearing down there too. I find that extremely suggestive."

On the surface, the Howiesons Poort is tantalizingly close to just what I was looking for: a sharp signal of behavioral change in Africa, occurring while the Neandertals and other archaics were still going about their Neandertalish ways. Even more intriguing is the possibility that this change in technology might actually *explain* the early emergence of modern human anatomy in Africa. In comes a more efficient tool kit, out go the robust proportions of African archaic *Homo*, because such muscle bulk would no longer be needed. But the story, as Deacon explained, is not quite so neat and simple. First, the modern human fossils at Klasies and Border Cave *predate* the Howiesons Poort by at least 20,000 years. Obviously, the new tools cannot account for a biological change that occurred long before the tools are known to have existed. Second, if the Howiesons Poort really does declare the emergence of a new, technologically superior kind of human, wouldn't these people be likely to survive into the future? One would expect the Howiesons Poort to carry on into the next technological stage, merging seamlessly into the Late Stone Age, with which it shares many traits. But it doesn't. At Klasies and elsewhere in South Africa, the Howiesons Poort deposits peter out as you go farther up toward the present. They are replaced, with stunning anticlimax, by the same old boring Middle Stone Age tools that had dominated the Old World for so many millennia. Ten thousand years later, Klasies River Mouth was abandoned by humans altogether, not to be reoccupied for another 50,000 years.

Deacon laid out the bare bones of this narrative in the deserted classroom in the archaeology building, his soft, slightly stuttered speech barely penetrating the darkened air. But there was nothing unsure about his answers. In Deacon's view, there is no doubt that the Klasies people were modern in every major respect, long before their Eurasian counterparts. Their anatomical form, the way they perceived their environment, the way they obtained food and related to one another were all essentially the same as those of Late Stone Age hunter-gatherers, and even more recent people. Indeed, Deacon believes that the human bones found at Klasies may well represent the direct genetic ancestors of the living Bushmen: an endemic South African lineage, traceable from the present moment back 100,000 years, to the point when humanity was born.

For Deacon, that point began when our ancestors took control of their own surroundings. Earlier populations like *Homo erectus* and the archaics of the Early Stone Age were more or less managed by their habitats—tied to valleys and water sources on the coastal plain, eking out a living from whatever resources were at hand in the area. The key change occurred at the beginning of the Middle Stone Age, when people began to manage the environment instead. Rather than take the surrounding landscape as a given, they molded it to fit their needs. The proof, he believes, is in the plants. Increasingly through the Middle Stone Age, the Klasies River Mouth region would have been taken over by open savanna landscape with few fruiting trees. The productivity of plants was concentrated underground instead, in "geophytic" buds and bulbs. Left to their own devices, geophytes are a very slowly renewing resource. In order to rely on them for sustenance, the surrounding vegetation would have to be systematically burned off to speed up new growth. This is exactly what African pastoralists do on the savanna today, to encourage the growth of new grass for their cattle.

Obviously, such management requires the ability to make fire at will. It also demands a perceptual leap: the sense that the habitat, and with it the very future, can be *designed*. We know that the Klasies people knew fire. We know they were depending on geophytes to survive, especially during Howiesons Poort times. "Putting two and two together," Deacon said, "we get a picture of people 'farming' patches of starch-rich plant foods with fire and supplementing this diet with meat from hunting and scavenging and from collecting shellfish when at the coast."

All this was very satisfying to my ears—a tiny glimpse of people who had crossed the gartel into that unprecedented "us-ness," in Africa, way ahead of their time. But there were other voices to listen to, people who think Deacon has put two and two together and come up with five, if not fifty. One such voice belongs to Richard Klein of Stanford University, a respected archaeologist who, like Deacon, has spent many years probing the behavioral depth of southern Africans in the Middle Stone Age. In contrast to his South African colleague, Klein insists that they had yet to cross the line. "The Middle Stone Age people at Klasies and other southern African sites were anatomically modern," Klein told me. "But that doesn't mean they were acting like us at all."

In his view, their sub-modern life-style reveals itself in the animal bones littering Klasies River Mouth, as well as other caves such as Nelson's Bay and Die Kelders. The only large mammal the Middle Stone Age people regularly hunted appears to have been the eland. While the bone record also includes Cape buffalo and bushpig, these more dangerous species are almost always represented by either young or very old animals—the individuals who could be scavenged or, in the case of the juveniles, captured with comparatively little risk. In contrast, the Late Stone Age hunters, who arrived on the scene some 20,000 years ago, regularly took on buffalo and bushpig in the prime of life. They could do so, Klein believes, because they had invented sophisticated weaponry—notably the bow and arrow—which enabled them to kill dangerous animals from a distance, greatly reducing the chances of getting gored or mauled in the process.

These Late Stone Age folk also exploited sea resources differently. The bones of fish and flying seabirds like gulls and cormorants are abundant in Late Stone Age sites, but virtually absent in the leavings of their Middle Stone Age predecessors. Along with these bones come bone points and other artifacts that may well be "fishing and fowling gear." Klein concedes that the Klasies people were burning geophytes and were probably advanced enough to use cooperative hunting techniques, such as cliff drives of eland. But their dependence on the sea was limited to what could be easily collected on the beach—scavenged seals and dead birds, along with the staple shellfish. By the Late Stone Age, humans had created a whole new realm of subsistence, actively and intelligently pursuing prey that would have been out of the reach of their predecessors. "The Middle Stone Age people just didn't do this," Klein told me.

"Why?" I asked. "Because they couldn't? Or because they didn't need to?"

"They didn't do it because they didn't have the capacity to. That's what I think. And they didn't have the capacity because they weren't bright enough."

"So there was a qualitative, intellectual difference between the people at Klasies and later hunter-gatherers?"

"Look, if you believe in human evolution, there have to be intermediate steps," he replied. "You could say the Middle Stone Age Africans were stumbling around the landscape, but they weren't stumbling around as much as Peking Man, and he wasn't stumbling around as much as *Homo habilis*. I would make the argument that each was smarter than the one who came before, and dumber than the one who came after."

What then does one make of the oddly precocious Howiesons Poort industry? On the basis of the stone-tool technology, the Howiesons Poort measures up to what early Cro-Magnons were producing in Europe many thousands of years later. But according to Klein, equally significant aspects of the European Upper Paleolithic "creative explosion" are missing. Tools made of antler and bone, a Cro-Magnon trademark, are virtually nonexistent in the African Middle Stone Age. Clear signs of art and personal ornaments do not show up in Howiesons Poort levels. "The Howiesons Poort isn't more sophisticated," he said, "it's just different. Okay, you find these geometrically shaped pieces that somebody has dulled on one edge. But the artifacts below and above them in time are often sophisticated too, if that's the term you want to use, in other ways. Maybe Howiesons Poort has some meaning, maybe it doesn't. Who knows?"

Hilary Deacon, perhaps. In sharp contrast to Klein, he contends that the mysterious Howiesons Poort makes sense only as a thoroughly modern response to crisis. If he is correct in setting the age of the industry at 70,000 years, its appearance in South Africa corresponds to the beginning of the last glacial cycle. Terrestrial mammals at the site appear to be those associated with cooler climates, and shellfish are scarcer, suggesting that by this time the ocean was farther away. The new, more standardized Howiesons Poort tools, with their backs prepped to receive spear shafts, might be a technological response to living in harsher, more demanding environmental conditions. This would explain why the industry suddenly disappeared as well. After the initial jolt of glacial conditions 70,000 years ago, the oxygen isotope record shows a brief amelioration of global climate. During this warmer phase, the Howiesons Poort way of coping with environmental stress was no longer needed, and the old Middle Stone Age way of life returned.

The Howiesons Poort tools are far more suggestive, Deacon believes, than any explanation based purely on their functional efficiency as weapons. "This whole idea of efficiency is a Calvinistic product," he said one night over drinks after dinner. "It's about Europe. It's about working harder, trying more. Well, that's just not the thing here in Africa."

"But I thought the whole point was that the Howiesons Poort enabled the Klasies people to survive through more difficult times," I said. "Isn't that by definition a matter of function?"

In response, Deacon excused himself and returned with a tray full of Howiesons Poort artifacts. "Just *look* at this," he said, plucking a small, triangular point out of the tray. It was a pretty thing, made from a rose-tinted translucent silcrete. No such raw material, he told me, exists within forty miles of the Klasies River Mouth site, where this one was found. He held it up between thumb and index finger.

"Do you really believe such a thing as this is *functional*?" he asked.

"If it's not functional, what is it?"

Deacon smiled. "I think these points are symbols in some way. I think they belong to the male realm."

Symbols? The male realm? By 70,000 years ago hominids had been making stone tools for well over two million years. Nothing found in all that time, however, suggested that a tool was ever anything more than a tool. But now, perhaps, something crucial had changed. A tool was a tool, and something else too. If Hilary Deacon is right, the hominids of the Howiesons Poort were finally getting around to acting like human beings.

Let's back up for a moment and look at the force driving change in the first place: the environment. One hallmark of humans is the ability to survive in a spectrum of habitats ranging from moist, dense forests to parched deserts and arctic barrens. The resources offered by habitats likewise range from abundant to scarce, from predictable to very uncertain. Some habitats might be ungenerous in resources most of the year but offer seasonal, highly predictable bonanzas of food—an annual salmon run up a river, for example. Other environments harbor dense patches of potential food—for example, nomadic herds of bison or reindeer— whose whereabouts in time and place are much harder to predict. What makes human survival possible against these very different environmental backdrops is our species' ability to alter the way we position ourselves on the landscape, adjusting the size of the social group, balancing the amount of movement needed to secure food against the energy costs of moving, developing new behavioral innovations to suck nourishment from food sources that would remain unavailable to a behaviorally less flexible species.

According to Hilary Deacon, the habitat at Klasies during Howiesons Poort times was cool, arid savanna. Such a habitat yields predictable but very sparsely distributed resources, rather like the central Kalahari now inhabited by the !Kung Bushmen. Regenerated by fire, geophytic plants could be relied on to provide an adequate, if hardly abundant, staple. Eland and other dry-country animals could be depended on for occasional rather than constant protein sustenance. In such circumstances, in order to feed themselves, human groups would have had to range over a larger territory than their Middle Stone Age predecessors, who could afford to "sit" on resources available locally, much as the Neandertals were doing in Europe at the same time.

This change from a sedentary existence to greater movement is echoed in the shift in raw-material use, from locally abundant quartzite to the silcretes that the Klasies people had to bring in from farther away for their Howiesons Poort points and blades. The possession of this finer-grained raw material was a prerequisite for making small, blade-based tools; quartzite is too coarse-grained to yield to the blade technique. Most archaeologists also agree that the Howiesons Poort represents a clear technological improvement at a time when the environment was changing for the worse and people needed new technology to help them adapt and survive. This does not mean that the switch to "foreign" flint was *caused* by the need for better tools. Rather, the people must have already been moving around more, extending their hunting excursions into silcrete-rich areas and only later discovering the technological promise sequestered in the stone. In the meantime, they were encountering something else in those farther-flung territories: other people. If Deacon is correct, this aspect of the altered landscape influenced the appearance of

their tool kits at least as much as the functional improvement offered by
the rock itself.

According to Deacon, his graceful little Howiesons Poort spearpoint
was meant for more than securing food. Projectile points have enormous
socioeconomic importance; they are the working edge of the hunt, the
material object that delivers meat to the group. But among modern
hunter-gatherers—or modern businessmen, for that matter—ostensibly
functional objects can quickly gather auxiliary importance as declarations
of social identity. The businessman's Mont Blanc pen is an indisputably
functional instrument, but nobody whips these sleek black powersticks
out of his vest pocket to sign a contract *solely* because they deliver ink to
the page better than other pens. The social message spoken by the pen
is far more important. Mont Blancs have *style*; they carry information
about the personal and social identity of the person holding one in his
hand: "I am a big player. I have arrived." Sometimes the social messages
embodied in an object completely preempt whatever functional value it
may once have had. Though most white-collar workers wear ties, those
ties do not fasten their collars anymore. They merely dangle there in front
of the shirt. But the ties are far from useless. Whether thin or fat, striped
silk, jazzy print, or heather wool, the tie is a self-imposed, unambiguous
advertisement of social class, occupation, and personality. When you put
on your tie, you know who you are and where you belong. So does everybody else.

Clearly a projectile point serves a more obvious economic function for
a hunter-gatherer than a tie does for an insurance agent or lawyer. Projectiles are nonetheless perfectly positioned to be loaded with social importance as well. If the hunt yields large game, the meat pierced and killed
by the spearpoint can be shared among members of a social group, binding
socioeconomic ties. Tellingly, among Bushman hunter-gatherers, it is not
the successful hunter who distributes the meat. That honor is reserved
for the person who made the fatal arrow. Since hunting among Bushmen
is largely a male activity, spearpoints naturally gather gender identity—
they "belong to the male realm," as Deacon put it—without resorting to
Freudian analogies. Polly Wiessner of the Max-Planck-Institut, who conducted a comparative study of Bushman arrowtip styles in the Kalahari,
was forced to work from photographs in some cases because Bushman
taboos forbade women to touch arrows. Perhaps most important, since
the hunting males by their movements define the limits of a group's ef-

fective territory, the spearpoint, the physical manifestation of the hunt, is a natural place to "materialize" the identity and physical reach of the group itself.

Wiessner's study offers a splendid example of how projectile points come to mean much more than their familiar function. Once widespread across southern Africa, the Bushmen hunter-gatherers (or San, as they are sometimes called*) have been reduced by Bantu and European invasions of their homeland to a few remnant populations in and around the Kalahari Desert. For the last few decades, the Kalahari Bushmen have fashioned the tips of their poison arrows out of hammered fencing wire. Wiessner compared the arrows of three Bushman language groups: the well-known !Kung and two neighboring, lesser-known groups, the !Xo and the G/wi. Within each group, the shape and size of the metal arrow tips were remarkably similar. But whereas the !Kung consistently fashioned small, triangular arrow tips, those of the other two groups were invariably twice as large. !Xo and G/wi arrows, meanwhile, were clearly distinguished from each other by shape, the !Xo arrows being notably blunter and thinner.

From her interviews with arrow makers, Wiessner knew that none of them had consciously set out to make arrows reflecting their linguistic affiliation. They said they made arrows "just as they pleased." Nonetheless, when she showed arrows made by one group to hunters from another, the social message embedded in the artifact rang loud and clear. "The !Kung reacted to the !Xo and G/wi arrows with surprise and anxiety," Wiessner reported. "A discussion ensued from one small group about what they would do if they found a dead animal with such an arrow embedded in it in their own area, saying that they would be worried about the possibility that a stranger was nearby about whom they knew nothing at all. Although afraid of !Kung strangers as well, they said that if a man makes arrows in the same way, one could be fairly sure that he shares similar values about hunting, land rights, and general conduct."

"You should hear Polly tell how they denigrated the arrows made by

---

*Both names, regrettably, are outsiders' terms for a people who have no name for themselves. When anthropologists balked at the pejorative overtones in the European "Bushman" (or "Bojesman" in Afrikaans), they began to use "San," the Nama peoples' name for their hunter-gatherer neighbors. It is hardly an improvement. In Nama, "San" refers to any poor people who live like wild animals, without property or livestock. Here, I will follow the lead of such writers as Elizabeth Marshall Thomas and David Lewis-Williams and revert to "Bushman," with the understanding that no negative connotations are intended.

hunters from another group," another anthropologist said to me. "There was a lot of scatological comment and references to the sexual organs of the maker."

In short, the distinctive style of each linguistic group's projectile points functions as a border between them. (A far more useful boundary, by the way, than the government-erected fences that provided the raw material for the arrows in the first place.) But the unity in arrow style *within* each linguistic group is just as essential. In a habitat where drought can quickly dry up local resources, it helps to have social contacts spread over a wide area, so that food and water from less affected areas can be shared. The Bushmen cement such alliances within their linguistic group through an elaborate system of gift-exchange, called *hxaro*. Arrows are an integral part. Wiessner found that almost half the arrows in the hunters' quivers had been received as gifts, rather than made by the hunter himself. It is interesting that among one sample of over two hundred !Kung arrow gifts, most were given either by hunters in the same camp or by people living as far as sixty miles away.

Back in Stellenbosch, after supper in the Deacons' living room, Hilary

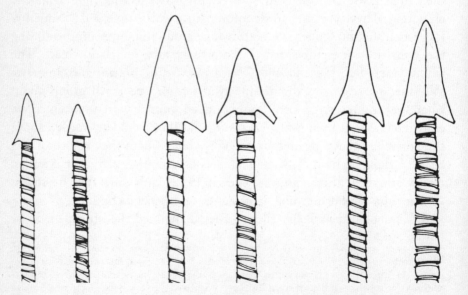

Group affiliation of Bushmen hunter-gatherers can be distinguished by the style of their projectile points. From left, typical arrows of the !Kung San, !Xo San, and G/wi San of the Kalahari POLLY WIESSNER, BY PERMISSION OF THE SOCIETY OF AMERICAN ARCHAEOLOGY, FROM *AMERICAN ANTIQUITY*, 1983

projected Wiessner's analysis back into the Pleistocene. Given the distinctive shapes and advanced workmanship of the Howiesons Poort tools, he surmised that the Klasies people were up to the same thing—investing their projectile points with style, group identity, and meaning. The exotic raw material they were using proves they had either expanded their foraging areas, which would presumably put them in touch with more people, or they had developed exchange networks that could deliver the silcrete stone through trade—more social contact. The Howiesons Poort projectile points are not nearly so standardized as the contemporary Bushman's metal arrow tips, but stone is a much harder material to work in the first place. It all comes back, Deacon believes, to the scarcity of resources in that long, long cold snap 70,000 years ago. This would be the time when alliances over long distances would pay the highest possible dividend: survival. If Deacon is right, it was a new and distinctly human way of dealing with a very old problem. And it wasn't happening back in Europe. At least, not yet.

"Tools like these just couldn't *be* in the Mousterian in Europe," says Deacon. "It would be like finding bobby socks in the Elizabethan."

Finally, the social content invested into silent stone by the Klasies humans would explain why the Howiesons Poort disappeared. Other anthropologists have documented that the stylistic flourish given to clothing, implements, and so forth in various societies rises and falls with the amount of socioeconomic stress the society is enduring. When the Pleistocene climate improved again, complex social networks and their material signals were no longer crucial. The Howiesons Poort disappeared, and a tool became, once again, just a tool. In a deeper sense, however, if the Howiesons Poort means what Hilary Deacon believes it does, perhaps the stress of that long cold time altered the human mental landscape in a way that could never be completely undone. Up to this point, our ancestors could look out on the grass and trees and the hard edge of the horizon and see only what was there to be seen. But a world in which stone tools— mere *things*—can become saturated with message and meaning is no longer a world that can be taken at face value. Looked at with human eyes, it is a world that has lost its innocence.

A couple of days later I was on my way to an appointment with Philip Rightmire in the South African Museum in Cape Town. He had promised to show me the original Klasies River Mouth fossils. I wanted to ask him whether the bones offered any further clues to the behavior of the people, as the skeletons of Kebara and Qafzeh do in the Levant. Some of Hilary Deacon's astonishments were still buzzing around in my mind. Among the most intriguing was his suggestion that the Klasies people might be linked to modern-day Bushmen by more than just analogy—that there in fact exists a direct regional lineage leading from the present people of the Kalahari straight back 100,000 years to Klasies River Mouth. For support, Deacon cited the geography of the African continent. Unlike Eurasia, which possesses "broad corridors for movement" running east to west, Africa is geographically aligned north to south, with no natural passage granting a free-flow migration of populations. "Any population that got to the south was effectively isolated," Deacon told me, granting the Bushmen a column of genetic purity running deep into time.

There is something uncomfortably symmetrical about this notion. The Bushmen, and especially the !Kung linguistic group, are the best-known, most intensively researched hunter-gatherers alive. As such, their lifeways are often carelessly used to stand in for those of Paleolithic people in general. It would seem entirely too neat if they turned out to be the "original modern humans" in a biological sense too. On the other hand, Linda Vigilant's study of !Kung mitochondria and other genetic evidence was suggesting that the common roots of the Bushmen go a long way back, perhaps as far as those of all living humans generally. Vitaly Shevoroshkin, a Russian linguist, believes that the clicked Khoisan languages spoken by Bushmen, Hottentots, and some other South African tribes make up the oldest language group on earth, a key to unraveling the "mother tongue" of all the world's languages. Such an astonishing assertion would gain credence if exclusively Bushman-like features also showed up in fossils 100,000 years old.

On my way to Rightmire's temporary office, I took a wrong turn into a storage area in the museum, and suddenly I was standing, dreamlike, face-to-face with a band of Bushmen. A couple of hunters were hovering about, while a young woman nursed her infant. All were naked, and their skin was chipped in places, exposing the dull beige plaster beneath. I later learned that the figures were life-size casts of Bushmen who lived early in the century. They have had a poignant history of display in the museum. First they were set up each on its own, like stuffed animals. By the 1930s they were granted a glass case and displayed as a group. By the late 1950s they had earned a naturalistic scene: Dressed in loincloths, they were given spears to hold and background scenery to enhance their verisimilitude. It was not until 1984, however, that someone pointed out that while the exhibits on white colonists and other immigrants to South Africa were displayed in the Museum of Cultural History, no one had thought to move the Bushmen from their place, among the birds and beasts, in Natural History.

I found Philip Rightmire on another floor, and he went to fetch the fossils. He brought them back in a dull-yellow cardboard carton not much bigger than a shoebox and plunked it down on a tabletop. I could see right away that looking for behavior in these bones would be a difficult business.

"It's a pretty scrappy lot," he said, arranging the fossils on the table, "but if you're careful and persistent you can get a good deal of information out of it." The information Rightmire is most interested in is whether or not the Klasies remains represent the most ancient anatomically modern humans on earth. The oldest fossils in the box included a scattering of teeth, a piece of elbow, and two fragments of upper jawbone, one little more than a morsel of jaw with a tooth stuck in it. "Nothing about this material suggests they *weren't* modern," said Rightmire. "On the other hand, you could argue that there just isn't enough there to tell."

The fossils that make Rightmire's case were found farther up in the sequence by Singer and Wymer, back in the sixties. In all, they pulled out five partial mandibles, a cheekbone, and a motley collection of skull fragments, teeth, and broken limb bones. Judging from these specimens, Rightmire believes that by this point human populations in South Africa had crossed the line into physical modernity. The limb bones are gracile. Most of the lower jawbones sport chins. In fact, one of the fragments I saw on the table amounted to little more than a chin alone, as if the individual was concerned that, should some piece of him be saved for posterity, let it be something that clearly announces, "I was modern." The same might be said of an irregular shard of skull, one that boasts part of a forehead and nose region devoid of any brow swelling or other archaic feature. The collection as a whole is not, however, a perfect match for human beings today. Many of the jaws are robust and big in the tooth, and one is clearly chinless. "You'd have to search long and hard to find something like that in a modern sample," Rightmire observed. "If there is evidence that this population retained archaic characteristics, this is it."

Such apparent holdovers from the primitive past have, in fact, led investigators like Milford Wolpoff and Fred Smith to deny that the Klasies River Mouth sample is fully modern. But most experts agree that Klasies River Mouth offers the best securely dated evidence for true modern human anatomy in Africa.

"What about Bushman features?" I asked. "Is there anything here that can directly link Klasies to the Bushmen today?"

"We have skulls twenty or thirty thousand years old that look pretty convincingly like Bushmen to me," Rightmire replied. "It's possible there is a connection even farther back. But the more I look at the Klasies material, the harder it is to see distinct resemblances."

Perhaps the best anatomical evidence on the table for a deep Bushman lineage was a tiny mandible with pearly little teeth, a jaw no bigger than that of a modern child of seven or eight. But this was no child. The molars in the jaw were badly worn, indicating that the individual, almost certainly a woman, was well along into adulthood when she died. The Bushmen, too, are childlike in their proportions, so perhaps this showed some link, although a very tenuous one. "It would be easier to tell if we had a complete skull," Rightmire said, a little wistfully. All the fossils from Klasies have been fragments, and the chances of finding a more complete specimen are remote.

Bushman or not, this slip of a jaw was intriguing for another reason. While its size was striking in itself, the jaw looked even stranger when placed up against other Klasies mandibles in the collection, which are large, robust, and presumably male; one of the jaws is some 50 percent bigger than the little female. Such "sexual dimorphism" is way off the scale of anything seen today. On the average, human males are now only some 3 percent larger than females in the same population. There is, of course, always the possibility that the little Klasies jaw belonged to an unnaturally small person, or that it represents a different population. But if the mandible is that of a fairly normal Klasies woman, there are only two known explanations for the yawning difference in sexual size.

One is nutritional. If females were being denied access to the protein-rich foods enjoyed by the males, their body size would end up disproportionately small. Rightmire told me of one such case among Paleo-Indians in Georgia, where the men were obviously eating meat at the same time that the women were living on cornmeal alone. The other explanation is social. It is not that the females were small, but that the males were disproportionately large. Among mammals generally, extreme sexual dimorphism is often a trait of polygamous species whose males must compete with one another for access to mates. Those with the largest bodies tend to prevail and pass on their genes for bulk. If this sort of social organization accounts for the dimorphism at Klasies, apparently modern humans there had not yet established the monogamous, pair-bonded pattern predominant in human societies today.

There was one other curious feature about the little lady with pearly teeth. The front portion of her jaw was discolored and blackened, as were many of the other bones in the collection. I had already heard Hilary Deacon's grisly explanation for these dark stains. Whatever else the Klasies people were eating, he told me, they were also eating each other. Deacon believes that virtually all of the hominids found at Klasies represent the remains of cannibalistic feasts. There is no doubt that the bones have been burned, and some of the skull fragments show signs of having been torn open. Tim White of the University of California at Berkeley has also discovered "indisputable cut marks" made by stone tools on the very modern-looking frontal bone from Klasies.

Late one evening, Deacon described just how the meal might have proceeded. "Let's say that you want to heat up some brains," he said. "Well, you'd place the head on the fire, which puts the mandible in contact with heat—hence the blackening that you see there now. When everything's nice and bubbly, you take it out. But you've still got to get the thing open, right? So you grab a hammerstone and give it a good bash. Naturally the bone is fresh and tears a bit as you pry open the skull. And that's where those tear marks come from."

Claims of cannibalism are not uncommon in the Paleolithic record. *Homo erectus* in China practiced it, Neandertals in Yugoslavia practiced it, Neolithic Frenchmen almost certainly did too—so why not the early African *Homo sapiens*? Deacon recited some historical examples, including a particularly unwholesome local morsel from Basutoland, where a famine led to systematic cannibalism in the 1820s. "After that, it became something of a habit," Deacon said. "They would break the legs of the victims so they couldn't run away—a sort of human larder. A number of accounts confirm this. There was a preference for children, and for certain body parts above others."

If the Klasies people were eating each other—and cannibalism seems the most reasonable explanation for how the bones ended up as they are—

does that help us decide how "human" they were? Not really. As Tim White pointed out to me, cannibalism has been reported in some seventy-five other mammal species, including fifteen nonhuman primates. A more studied answer would depend on *why* human flesh was on the menu in the first place. Most experts agree that people eat other people either because they have nothing else to sustain themselves, or as a ritual act of some sort. The belief systems of many tribal societies include ceremonial eating of selected parts of the deceased, in the communal hope that some quality of the departed—his spirit, wisdom, strength—will thereby be passed on and preserved. The fierce Yanomamö Indians of the Amazon consume the ashes of their comrades who have been killed in raids. A pinch of Fallen Comrade is added to boiled plaintain soup at the funeral ceremony, and served up again whenever a revenge raid is in the offing. "This puts them into the appropriate state of rage for the business of killing," writes anthropologist Napoleon Chagnon, who has spent many years with the Yanomamö.

By its very nature, ritualistic cannibalism would seem to demonstrate a layered mental landscape, a sense that things—in this case, the ashes of a corpse—embody powers, values, or meanings beyond their physical opacity. Eating your own kind merely to stay alive, on the other hand, demonstrates nothing but an extreme state of starvation. Unfortunately, the line between the two motives is not always clear. "If you are starving to death and you eat Uncle Harry, there is probably going to be some ritual involved," says White. "Even so, your motive in eating him was still starvation. On the other hand, there are societies where they eat Uncle Harry purely because they thought he was a neat guy and want to be a little more like him."

And at Klasies? It is impossible to tell whether the people eaten there, if eaten they were, went down with ceremony. "I don't know whether Hilary is right or not about the cannibalism at Klasies," said Philip Rightmire in Cape Town. "But if it takes him back to the site to look for more specimens, I'm all for it."

A few days later, I was lying on my back on a slab of stone, looking up at a man with the head of an elephant. Or perhaps it was an elephant with the head of a man. The body parts were so jumbled up, it seemed futile to try to sort them out into categories. The figure had four legs, but each fanned out into a flurry of digits. Though the head in profile was manlike, a tapering trunk dangled out of its forehead and appeared to pour liquidly out of the interior of the man's skull, trail along the ground briefly, then launch back up into the air like a dancing cobra.

People who study Bushman rock art call this sort of image a therian-thrope, a man with the features of a beast. Before they were driven extinct by Europeans in all parts of southern Africa except the Kalahari, the Bush-men embellished great tracts of the landscape with their delicate, effer-vescent creations—prancing elands, antelope, and buffalo, lithe hunters and transported shamans, geometric patterns, fantastic hybrid creatures, some painted in polychromes of white, black, and red, others engraved into the rock itself.

Barnes Shelter, the little enclosure where I had just woken up from a nap, is one of dozens of caves in the vast Giant's Castle Game Reserve in the Drakensberg mountains of western Natal. Virtually every one of those caves holds paintings or traces of paintings all but erased by time. I had chosen Barnes Shelter as a destination for a hike because it was accessible but not publicly known. Tucked in a grassy ravine above the camp, it was pervaded by the smell of wild sage and a deep, restoring silence. There was just the Elephant Man, myself, and my thoughts.

What, in the end, had I learned from Klasies? The cannibalism was a sort of human-interest snack—fun, but not fulfilling. Similarly, the con-nection between the Middle Stone Age people there and living Bushmen was intriguing, but barely more than a speculation. The sexual dimor-phism witnessed in the size of the jaws was possibly meaningful, but what it meant remained a mystery. The evidence from the site for an early appearance of modern anatomy seemed solid enough, but there was no attendant behavioral testimony, no other lead to follow that could answer the key question *Why? Why in Africa? Why at that point in time?*

"I don't know the answer, and nobody else does either," Richard Klein of Stanford University told me later. "Maybe the anatomical change is purely coincidental. Maybe it's linked to some neurological development. Maybe, maybe, maybe." Then he added, with a hint of wistfulness that I had also seen in Philip Rightmire's doleful countenance, "Trouble is, you get too many maybes, and people stop listening."

In my search for a human-defining behavioral shift, the richest lead yet was Hilary Deacon's explanation of the Howiesons Poort industry as a social phenomenon, rather than simply a functional response to an en-

vironmental problem. It seems to me that only a doubly-wise consciousness could endow objects with such double purposes and negotiate the
layered intricacies of the new world it had thus invented. Much as I liked
the *idea*, I wondered if the evidence of the Howiesons Poort was strong
enough to bear such consequence. Deacon's argument depends heavily
on the new movement of raw material across the region. But even if the
Klasies people were making a concerted effort to obtain higher-quality
stone, you can't rule out the possibility that they were primarily interested
in the functional value of its cleaner cleavage planes. According to Richard
Klein, raw material in Africa doesn't really start to move often and far
enough to suggest a pattern of social contacts over distance until later, in
the Late Stone Age.

And there is the problem of time. The Howiesons Poort appears around
70,000 years before the present. The first great flourish of symbol and art
does not occur until the European Upper Paleolithic, some 30,000 years
later. If the key to being human lies in an awareness of the power of
symbol, born in the deep time of the African Middle Stone Age, why does
it take so long to bear fruit? I looked at the painted image inches from
my face. Where in time and place are the intermediate steps, the stages
between a tool passively carrying a measure of social meaning and this
deft engagement of pigment and line before my eyes, the interlaced essences of man and elephant, set dancing on a shaded rock? Perhaps the
Howiesons Poort did mark a fundamental change in the human mental
landscape. Perhaps any difference between the thoughts of a 70,000-year-
old man or woman in South Africa and one today is only a difference in
degree, not in kind. But no direct link between that past and this present
has come forth from South African ground.

It probably never will. The link breaker, the destroyer of continuities,
was a blanket of ice creeping down from the opposite pole. As the glaciers
cooled the global climate and sucked water from its cycle, the southern
edge of the African continent dried up. The people of Klasies River Mouth
and the other caves either migrated northward or dwindled down to numbers so small that their traces have yet to be discovered. Human populations throughout much of the rest of the continent apparently suffered
the same fate. There is one region, however, that remained inhabitable
through all the glacial perturbations of the Pleistocene. If there is one
place that can bear witness to the emergence of truly modern human
culture across that chasm of time, it is East Africa. For the most part, the

Middle Stone Age there remains uncharted, an enormous hanging question mark.

A couple of American archaeologists recently ventured into the little-known territory, however, and came back with a sensational discovery. The only trouble is, nobody believes them. The find is so startling that sometimes they can't quite believe it themselves.

The surprising harpoons from Katanda, Zaire ELIZABETH SHREEVE

Chapter Nine

A MATTER OF TIME

"Everybody is anomalous." —HILARY DEACON

Near the border of Ethiopia and Kenya, the giant tectonic scar known as
the East African Rift Valley splits into two long, curving forks, holding
Lake Victoria between them like a sapphire between thumb and forefinger. The younger, eastern fork, running through the airy savannas of Kenya
and Tanzania, is the most fertile source on earth of information about
human origins. Here, the natural forces of volcano, earthquake, erosion,
soil chemistry, and climate have conspired to create an ideal environment
for growing and mining fossils, as the famous discoveries of the Leakeys
and others attest. The western fork, mountainous and heavily forested,
has never been so generous. A land of dark corrosive soils and jungled
impenetrabilities, it clutches its ancient human treasures possessively. If
the eastern fork is knowledge and light, the western is all murkiness and
mystery. Few expeditions go there anymore, and those that do mostly
return empty-handed.

   In the northeastern corner of Zaire, a few hundred yards from the
equator, there is a deserted strip of cleared turf, just long enough to land
a bush plane. On a July morning in 1990 I was sitting on my duffel bag
at the edge of the clearing, listening to the fading drone of the Cessna
that had just dropped me off. I was late for a rendezvous by exactly one
day, so I wasn't surprised that no one was there to meet me. All I could
see was tall grass and acacia—high savanna terrain. The plane's lonely
hum thinned to silence, the only sound a hot breath of wind in the grass.
Just before landing I had seen a collection of huts resting on a bluff above
a lake, but at ground level they were nowhere in sight. Without an idea

of which direction I should take, I stayed put at the landing strip, watching a dung beetle trudge toward a pile of fresh feces.

The place was called Ishango. Built for a vacation of the queen of Belgium, it had been converted into a remote station in Virunga National Park, a narrow, wild tumult of contrasting habitats along the Ugandan border. It overlooks the source of the Semliki River, the western Rift Valley jewel once called Lake Edward and later Lake Idi Amin Dada. When Amin lost his job as Ugandan despot, and with it the privilege to name bodies of water after himself, Lake ex-Edward, ex-Amin became known as Lake Rutanzige, or simply Lake Ex. Its waters drain at Ishango in a gorgeous union with the young, free-flowing Semliki, which runs a hundred increasingly turbulent miles northward along the rift floor before spilling into Lake Mobutu Sese Seko, the former Lake Albert, the legendary headwaters of the Nile.

I had come here at the invitation of archaeologists Alison Brooks and John Yellen, a wife-husband team from Washington, D.C. For years, beginning long before the Eve Hypothesis refocused attention on the mystery of modern human origins, Brooks has been quietly insisting that sophisticated culture emerged in Africa, thousands of years before the Cro-Magnons in Europe. If conclusive evidence hasn't turned up, Brooks believes, it is because few people have bothered to look for it, and those who have refuse to credit what they have found. It is a widely held notion, for instance, that people of the African Middle Stone Age could not, or did not, make tools out of bone. Brooks points out that although bone tools are rare in the African Middle Stone Age, so, too, is archaeology. "In the whole continent, there are a couple of dozen Middle Paleolithic sites that have been excavated, most of them in South Africa," she says. "In France alone, there must be three hundred. You need a magnifying glass just to read them on the map."

Among the handful of African sites, Brooks maintains that over 25 percent contain worked bone, etched shell, or some other artifact showing that more was going on in the Middle Stone Age mind than the numbing redundancy of crude flake tools would suggest. None of the previously known MSA sites, however, come close to expressing the level of competence revealed in the discovery she and her husband believe they have made here in Zaire. In 1988, while excavating at a place called Katanda, a few kilometers downriver from Ishango, Yellen uncovered a clutter of river cobbles and rough flake tools, lumped together with the bones of enormous catfish and other fauna. Both the concentration's density and its sharp edges suggested that a human hand had been involved in its formation—much like the "Neandertal garbage" found in the center of Kebara cave in Israel. But the last week of the field season yielded what no one would dream of finding in a Neandertal cave: half a dozen superbly crafted bone points, barbed along one side, with a circular groove cut into the butt end, suggesting they had been tied to a shaft like fishing harpoons. Barbed points in Europe first appeared during the late Upper Paleolithic Age, about 14,000 years ago. Brooks and Yellen believe their equally sophisticated African harpoons are three times as old. To put this in perspective, imagine discovering a prototypic Pontiac in Leonardo da Vinci's attic.

"If these things are as ancient as we think they are," Alison told me back in Washington, with the harpoons laid out on a table for my benefit, "it could clinch the argument that modern humans evolved first in Africa."

By listening to such talk I had ended up sitting on the edge of a deserted landing strip in Zaire, watching a dung beetle crawl through the heat. The beetle had covered only a few inches before shouts and giggles confirmed that my arrival had been noticed. Half a dozen children scampered out of the grass. We greeted each other in the universal language of chocolate (I give, they eat). I asked them where I could find Leo Mastromatteo, a European Economic Community staff official who was supposed to know of my visit, and followed them back to the compound at Ishango, entering the gate under the menacing glare of a young man wearing bits of an army uniform and swinging a semi-automatic rifle under his arm. He motioned me farther on, where a pair of denim legs stuck out under the chassis of a truck. As I approached, a barrel chest appeared, then a massive black mustache, and finally the rest of Mastromatteo's grease-spattered face. He greeted me with a great deal more spirit than the soldier.

"Don't let these guys worry you," he said. But they clearly did worry him. The army was in the area to protect the Ugandan border, he explained, but with their already meager pay embezzled by their own officers, most of the time they hung around in the park, extorting food from the locals and getting drunk on *kidingi*, home-brewed banana liquor. "These aren't soldiers, these are drunkards with guns," he said. "Their officers have no control. They're drunk too."

He assured me that Brooks and Yellen must have seen my plane landing and would be along to pick me up. In the meantime, he suggested a swim in the river to cool off. From the path down the bluff the water was a gorgeous, inviting sight, rimmed with egrets and pelicans and so clear that the hippos looked like submerged heaps of jade. Leo explained the Ishangan food chain. The abundance of bird species are drawn here by the ample fish population—barbus, Nile perch, and dozens of other species. The fish, in their turn, congregate for the wealth of plankton in the water. "And the plankton," Leo continued, "live off all the hippopotamus droppings. Beautiful, isn't it?"

"Well, why do *they* come?" I asked.

"The hippos? Oh, they've *always* been here."

On the river shore, we stripped and glided into the biology, scarcely disturbing the egrets, storks, and cormorants flocked on the far bank. A pod of hippos floated by with the heavy grace and tempo one would expect from animals that had *always* been there. In the full afternoon light, the sandy bluffs glowed like polished gold. The water was pure and blessedly cool. I leaned back and closed my eyes, listening to the calls of the birds and the serene exhalations of the hippos. If there once was an African Eve, I thought, surely this was her paradise.

Back in the 1950s, when Zaire was called the Belgian Congo, a Belgian geologist named Jean de Heinzelin came to Ishango and discovered an Eden of sorts. He uncovered the remains of what he called an "aquatic civilization"—a culture that had for thousands of years hunted and fished with barbed spears and harpoons, before their presence on earth was obliterated by a volcanic eruption. Among the artifacts they left behind were bones with parallel, grouped lines etched into them, which de Heinzelin took to be evidence for an early numerical system. He speculated that from its source at Ishango, this number system had later spread to dynastic Egypt, and beyond.

"Because the Egyptian number system was a basis and a prerequisite for the scientific achievements of classical Greece, and thus for many of the developments in science that followed, it is even possible that the modern world owes one of its greatest debts to the people who lived at Ishango," he wrote in *Scientific American*.

Using the then-new technique of radiocarbon dating, de Heinzelin originally dated the Ishango civilization at 21,000 years before the present. Later, knowing that his radiocarbon dates had been gleaned from iffy material, and fully aware that such an age predated the appearance of harpoons in Europe, de Heinzelin revised his estimate. The Ishango civilization, on second thought, was only some 8,000 years old, securely within the Holocene period, well after European harpoons and at a time when similar weaponry was spreading throughout north and east Africa. The entrenched belief that white, European, Cro-Magnon Man represented the *avant garde* in matters technological remained unchallenged.

That is how matters stood until 1985, when Alison Brooks and a colleague opened a new dig at Ishango. They found several more harpoons, as well as fragments of a human skull and a wealth of rather crudely made flake tools. But the most significant find proved to be a triangular bit of ostrich-egg shell. Brooks tried two different dating methods on this seemingly insignificant morsel, trying to establish the true age of Ishango. The results convinced her that de Heinzelin had been right the first time around. The harpoons, notched bones, and other artifacts were close to 25,000 years old. Nobody believed her, of course, least of all de Heinzelin. There was little reason why anyone should. De Heinzelin was the more experienced investigator and far more familiar with the geology of Ishango. Moreover, one of the techniques Brooks used to date the shell, amino-acid racemization, had performed very poorly in the past. And of course, everybody *knew* that Africans 25,000 years ago couldn't even conceive of barbed harpoons, much less make them.

In the face of such skepticism, it would be understandable if Brooks had revised her date and cut the age of harpoon-literacy in half. Instead, she doubled it. Although she was careful not to mention precise numbers, she hinted broadly that the harpoons her husband had found at Katanda were perhaps twice as old as those at Ishango. The implications are astonishing. Anyone can learn to prepare a rock core and knock off a sharp flake. But working bone in a manner that produced those barbed beauties would have required considerable training and dexterity, not to mention the mental capacity to conceive of the object and its complex function in the first place. "To be honest, I'm not sure *how* they made them," Yellen told me. One thing is sure: no other technology remotely as sophisticated as this is known to exist anywhere in the world so far back in the past.

Not surprisingly, there has been no shortage of skeptics among Brooks's colleagues. "Alison's harpoons are totally anomalous," says Stanley Ambrose of the University of Illinois. "I don't see how they could be as old as she says." "It's Calico Hills all over again," sniffs another archaeologist, referring to Louis Leakey's embarrassing debacle in his waning years, when he supported the claim that some stone tools in California were a million years old. Even Brooks's strongest critics, however, admit the hint of a "what if." "What she's got in Zaire is noise, not pattern," says Stanford's Richard Klein, who puts as little faith in Middle Stone Age mental abilities as anyone. "But who knows? She may be right. If so, she has made a very remarkable discovery indeed."

Rejuvenated by my dip in the Semliki, I climbed back up the Ishango bluff, reaching the top just in time to hear Brooks's Land Rover labor into the compound. I had met her several times before, but always in her office at George Washington University or some other academic setting. The daughter of an undersecretary at the Smithsonian Institution, a *magna cum laude* Harvard graduate who dabbled in classics, medieval literature, and astrophysics before settling into anthropology, she had always impressed me by her calm, scholarly presence and the melodious assurance in her voice. Now I found her in the driver's seat, dressed in dusty khaki shorts and shirt, hair bursting out from under a safari hat, looking like a blond adventuress from another era. I threw my gear in the back and climbed in. Before we could leave, the young soldier approached and leaned into the car, demanding to know why Brooks had not yet provided Semliki Research Expedition T-shirts for himself and his buddies, as promised. She put him off, delicately, and as soon as he backed away threw the engine into reverse and wheeled around. "I have no doubt that guy would shoot me in a second if he thought he could get away with it," she said.

We bumped along the path to Katanda. I was eager to know what had developed at the site, especially concerning the date of the harpoons, but Brooks was uncharacteristically elusive. After a few kilometers, the Land Rover turned into a woody area sprinkled with an assortment of tents on the bluff above the river. Nobody was in camp. The expedition staff was down on the riverbank, sudsing up after the day's work. Brooks introduced me. I was surprised by how few people there were to be introduced to. David Helgren, a burly, bearded geomorphologist from San Jose State University, was handling the geology. The only archaeologist in residence, except for Brooks and Yellen, was a young Belgian named Els Cornelissen. A handful of students completed the scientific staff. Everyone seemed subdued. The only people clearly enjoying themselves were the three children in camp this summer: Elizabeth and Alexander Yellen, and Karin Helgren, who were taking turns cannonballing off an old raft found among the reeds. From a distance, a trio of hippos watched the commotion with gargantuan indifference.

John Yellen, director of the Archaeology Program at the National Science Foundation, was standing by the water's edge, wearing only a bathing suit and a thick, rabbinical beard. A native Brooklynite, Yellen exhibits the ectomorphic angularity that often comes from being raised in city air. But he was tan and leanly fit, with deep brown eyes worrying out over the beard. Well known for his own studies of Bushman archaeology in the Kalahari, he has spent much of the last decade administering to the dreams of other investigators, relishing his limited time in the field. This afternoon, he seemed burdened by his thoughts. I was beginning to wonder whether something had gone seriously out of kilter with their hopes.

"The problem is right here," Yellen said, as if reading my mind.

He was pointing down to where his long toes lay buried up to their first joints in the black-flecked sand of the riverbank. The problem concerned those little black flecks—volcanic ash, which plays an essential role in the dating of Katanda. Brooks and Yellen were convinced that the Ishango site up the river is about 25,000 years old. They had several clues hinting that Katanda is older still. The crude quartz tools appear to be Middle Stone Age, while Ishango's contain small, crescent-shaped artifacts and other Late Stone Age elements. The mammal species at Katanda appear to be from cooler, drier times, which would put them back into the last glacial period, rather than the warm Holocene that followed. But the strongest evidence was geological. The glittery black flecks in the sand around Yellen's feet were bits of a volcanic ash called perovskite, which is traceable to eruptions of Mount Katwe, twenty miles across the Ugandan border to the east. Perovskite is common all over the surface in the region and can be found embedded in the archaeological deposits exposed at Ishango. In fact, the whole Ishangan sequence ends in a great blanket of ash, the result of a massive eruption of Katwe.

Brooks and her colleagues had found perovskite here at Katanda too, but only high up in the deposits, well above the level where the harpoons were exposed. They believed the absence of perovskite at that level could mean only one thing: Whoever made the harpoons lived before Katwe erupted. Judging by the amount of soil that had accumulated over time above the site, they lived a *lot* longer before. Jean de Heinzelin was maintaining quite the opposite: that Katanda is much *younger* than Ishango, and he had not budged on his conviction that Ishango itself is only around

8,000 years old. If de Heinzelin was right on either count, the Katanda harpoons would sink in importance from astonishing discovery to mere footnote, and not a very interesting one at that.

John Yellen, full of consternation, stood sole-deep in the Semliki bank. The fact that perovskite darkened the sand around his feet was not the problem; throughout the region perovskite is everywhere on the surface. But the day before my arrival, David Helgren had discovered traces, horribly uncomfortable traces, of the black Katwe ash where it was not supposed to be: deep in the Katanda geological beds that also harbored the harpoons. Rather than support Brooks's ancient age for the site, the ash was threatening to turn against her, lending support to de Heinzelin's blandly youthful date instead. Yellen told me they had photographs of Katwe showing multiple craters, so perhaps there had been an earlier, Middle Stone Age eruption of the volcano that could account for the perovskite in the harpoon layer. But he admitted that this kind of reasoning smacked of an attempt to explain away data that didn't fit their theory. "It's unfortunate that we didn't think of that possibility before," he said. David Helgren, the expedition's own geologist, was taking an even dimmer view. He told me privately that he thought de Heinzelin might be right after all.

Having traveled several thousand miles to seek the birth of modern culture, I was not exactly thrilled by this news. I fretted. My notebook reads, "First day at Katanda, and it's a bust. No answers here. Just stale water, heat and ants." The next day Helgren took me up and down the sweltering Katanda ravines, jovially poking at the exposures and explaining the geology all the way back to when Africa was part of the supercontinent Gondwanaland, a hundred million years ago. But there was a sadly disengaged tone to his *bonhomie*, as if it didn't matter quite so much anymore. The students were livelier, but I was worried by the inverse relationship in camp between experience and enthusiasm. You often hear about the great discoveries made by mavericks who bucked the conventional wisdom. What you don't hear about are the many occasions when the conventional wisdom was right all along.

The next morning Yellen and Cornelissen took a crew down to Katanda 9, the site where the harpoons had been found. Helgren went with them, and the graduate students dispersed to other sites, leaving Brooks behind to prepare the payroll, arrange for the workers' safe escort, and plan a supply trip into town. Everything on her shopping list would have to be bargained for—charcoal, chickens, tomatoes, leeks, manioc flour, one hundred eggs, "each wrapped in a palm leaf and separately negotiated." I began to develop a deep respect for her organizational talents. Later in my visit, one of the workers developed a bulbous infection under his arm, resulting from a burn. Brooks cleaned it up and put him on an antibiotic. When a grad student lost a filling, she poked around in the clutter of the supply tent and returned with a dental repair kit. At the same time, she was mother to two of the camp members, one with a nine-year-old boy's restlessness and energy. There were graduate students who needed guidance, an official visit from the American ambassador to prepare for, and last and deservedly least, fretting writers to show around—all in addition, of course, to the work she had come to Katanda to accomplish.

Brooks completed the payroll and got her son started making brownies on the camp wood stove, and she and I headed over to the original Katanda excavation (Katanda 2), located a hundred yards across a plunging ravine running perpendicular to the river. Alison took me there the long route around, away from the river and then back again. On the way I asked her how she and John met up. At the time she was working summers with the notoriously demanding Harvard archaeologist Hallam Movius, excavating the Aurignacian site of Abri Pataud in the French village of Les Eyzies, a few doors down from the original Cro-Magnon cave. She knew Yellen from classes at Harvard. Then John spent a summer in Les Eyzies too. It must have been the romance of the place. When she told Movius she was getting married, he was furious. "You might as well go teach kindergarten now," he told her, turning on his heel. A short time later he suffered a stroke.

"I still feel partly responsible," she said to me.

Of the three principal investigators at the site, Alison seemed the least perturbed by the perovskite problem, pinning her hopes on other, more dependable ways of dating the harpoons. At the end of the 1988 field season, the team had taken soil samples back to the United States, to test by a new thermoluminescent method, which can reveal when buried soil was last part of the open landscape. The initial results were promising. "Either the harpoons are very, very old, or they were sitting on a uranium mine," she said.

She conceded one other possibility. The samples might have been put through the X-ray machine at the airport by the excavator charged with bringing them home. The extra dose of radiation from the machine could falsely balloon up their apparent age. New samples would have to be gathered. Meanwhile, electron spin resonance studies were also under way, using hippo tooth enamel found in the harpoon level. These experiments were still in their preliminary stages, but initial indications were giving her confidence.

In a few minutes we reached Katanda 2. At ten o'clock in the morning the site was already a windless oven. A half dozen Zairian workers and two students were squatting in an opened square of earth at the top of a step trench running like a staircase down the slope. One of the workers had uncovered a sooty cylindrical object, possibly charcoal. At Katanda, datable morsels are more precious than fossils, and if this was ancient carbon, it could be targeted for radiocarbon dating. Everyone was silent as Brooks unwrapped the sample and tested it with her finger. It smeared. "Looks like it might be something," she said. "Congratulations." She was squatting with the students to discuss details of the next step in the excavation, when a shout carried over from the distant opposite bluff. It was her son, Alexander.

"Hey, Mom! How do I know when the brownies are done?"

Brooks stood, walked to the edge of the ancient ravine, and cupped her hands over her mouth. "If they're crusty on top," she called back, "take them out!"

We headed back toward camp. Out of earshot of her students, she confided that she was more skeptical about the black cylinder of material than she had let on; it could turn out to be just a decayed bit of tree root. Dating the top level of Katanda 2 would be only a glancing blow on their target anyway. To make their case for the harpoons, they needed something from deeper down in time. "Two things could save our necks," she said. "One would be a geological horizon directly linking deposits here with those at Ishango. The other would be a nice fat piece of ostrich-egg shell coming out of John's site. Hell, I'd settle for a little mollusc."

Her first, geological hope was simple and pure. If you know how old one place is and you want to prove that another nearby is even older, find a common geological thread that runs between the two. Chances are good that everything *below* the linking thread at the second site will be older than everything *above* the thread at the first one. A stratigraphic marker like this wouldn't pin an age on the harpoons in calendar years, but it would prove once and for all which of the two sites was the older.

Her longing for ostrich-egg shell requires a little explanation. For some time, Brooks has been championing the revival of a dating method called amino-acid racemization. Until recently, saying these words to an anthropologist was like whispering "Edsel!" into the ear of an automotive engineer. In theory, the technique is simple. The process of fossilization turns bone and shell into inorganic matter, but there are still bits of protein left long after fossilization has occurred. Like all proteins, these are made up of amino acids, which exist in one of two chemical configurations that are mirror images of each other, like two hands. In nature, the amino acids in living organisms are all "left-handed." But when the organism dies, its proteins break down into their constituent amino acids, and some of them flip around and assume the "right-handed" configuration instead. This flipping phenomenon, called racemization, is known to occur at a slow, steady rate through time. Thus counting the proportion of left-and right-handed amino acids in a fossil should tell you when that bone was part of a living animal.

When amino-acid dating made its debut in the 1970s, it was hailed by anthropologists as a godsend, giving the ability to date fossils far beyond the range of radiocarbon. Museum curators in South Africa eagerly handed over many of their most prized treasures to a young German investigator named Reiner Protsch. His aim was to test each with AAR, extracting a hefty piece of bone in the process, and return with a chronology for modern origins in Africa. Meanwhile, other anthropologists were applying AAR to Paleo-Indian skulls from California. According to their results, the skulls were an astonishing, not to say perplexing, 70,000 years old—more than 50,000 years older than any other trace of humans in the Western Hemisphere.

Unfortunately, Protsch and the others did not understand how sensitive the racemization process is to moisture. Over time, groundwater seeps into a fossilized bone, causing the proteins to break down more quickly than normal, washing out the resulting amino acids and generally playing havoc with the notion of a steadily ticking clock. The 70,000-year date for the North American specimens eventually proved to be off by one full goose egg (they were really only 7,000 years old). In South Africa, all that was left from Protsch's work was a trail of irreplaceable human skulls that now had quarter-sized holes drilled through their crania.

After such an embarrassing opening act, it is no wonder that amino-acid dating was shunned. Then one day in 1982, Alison Brooks happened to ask geochemist Ed Hare of the Carnegie Institution, one of the originators of the amino-acid racemization technique who was still intent on improving its capability, if she could bring her graduate class in laboratory methods for a look around his laboratory. Hare suggested she bring a sample of something to test. "I had some ostrich-egg shells in my lab, so we brought them along," she told me. "We ran some tests, and much to everyone's amazement, they showed that the ostrich shell was virtually impervious to water. Nothing was getting in, and nothing out."

Since that discovery, Brooks and Hare, along with Gif Miller working independently at the University of Colorado, have led a small, stubborn movement to redeem racemization dating from its early ignominy and establish it as a viable instrument for dating the origin of modern human beings. Their hopes rest on eggshells. Completely "waterproof," ostrich-egg shells and the shells of other big rattite birds are not only immune to moisture's effects but also ubiquitous in the geological record of time, offering to rescue from chronological limbo sites dispersed throughout the Middle East, Asia, Australia, and of course, Africa. (Mollusc shells are even more common, but not quite so reliably watertight.) Miller has dated the Border Cave modern humans at 80,000 years. Brooks pinned a shell-age of 105,000 on the hominids of Klasies River Mouth, a date that agrees nicely with ones garnered by other methods. The AAR method is still far from perfect. Using ostrich-egg shell instead of bone eliminates moisture as a variable, but the rate of racemization is also affected by the temperature of a site through its long history, which can be extremely difficult to estimate. "Temperature is a real problem," said Miller, when I saw him at a meeting in London. "If you screw up there, you are really screwed."

Miller believes that amino-acid dating works best if you piggyback it onto a more trustworthy dating method, like the tried-and true radiocarbon method. Radiocarbon is very accurate, but it extends back only 40,000 years under the best conditions. If you can find some shell within that range, and date it with *both* methods, you can extrapolate downward and also date ostrich shell farther down in the same region, where radiocarbon cannot reach.

And that explains why Alison Brooks was longing for an ostrich-egg shell at Katanda. Her date of 25,000 for the Ishango "aquatic civilization" rested mainly on the shell fragment she had found there and dated with both amino acid and radiocarbon. Using that shell as a reference point, she might be able to come up with a solid age for a much older piece of ostrich shell here at Katanda too. But she had to find one first.

*   *   *

The next afternoon I visited Yellen where he spent his days, down on his knees in the cryptic mélange of Katanda 9. If Katanda 2 was an oven, Katanda 9 was an open broiler, cut into the east slope above the river where it takes the full brunt of the post-meridian sun. Yellen was thoroughly roasted, but it didn't slow him down; he hopped agilely up onto the rim of the excavation to greet me, eager to show off his site. In contrast to established, intensive excavations like Kebara and Klasies River Mouth, this was a tiny enterprise, a mere nick in the bluff, not more than a dozen feet on a side. Most of the surface was covered with the compacted clutter of fish bones, animal bones, rounded river cobbles, and crude tools that Yellen began excavating in 1988. The accumulation was so dense that a single foot-square plot was yielding hundreds of bits of bone and stone. Just that morning, the work had turned up giant catfish skull plates an inch thick, an antelope's foot bone, and assorted bits of its pelvis and scapula.

"What you'd like to find is evidence that people were camping on this spot and left all this stuff behind, just as you see it," Yellen said, waving a bony arm over the debris. In this ideal scenario, the whole site could be read as a cultural text, a moment of human behavior frozen in time. Unfortunately, ancient life rarely shines through an archaeological accumulation in such unfiltered purity. After humans move on, their Paleolithic trash can be jostled about in any number of ways—kicked by passing bovids, carted away by hyenas, rolled downslope or downstream, nudged around by burrowing insects, and so on. The *worst* scenario is that humans didn't have anything to do with accumulating the material at all.

"Lots of stuff could just get lumped together," Yellen explained, "the products of people, hyenas, whatever, all ending up in this place as a result of a flash flood or something like that." Back in 1988, when the site was first opened, a visiting geologist thought they'd simply found a little potholelike depression where debris from all around the area had naturally concentrated as rainwater washed it down. This bleak hypothesis now seems unlikely. Most of the artifacts are positioned on the upward slope rather than in low pockets. Those that *have* come to rest downslope or in depressions tend to be the larger bits of bone and stone. This is exactly opposite to the picture one would expect if water action had sorted things out, because bigger pieces are harder to move.

"I'm certain now it's not been washed down a hillside," Yellen said. On the other hand, Cornelissen had convinced him that the accumulation had been at least gently disturbed by running water: The artifacts, especially the long bone pieces, tended to be oriented in two, perpendicular directions, a pattern known to be caused by stream action. Taking all the data gleaned from the site so far, Yellen favors a compromise. The clutter represents an ancient occupation site of humans, most likely on a raised riverbank, but the artifacts were probably moved around a little by water before getting buried, blurring their exact relationships. It is as if you painted a portrait and then dragged a piece of paper across the surface while the paint was still wet.

The question is who, and what, that blurred portrait reveals. In the best of all possible archaeological worlds, the stone tools in a site have the clear defining characteristics of a particular era, style, or industry, and the animal bones bear cut marks and other signs of being worried by human hands. The debris at Katanda 9 has neither. With a few exceptions, Yellen told me, the stone tools were made of "extremely crummy" local quartz, producing rough flakes hard to pin down even as Middle Stone Age. "Most of the cores look like people were just picking up a rock and thinking, 'How do I get a flake off of this'?" he said. The animal bones are also disappointing. Their surfaces have been scratched by movement against the river sands, obliterating any cut marks. The only artifact that has any distinction at all is the occasional serenely elegant bone harpoon point, perched among the paleo-rubbish like a Rolls parked in an auto wrecker's lot. It was this jarring juxtaposition of technological virtuosity and rank obscurity that stretched belief beyond what most investigators would allow.

"How can you be so sure the harpoons and this other stuff belong together?" I asked. "Suppose fifty thousand years ago one group of hominids came along and used this place to bash up catfish bones. Thousands of years later, their garbage was exposed by erosion or something, and another, much more modern group came along and left behind some nice harpoons. Wouldn't you end up with the same result?"

"We can rule that out with almost absolute certainty," Yellen replied. "Everything in the site was embedded in and covered by a distinct layer of fine gravel, including the harpoons. Think of it as if a truck came along and dumped a load of sand for a playground. Everything buried by that sand had to have been there before the truck arrived."

"But why are the bone harpoons so much more sophisticated than the stone tools?"

"Everybody seems to think that's really weird," he said. "But if you view the people who made them as human beings, instead of something not quite yet arrived, it's not really unusual. In western Australia today there are aborigines who make very crude-looking stone tools. But their wooden implements are very elaborate, with fancy painting on bark, and beautiful spearthrowers and shafts. They also have extremely complex social systems, cosmology, and narrative traditions. If you were to dig up one of their sites a thousand years from now, however, all you would see would be the clunky stone tools. Does this mean those aborigines were technologically inferior? Not at all. They were simply relying on perishable materials."

"But isn't it still weird that they were making these bone tools so much earlier than anywhere else?"

"Maybe the weirdness people see is just the distorting effect of working from the present backward. We look at ancient humans and say, 'We know they were more stupid than we are, but just *how* stupid? How smart?' Maybe that's not the right approach. What these harpoons tell me is that people in Africa at this time could exploit a particular resource. They must have had the capacity to conceptualize, to socialize. I can't tell you just how much like us they were. Let's just say, the soil was fertile."

"But why here? Why don't you see these harpoons anywhere else?"

"I don't know," John said. "It does seem to be a purely local tradition. But isn't it interesting that the next time you find harpoons in the archaeological record, they are only six kilometers away, at Ishango? However isolated the tradition was, it seems to have lasted for a very long time."

After a few days at Katanda, I had heard and seen much to suggest that the harpoons *could* be as old as Brooks and Yellen were maintaining, but nothing that convinced me that they *are*. Given the heat and all the emerging ambiguities, I was amazed that the two did not wilt into puddles of doubt and despair. Brooks moved forward with the inner calm of the fanatic. She *knew* she was right about the dates, Katanda and Ishango too, and with patience the decisive evidence would emerge and all the problems would reveal themselves to be silly passing trifles. Yellen's approach was the opposite. He pre-critiqued each new development from all angles, cushioning it with a blanket of worry so thick that no sharp, uncomfortable fact could possibly poke through. The next day, he found a new harpoon at Katanda 9.

"This is *it*," Brooks said. "I'm glad this happened. I'm happy for you."

But the new point left Yellen unsatisfied. Unlike the other Katanda harpoons, this one looked suspiciously like the ones from Ishango. Were there clearly two different styles, separated by thousands of years? Or could all the harpoons belong to the same time period after all? Without the perovskite ash to guide them, what proof did they have that Katanda really lies below Ishango in time? The uncertainties kept reappearing. He would feel better when he had some Ishangan harpoons to compare the new point to directly. They would all be more certain of the harpoon dates once their fish paleontologist arrived to identify the species of catfish in the Katanda deposits. Everyone would rest a little easier once the TL dates were finished, or the ESR, or when a piece of ostrich-egg shell turned up.

The one certainty for me was the day the bush plane would return to take me out. It arrived on schedule, and I flew to Goma, at that time a dusty little town in the Rift, impossible to imagine flooded with millions of refugees, and traveled overland to Kigali in Rwanda, where I got a commercial flight to Nairobi, and home. For several weeks I was too preoccupied with other matters to think much about Brooks and Yellen and their Katandan wonders, but when I did get in touch again, they were full of good news. The day after I left, yet another harpoon had appeared— not at Katanda 9, but in the Middle Stone Age deposits at a new site a few hundred yards farther away from the river. Finding harpoons and MSA stone tools mixed together *once* could be a geological fluke; finding them *twice* makes it very difficult to argue that the two kinds of tools belong to separate time periods. (Later, another bone point turned up in the material collected from Katanda 2.) Then Kathy Stewart, the anticipated fish specialist, arrived at the site. She took a long look at the giant catfish fossils at Katanda and declared immediately that they were older types than the ones she had previously studied at Ishango.

The best news of all came at the end of the field season. Brooks and David Helgren had been making boat trips up the Semliki from Katanda toward Ishango, scouting the cliff exposures for geological clues, and one day they came back to camp and matter-of-factly said, "We solved it." At a place called Kabali where the river takes a big bend to the west, they had hiked into a little ravine and found a layer of Ishangan gravel, identifiable by the species of mollusc it contained. Below that deposit was a layer of yellow sand, and below the sand a layer of the soil that also covered the harpoons at Katanda. They had found the common geological thread tying the two sites together, Katanda clearly residing deeper in time. David Helgren, nearly convinced by the Katwe ash that Katanda was the younger site, abruptly switched his allegiance back into Brooks's camp.

Given the importance of such a find to the modern human origins debate, which had not lost momentum since mitochondrial Eve first showed her face, I expected to read something in the academic press about the amazing harpoons of Katanda. But months went by with no word. Like other scientists, anthropologists do not formally acknowledge a discovery until it is described in a publication. You can drop hints about the fossil or tool you've found, you can lecture about it openly in meetings and even bring it along in a box to show. But in a very real sense the thing does not really exist until you commit yourself to it in print. For Brooks's critics, her silence spoke volumes.

"She's not publishing because she knows that she'll get stepped on," Richard Klein told me.

A year after leaving Zaire, I ran into Brooks and Yellen again at a conference in Bratislava. They cheerfully filled me in on all the latest news from Katanda, including some new amino-acid dates on molluscs from the site that were further bolstering their confidence that the harpoons are "twice as old as Ishango." But I was getting impatient.

"You say you're certain that Ishango is twenty-five thousand years old," I interrupted. "That means the harpoons are around fifty thousand, right? So why don't you come out and say it? Why don't you publish?"

"It's because no one believes us that we need the case to be airtight," Alison replied. "We need dates confirming dates confirming dates."

What they needed most were the magic letters TL and ESR. After their success in the Israeli caves, thermoluminescence and electron spin resonance have both been promoted—too quickly, some researchers were protesting—from experimental techniques to the *de rigueur* method for anybody tinkering around with a modern human origins site. Mention ostrich-egg shells, or some relative dating method, and people nod thoughtfully; mention a TL number for your site, and suddenly they are all ears. Brooks told me the TL dates for Katanda were due in October. October passed. Later I heard from Yellen that no less an authority than Henry Schwarcz, the much-sought-after chronologer, was now at work squeezing an ESR date out of the hippo teeth at Katanda 9. Weeks went by. I called Schwarcz. He had some "preliminary numbers," but wouldn't tell me what they were.

"This site is such a touchy issue, I can't say anything until the work is done," he explained. I called Brooks back. "Don't worry," she said, "we'll have the dates by the spring." Then it was by summer. By fall. But already years had gone by since the harpoons first came to light. Gradually, a little guiltily, I felt myself slipping into the ranks of the skeptics.

In the meantime, I moved my family to France and plugged in my computer and fax machine, ready to reveal the oldest mystery in human evolution to the general reader. The fax promptly blew up, and the computer keyboard grew stiff with uncertainty. I had now talked to one hundred and fifty scientists—archaeologists, anatomists, geneticists, geologists, dating experts—and sometimes it seemed I had come away with one hundred and fifty different points of view. Early modern humans appear first in Africa. No, they don't. Or it depends on what you call early. Or how you define modern. Or what you really mean by human. I indexed my notes, and indexed the indices. A city of Post-it notes grew on my office wall, each with a revelation scribbled on it. Arrows of blue chalk sprang up to link brainstorm to brainstorm. But the arrows sprouted question marks. The Post-it notes lost their sticking power and fell to the floor.

Awash in conflicting perspectives, I took an almost religious reassurance in the one gross certitude I could cling to: The past did, in fact, *happen*, and in only one way. *Homo sapiens* is not the product of an amalgam of viewpoints, but a living proof that only one view can be right. Which one? I drew charts, weighed the evidence, played out scenarios. In every case, the story seemed to hinge on the age of things. So long as the modern humans of Qafzeh and Skhul were believed to be 40,000 years old, their fossils spoke for a seamless evolution from Neandertals to us. Redated to 100,000 years, they were instantly transformed into the best evidence around for the opposite point of view—that Neandertals were off the modern human line entirely. But how sure were those dates? The same was true for the genetic evidence. If our genes trace back to a common ancestor a million years ago, then the human family has been evolving gradually in all regions of the Old World since the *Homo erectus* migration. But peg an age of 200,000 on Eve instead, and only one region on earth could be our common homeland.

So much hinges on simply knowing how to tell time. Early in 1992, Chris Stringer and Paul Mellars organized a meeting at the Royal Society in London, devoted specifically to the new ways of dating the origins of modern humans. Even before the meeting began, the lobbies and corridors were buzzing with scandalous news about Eve. Six months before, Linda Vigilant, Mark Stoneking, and their colleagues had published their more detailed comparison of living mitochondrial DNA, offering the most powerful arguments yet for a recent African origin for all living humans. But the Berkeley triumph was short-lived. Just a couple of weeks before the dating meeting, a geneticist named Alan Templeton at Washington University in St. Louis, and a team led by David Maddison of Harvard, independently demonstrated that the mitochondrial tree rooted in Africa suffered from a fatal inner flaw.

The underlying method of the Berkeley geneticists was to sample mitochondrial DNA from women in various parts of the world and discover how different and how much the same each was compared to the others. Invoking the scientific principle of "parsimony," the team sought to reconstruct a genealogical tree that would account for the global pattern of relatedness in the simplest possible way. "The problem has always been that there are literally billions of parsimonious trees that could be drawn from the data," Stoneking told me in London. "We knew that we couldn't sample this whole universe to find the best one. Instead, we used a computer program that narrows the search, and worked from it. We now know that was a mistake."

The computer software program the Berkeley team used is called PAUP, short for Phylogenetic Analysis Using Parsimony. The researcher feeds the raw data gleaned from the mitochondrial samples into the computer, and at the push of a button it begins to spit out random sets of possible trees relating all the mitochondrial types. In both the original 1987 study and the more detailed 1991 work led by Linda Vigilant, the Berkeley group selected the most parsimonious trees generated by one run of the computer program. In both studies, the trees were rooted in Africa. But the team should have spent more time at the computer terminal. Templeton and Maddison demonstrated that if the mitochondrial types are entered in a different order, another run of the PAUP computer program produces a completely different sample of trees, including thousands that are more parsimonious than the original Out of Africa version. A third run generates yet another batch, and so on. Among these competing "simplest" trees are many with African roots, but many others place the last common ancestor of all humans in Europe or Asia instead. In short, as Templeton made clear in *Science* magazine, it is statistically impossible to pinpoint the geographic location of the last common ancestor of humanity by comparing living DNA and constructing genealogical trees from them. For those who had been fighting Eve all along, this new development was manna from heaven.

"Templeton's pulled the plug on the whole thing," Milford Wolpoff was quoted in *The New York Times*. Unlike his own attacks on the Eve hypothesis, this one did not question the Berkeley team's interpretations of their results. Instead, it pointed out a clear mistake in generating those results in the first place. Even if modern-looking people evolved first in Africa and spread out to other regions, the mitochondrial evidence could no longer rule out the possibility that they interbred with the residents they encountered. Such intermingling between Africans and residents could account for the regional characteristics that Wolpoff and his colleagues see persisting in Eurasia. Their claims that "Eve is dead" are, however, overstated.

"The bottom line is that an African origin still provides the best explanation of the data," Mark Stoneking told his audience in London. He conceded that constructing parsimony trees could no longer root Eve in Africa beyond a doubt. In a display of evenhandedness rare in science, he even added his name to a paper in *Science* pointing out the error in his own work. But other methods of analyzing the same data were still coming up with a tree rooted in Africa, though with less statistical confidence. Meanwhile, every study of nuclear genes, as well as comparisons of the "classical" markers like blood types, were continuing to demonstrate that Africans are genetically much more different one from another than people elsewhere. If our genes are diverging away from one another at a steady rate, this inflated diversity suggests that humans have lived in Africa longer than anywhere else.* And no statistical glitch discovered in a computer program affects the fossil evidence, which still shows the clearest transition to modern anatomy occurring in Africa. The key question, once again, is when.

A possible answer emerged in London—not a single dramatic revela-

---

*In 1994, a new study of nuclear genes underscored the likelihood of a recent African origin. Using a method that traces the pattern of inheritance of two neighboring segments of DNA in populations around the world, Sarah Tishkoff and Kenneth Kidd of Yale showed that there was tremendous variation in the pattern among sub-Saharan Africans. Northeast Africans—Ethiopian Jews, Somalians, and Egyptians—showed less diversity, while people from other parts of the world almost all carry one particular pattern. The most likely explanation is that modern humans arose in Africa; then one population bearing that particular pattern migrated out so recently that there has not been time enough for random mutations to reintroduce variability into this stretch of DNA. "It's very consistent with a recent spread of modern humans out of Africa," says Kidd. "It's hard to imagine anything else that could explain the data."

tion but rather a gathering congruence of several lines of research. In his talk, Stoneking took the offensive and unveiled a shocking new age for Eve. For several years, he had been examining mitochondrial evolution within the populations of Papua New Guinea. According to the latest thermoluminescence dates, New Guinea and Australia were first colonized around 60,000 years ago, when they were still joined to the greater continent of Sahul. Given the sheer physical difficulty of reaching New Guinea before modern times, it is unlikely that there was much gene flow back and forth to the mainland after that original colonization. Most native New Guineans can therefore trace their common ancestry back to a single point in time. Equally important, they are about as far off the human mainstream as people can be. The rugged, jungled mountains isolate different tribes—almost a thousand separate languages, or one-fifth the total on earth, are spoken there—and the sea isolates them from everyone else. This isolation, combined with the likelihood of a single, known point of common origin, makes New Guinea the best possible natural laboratory for studying the rate of human molecular evolution.

Stoneking explained how he and his colleagues had sampled the mitochondrial DNA of fifty Papuan New Guineans from various highland and coastal tribes. He compared their mtDNA to that of individuals from Indonesia, the nearest part of Asia and hence the likely jumping-off point for the New Guinean gene pool. Assuming that all the genetic diversity they witnessed among the samples had accumulated in 60,000 years, and that New Guinean mitochondria evolve at the same pace as other human populations, he figured that it would take roughly 133,000 years to account for the amount of mtDNA diversity seen among humans from all over the world. Thus the mitochondrial mother of all humans, whether she is African or not, lived around 133,000 years ago. "These results provide the strongest evidence yet for a relatively recent origin of the human mtDNA ancestor," Stoneking concluded.

The new date nudged Eve astonishingly close to the present. But the number gathered even more intriguing overtones as the meeting continued. French paleoanthropologist Jean-Jacques Hublin presented new information on the Jebel Irhoud skull from Morocco, including a new date for the fossil. Once thought to be only 40,000 years old, it is now pegged closer to 125,000 by new ESR studies of the site. Henry Schwarcz mentioned in his talk that uranium-series dating tentatively puts the nearly modern Omo skull from Ethiopia around 130,000. Using the same method, the Ngaloba skull (Laetoli 18) from Tanzania appears to be about the same age. So do some quasi-modern teeth from a nearby rock shelter called Mumba Cave. The Florisbad fossil in South Africa, another nearly modern human skull, could date from the same period as well.

In short, the age of the alleged common mitochondrial mother and the age of the transitional African fossils—the ones no longer totally archaic, but not yet fully modern—seem to coincide. Eve could thus have been a member of the first anatomically modern population, an original modern woman. The agreement in time between the genes and bones gathers potent implications when one considers what was happening to African habitats at the same time. According to the global oxygen isotope record, the glacial cycle just before the last one—called Stage 6—reached its coldest extreme between 135,000 and 130,000 years ago. Africa was far from immune to the climatic transformations wrought by the glaciers descending from the opposite pole. While Middle Stone Age Africans would not have fallen under the near-Arctic conditions endured by their Neandertal contemporaries to the north, their environment would have become several degrees cooler, and a great deal drier. Where there had been dense forest, there would now be open woodland, and where the woodland once grew, nothing but parched grass and a few scattered trees. Arid grassy plains, including much of what is now the Sahara, would have deteriorated into uninhabitable desert.

Such a transformation in the environment would have had a serious impact on all the inhabitants of Africa. As resources diminished, populations would shrink. Landscapes of continuous habitation would be broken up into separate regions, and at the glacial apogee, those regions may have fragmented into a mosaic of isolated local breeding communities of a few thousand individuals, or even a few hundred. Upon such raw material natural selection and random genetic drift work their most powerful transformations. "If there is any element of the theory of speciation that is likely to be generally true, it is that geographic isolation and the severe restriction of genetic exchange between populations is the first, necessary step in speciation," the distinguished Harvard evolutionist Richard Lewontin writes.

So long as one is talking about *other* species—a kind of marsh grass, or a jackal, or even an intelligent primate like a chimpanzee—most evolutionary biologists would agree. But if you attach the adjective "human," the anthropologists start to squirm. Speciation is a drastic, last-ditch response to environmental pressure. Some evolutionists would argue that, like extinction itself, speciation occurs only when a population cannot adapt to a change in its habitat through the gradual adjustments of anatomy and behavior that continually take place within the population. Hominids, buffered against the vagaries in the environment by their clever cultural props, are the natural world's masters of adjustment. If marsh grass cannot obtain water, it quickly dies; hominids use their wits to find more, or pack up and look for a new place to live. If a jackal gets cold at night, it huddles and hopes; hominids build a fire. When the chimp finds its forest range shrinking, it makes anxious excursions into the surrounding savanna; the hominid marches out and claims the savanna as its new domain.

Leaning on its big brain, the *Homo* lineage had endured a dozen ice ages before Stage 6, with only one morphological shift grand enough to be deemed a clear species break. This was the evolution of *Homo habilis*-like hominids into *Homo erectus* some million and a half years ago, an event marked by, among many other traits, enormous increases in brain and body size. The emergence of modern anatomy was not so dramatic a change from the status quo. From the scattered fossil evidence in hand, one could argue that the small faces, the distinct chins, the vertical foreheads, and other modern anatomical grace notes represent a new kind of human altogether, but it would be just as valid to maintain that these developments were part of a continuing trend: a gradual reduction of robustness as new technological innovations took over the functions previously served by thick bone and heavy muscle. It all depends on how you like your hominids: wrapped up in species with nice clean edges around them, or blending one into another through time.

Either way you look at it, the most humbling aspect of our physical shape's evolution is how little it seems to have mattered. Immediately after the glacial maximum of Stage 6 came a brief, balmy period, the last interglacial period before the one we now enjoy. If glacial maximums are times of environmental fragmentation, interglacials knit habitats back together again. Genes flow more freely across former boundaries, populations move, mingle, and expand. By 100,000 years ago, humans who looked modern were spread out from the Middle East to the coast of South Africa. But as we have seen, they were no more "modern" in their culture or behavior than either the Neandertals in Europe or their own glacial ancestors. Qafzeh, Skhul, Klasies River Mouth, Border Cave—Mousterians all. If there is a single defining line in modern origins, it had not yet been crossed.

Still, such a decisive moment may have come, spurred by climate, long before the traditional explosion of humanness of the European Upper Paleolithic. Around 75,000 years ago, the planet entered the last Ice Age, identified in the oxygen isotope record as Stage 4. Shortly afterward, the earth experienced the single most devastating climatic event in the last 450 million years: A massive volcano called Toba erupted on the island of Sumatra. According to Michael Rampino of New York University, in two weeks Toba spat a billion tons of ash as high as twenty miles into the atmosphere. The largest eruption in modern times, Tambora in 1815, deposited 50 cubic kilometers of magma; Toba left behind 3,000 cubic kilometers on Sumatra and the surrounding ocean floor. Within a few months, most of the ash had fallen back down to earth. In the meantime, however, sulfur gases exhaled by the volcano were converted by water vapor and sunlight into long-lasting droplets of sulfate. According to Rampino and his colleague Stephen Self of the Goddard Institute for Space Studies in New York, the dispersal of these aerosol particulates through the stratosphere blocked sunlight around the world, plunging the globe into a "volcanic winter."

"The global temperature was already on its way down, but Toba shoved it through the floor," says archaeologist Stanley Ambrose of the University of Illinois. "There must have been a really hellish ten years."

If Toba's effects had lasted only a decade, it would not have been a total catastrophe. But Rampino and Self point out that the volcano could have triggered a "positive feedback mechanism" lasting long after the atmosphere had cleared. During the volcanic winter, snow cover and sea ice would have expanded their reach over the surface of the globe, reflecting sunlight back into space and further cooling the planet. The consequent lowering of temperature would cause more snow and ice to accumulate, depressing temperatures even more, and so on in a cascading downward spiral. Within five thousand years, sea surface temperatures in the North Atlantic had plunged as much as ten degrees centigrade. The glaciers had sucked enough water out of the ocean to drop sea levels a hundred and fifty feet. "Increased sea ice and snow cover following the eruption may have provided the extra 'kick' that caused the climate system to switch from warm to cold states," according to Rampino and Self.

Against this dramatic change in environment, human populations everywhere were faced with the same choices as other animals: adapt, migrate, or die. Stanley Ambrose believes that human populations in Africa were devastated by Toba's effects, squeezing our lineage through a "population bottleneck" that left perhaps as few as ten thousand individuals alive at one time. But obviously, some populations made it through. There is no indication that anatomy changed drastically at that point, either in Africa or anywhere else. How did they adapt? Did they survive by adopting new patterns of subsistence, new technologies? Unfortunately, the sparse archaeological record in Africa offers no hint.

Or maybe just one. "So what do you do when it gets cold and the food gets scarce?" Alison Brooks said to me one day in Washington. "Simple. You go fishing."

By December 1992, I had long since given up hope of using Katanda as a showcase for an early, African blooming of modern culture. I called Brooks on another matter, and out of the blue she mentioned that the dates were in and "everybody is happy." Those barbed harpoons were not, as she and Yellen had hoped, 50,000 years old. They were older. "Twice as old as Ishango" turned out to be an *under*estimate. Thermoluminescence dating of soil samples, completed by Bill Hornyak and Alan Franklin of the University of Maryland, yielded an age of 82,000 years before the present, give or take 8,000 years. Henry Schwarcz's electron spin measurements bracket the TL date, giving a range from 65,000 years to 105,000.

"So what do you do now?" I asked her.

"We publish, of course," she said casually, as if there never had been any doubt in the first place. She was preparing one paper on the geology of Katanda for *Science*, while Yellen was submitting another on the archaeology. "That way, we get the maximum number of pages."

Still wary, I called the daters themselves. Schwarcz was cautious. He had completed the ESR on only one hippo tooth, with two others yet to be tested. There could be problems in the way the uranium content of the soil at Katanda was determined, which could have skewed the results. He regretted that the political situation in Zaire was worse than it was in 1990, so there was no chance to go back and collect new samples. But yes, if I *had* to know, the site did seem to be alarmingly old, way beyond the 40,000-year range of radiocarbon. "There's no two ways about it," he said. "If these preliminary results are supported, this would seriously revolutionize the time frame for the emergence of modern behavior."

Bill Hornyak was further along in his TL work, preparing a detailed technical account of his methods. "This is going to raise a real ruckus," he said, his delight at the prospect palpable even over the telephone. "The date is many times older than this style of bone work has a right to be."

On the surface, words like these should make an Out of Africa advocate shout with glee. This could be the missing piece to the puzzle, the behavioral link in the chain of evidence pointing to Africa as our common homeland. Even with the tarnish on the reputation of the original mitochondrial Eve, the genetic testimony still favors Africa over other regions. African fossils show the smoothest, earliest transition to fully modern anatomy. And now, if the dates hold for Katanda, there is stunning proof that at least some African Middle Stone Age people were ahead of their contemporaries in Europe and the Middle East. The intelligence embodied in those harpoons could be the wedge that would explain how one kind of human being could spread out from its place of origin and replace the residents of the rest of the world.

Why, then, isn't everybody shouting? The simplest answer is that archaeologists, like other scientists, hate anomalies. The African Middle Stone Age makes a lot more sense *without* a few fancy harpoons cropping up in one isolated place. The MSA record everywhere else plays nothing but a monotonous theme of flake tools and more flake tools, dribbling out of every known site dated between 200,000 and 40,000 years ago. Of course, there is the odd artifact here and there, a pierced shell or a bit of bone with some curious scratches along one edge. And there is the brief incursion of the Howiesons Poort, with its precocious blade tools and hints of social alliance to explain. But the Howiesons Poort dies out, and couldn't the rest of these scattered baubles be the exceptions that prove the rule? Without Katanda's collection of bone points, the great dividing line between us and the rest is kept comfortably intact. Nothing changes until *everything* changes, in the great cultural explosion called the Upper Paleolithic in Europe, and the Late Stone Age in Africa. Brooks's harpoons poke a hole through that defining barrier, leaking confusion all over the pattern.

"It is such a *peculiar* thing," Stanford's Richard Klein says, his skepticism as tangible over the phone as Hornyak's excitement. "I've looked at thousands of artifacts from dozens of other sites in that age range, and I've never seen anything that comes remotely close to those harpoons." The most persuasive defender of the Out of Africa hypothesis among archaeologists, Klein believes the African origin model works a lot better *without* Brooks's precocious bone workers. "If it turns out that people had what it takes to make bone harpoons eighty thousand years ago, then why did they stay confined to Africa for another forty thousand years?" he asks. "The only thing that could explain it is that, however modern they looked, they weren't behaving modern. There is certainly no evidence for it anywhere else."

To which Brooks would respond, that's because we haven't looked hard enough. Both scientists agree on one point: Ultimately, the significance of Katanda rests in the future. Either more sites will confirm an early jump on sophisticated bone technology in Africa, or the lack of such support will make Katanda stand out more and more as a red herring.

My bets are on the first option. I think that the Rift Valley Africans of the Middle Stone Age were capable of all sorts of impressive ingenuities if a situation developed that demanded them. The sudden descent of the last glacial age provided that spur. But human groups scattered elsewhere were equally capable; they simply responded in different ways. If Hilary Deacon is right, the Howiesons Poort people relied on distant social contacts. Brooks, meanwhile, has evidence that at the beginning of the last glacial period, people at a site called ≠Gi in Botswana were killing zebra, giant buffalo, and other prey supposedly too dangerous for Middle Stone Age hominids to handle. In this one site alone, she has uncovered some four hundred small stone points, more points than there are in all of the South African MSA cave sites combined. Could ≠Gi be some sort of aggregation spot where human groups came at a fixed time of year to hunt game and exchange information? If so, they were acting not like Middle Stone Age hominids are supposed to but like modern hunter-gatherers.

Such localized expressions of human ingenuity were not confined to Africa. British archaeologist Clive Gamble, attempting to show that Neandertals lacked the ability to hunt the mountain goat in high rugged terrain, came up against the Uzbekistan site of Teshik-Tash, on the far eastern edge of the Neandertal range. Here, fifteen hundred meters up, in a landscape of great limestone gorges, where Neandertals have no right to be, is the grave of a young Neandertal, surrounded by a litter of mountain-goat bones, horns, and Levallois tools. The Neandertals may not have exhibited this facility elsewhere, but clearly they were capable of it.

"What am I to do with *that*?" says Gamble. "I have to concede that Neandertals had the capacity for fully modern behavior."

A more telling display of Neandertal hunting prowess has emerged through the work of Mary Stiner, a Lew Binford student now at Loyola

University in Chicago. She analyzed the bones of red deer, aurochs, and other ungulates from four Middle Paleolithic sites in west-central Italy spanning the duration of the Middle Paleolithic, and discovered that Neandertals in the early part of that time period, roughly from 110,000 to 65,000 years ago, were hunting much like other large carnivores: ambushing whatever prey they happened to come across and killing only the easily captured weak and young. Much of their meat may have been scavenged. But as the climate turned colder with the glacial advance, *these* Neandertals abruptly shifted their predatory pattern and began killing prime adult prey instead. Stiner speculates that this fundamental shift was occasioned by a greater metabolic need for fat, triggered by the increasing cold. But no matter what the cause, they were capturing a more dangerous, difficult prey with the same old Middle Paleolithic tool kit, implying a greater reliance on something else to make the kill—probably increased social interaction and cooperation. In this case it was not the technology but the hunting strategy that was modernized.

Stiner's revelations are a new addition to Neandertal archaeology. As the work continues, I believe even more evidence for the fully modern capability of Middle Paleolithic humans will pop up all over the Old World. There is nevertheless a missing element that will stay forever missing, no matter how much digging is done. It doesn't have so much to do with the way we think as it does with what happens to a thought once it has been conceived. Among modern humans, thoughts *move*. If an idea or innovation has real adaptive value, it should, by definition, spread from one brain to another. Individuals who are capable of receiving the new idea, either by exposure to it directly or by imitation of its material product, should have a competitive advantage over those who fail to pick up on it. Similarly, human groups that contain such savvy individuals will be more likely to survive than those composed entirely of less endowed intelligences.

Now look again at those pockets of Middle Paleolithic modernity. Take Katanda. Assuming that the dates for the harpoons are right, somebody (or a few somebodies) deep in the African Middle Stone Age thought of a way to fashion bone into a point that would not only spear a fish but prevent a speared fish from wriggling away. Viewed in the context of 100,000 numbing years of crudely flaked stone tools, the *idea* of a harpoon—and more generally, the recognition that bone could be crafted in the first place—gives new meaning to the word *originality*. Its adaptive value is obvious. Those who could make harpoons would have benefited from a great deal of fish protein just when resources were growing scarce, while those who couldn't would have to go without. If John Yellen is right, the bone-crafting idea was such a dandy one that it persisted, with only minor modifications, for the next 40,000 years. *But in all that time, the idea never moved.* Apparently it remained a discrete, local tradition in spite of its inherent usefulness to any other lakeside people, and in spite of any evidence that those harpoonless neighbors were qualitatively less intelligent.

In the end, the mystery of the Katanda harpoons may not be how Middle Stone Age humans in one isolated region managed to make such glorious objects at the very beginning of the last Ice Age, some 75,000 years ago. The sharper enigma is why, once the tools were introduced onto the human landscape, they never went anywhere else.

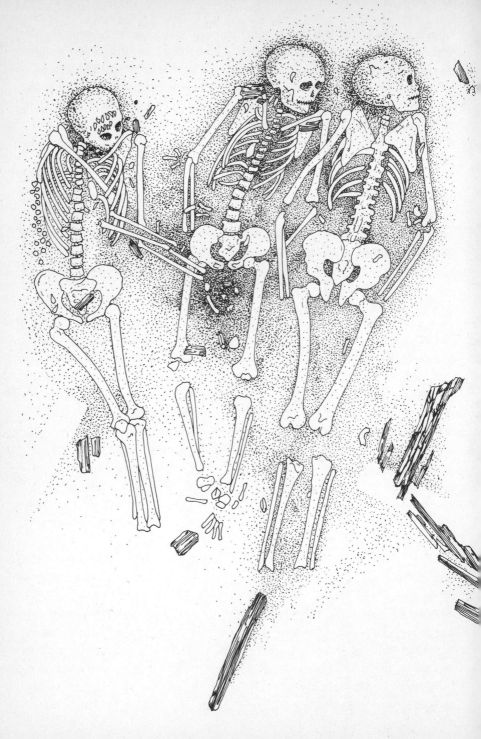

The triple burial at Dolní Věstonice   DAVID DILWORTH, REDRAWN FROM *THE JOURNAL OF HUMAN EVOLUTION*, VOL. 16, 1987, BY PERMISSION OF ACADEMIC PRESS LTD.

Chapter Ten

HOW DO YOU MAKE A MODERN?

Men begin as men, in distinction to other animals, precisely when they experience the world as a concept.

——MARSHALL SAHLINS

In the spring of 1986, near a village called Dolní Věstonice in the Czech province of Moravia, the bodies of three teenagers were discovered in a common grave. A specialist was immediately summoned from Brno, some twenty-five miles to the north, and under his care the remains were exhumed and faint remnants of the youths' identities revealed. Two of the skeletons were heavily built males. By its slender proportions, the third was judged to be female, aged seventeen to twenty. A marked left curvature of the spine, along with several other skeletal abnormalities, suggested that she had been painfully crippled. The two males had died healthy, in the prime of their lives. The remains of a thick wooden pole thrust through the hip of one of them hinted that his death might not have been entirely natural.

The bodies had been buried with curious attention. According to the expert—Bohuslav Klima, of the Czech Institute of Archaeology in Brno—both young men had been laid to rest with their heads encircled with necklaces of pierced canine teeth and ivory; the one with the pole thrust up to his coccyx may also have been wearing some kind of painted mask. All three skulls were covered in red ocher. The most peculiar feature of the grave, however, was the arrangement of the deceased. Whoever committed the bodies to the ground extended them side by side, the woman between her two companions. The man on her left lay on his stomach, facing away from her but with his left arm linked with hers. The other male lay on his back, his head turned toward her. Both of his arms were reaching out, so that his hands rested on her pubis. The ground surrounding this intimate connection was splashed with red ocher.

The skeletons lean into each other, like nestled question marks. In his written report, Klima speculated that the arrangement of the grave might reflect "a real life drama which precipitated the burials." His drama revolved around a young woman who had died in childbirth. The two male skeletons where those of her husband and a medicine man—the man wearing the mask. Held responsible for her death, the men had been compelled to follow her into the afterlife. One can imagine other "dramas," other characters, leading to the same denouement. A tragic love triangle? A young queen and two consorts? A sacrificial surfeit of youth? Or is the semblance of urgency in that morbid tableau mere accident— the reaching of hands the meaningless effect of rigor mortis, or the cynical toss of the body into a hole in the ground? Whatever the truth, it was covered over by a layer of burning branches, then soil, and finally by 27,640 years of future, give or take a few decades.

At that age, the triple grave at Dolní Věstonice belongs to a period known as the Gravettian, the second in the succession of cultural industries that define the Upper Paleolithic Age in Europe. If human evolution were an epic, the Upper Paleolithic would be the chapter where the hero comes of age. Suddenly, after millennia of progress so slow that it hardly seems to progress at all, human culture appears to take off in what the writer John Pfeiffer has called a "creative explosion." New types of stone tools proliferate, taking on regional style where before there was global monotony. In Africa as well as Europe, elegant implements carved from bone, antler, and ivory appear in abundance. Change replaces stasis. In France, new industries rush in and disappear again like Paris fashions— the Aurignacian, the Gravettian, the Solutrean, the Magdalenian—each with technological styles and innovations all but unknown in the period preceding it. From Spain to the Urals, site lists begin to read like proto-Sears catalogs: sewing needles, barbed projectile points, fishhooks, rope, meat-drying racks, stone lamps, temperature-controlled hearths, complex dwelling structures.

Dolní Věstonice can boast its share of such functional gear, but it is also a generous witness to a wholly original aspect of the Upper Paleolithic. For acres around, the fine, fertile loess soil is seeded with carved and molded images of animals and women, strange engravings, personal ornamentations, and decorated graves. A few thousand years before, such fanciful objects did not exist. Representational art especially seems to blossom, fully refined, around the beginning of the Upper Paleolithic, without a gallery of clumsy, childlike renderings from earlier periods to serve as precedents. At a German site called Vogelherd someone picked up a piece of ivory 32,000 years ago and carved an exquisite horse in miniature—mouth, flared nostrils, jowls, curved haunches, and swollen belly, all breathlessly realistic. Before Vogelherd, there were no representational horses. Before Vogelherd, all horses were horses.

It is hard to overestimate the significance of this transformation. By all appearances, the people of the Upper Paleolithic came into an innocent, unexamined world and galvanized it with symbol, art, metaphor, and story. They did not simply invent better means of surviving. They invented meaning itself. "The Upper Paleolithic," according to Stanford's Richard Klein, "signals the most fundamental change in human behavior that the archaeological record may ever reveal, barring only the primeval development of those human traits that made archaeology possible."

Fundamental changes call for big, bold, fundamental explanations. The old one, still viable even a decade ago, filled the bill perfectly. The enormous elaboration of human culture in the Upper Paleolithic happened so suddenly because a new, modern kind of human being entered the European continent: Cro-Magnon Man. New behaviors explained new anatomy, and vice versa. This cozy synchrony has collapsed under the weight of the bones of Klasies River Mouth, Border Cave, Qafzeh, and Skhul. The old answer is dead. With a 60,000-year time lag between the appearance of modern human skeletons in one part of the world and the florescence of modern behavior in another, the most astonishing advance in the history of human culture has been orphaned of a cause.

The most popular way to cope with this crisis, like any other, is simply to deny that it exists. Many archaeologists continue to account for the cultural events of the Upper Paleolithic by tying them to the emergence of a more modern, intellectually superior form of human being. If the appearance of anatomically modern skeletons in Africa and the Levant comes too early to coincide with this crucial transformation, so what? There must have been a *second* biological event later on, something that did not leave its mark in skeletal shape, but was nonetheless far more important.

"There is a developing consensus now that specimens like Qafzeh and Klasies River Mouth are better described as *near*-modern," says Klein, a

The exquisite Vogelherd horse, carved in ivory 32,000 years ago   DAVID DILWORTH

leading proponent of the "Biological Event" solution to the Upper Pale-olithic. "They tell us that Africa is the place where modern people origi-nate, although they are not themselves fully modern. But sometime around fifty thousand years ago, a breakthrough occurred in this African lineage, a neurological change that allowed them to develop all these new cultural behaviors. It could just as well have happened to the Neandertals instead. But it didn't."

In other words, the Upper Paleolithic did not happen before 40,000 years ago because there were no brains around developed enough to *make* it happen. Once the "neurological change" occurred in one human line-age, there was no stopping these well-endowed folk—our ancestors—from replacing archaic populations less favored by selection. This scenario has all the seductive simplicity of the old one and at the same time neatly folds in more recent theories like the Out of Africa hypothesis. But it has some nagging problems too. First of all, it is disturbingly short on evi-dence. There is no increase in brain size, no change in the markings on the inside of skulls that hint at brain structure, no trace whatsoever in the fossil record of this enormously important neurological development. Klein and other like-minded archaeologists cite the array of new, more sophisticated behaviors in the Upper Paleolithic as indirect support for their view. But since those impressive new behaviors are precisely the phenomena supposed to be explained by the change in neurological capacity, this argument sounds suspiciously circular.

Furthermore, the cultural evidence can be used to argue *against* the Biological Event solution too. If a neurological change in a discrete population of moderns was responsible for the Upper Paleolithic, one would expect the whole behavioral package—the art, the new tools, and so on—to begin in one geographic location and spread out, as the moderns themselves move into new territories. No such pattern can be seen, even in Europe, where most of the looking has gone on. The pattern is a complicated mosaic of mini-explosions that resemble one big explosion only when you stand back and take a long look at the whole. Different aspects of the Upper Paleolithic appear early in some areas and later in others. The Middle East is the last area where art appears, even though it is the first where people begin to rely on blade tools. In Eastern Europe, fully modern-looking fossils turn up thousands of years before any Upper Paleolithic artifacts. Furthermore, at the end of the Upper Paleolithic, around 12,000 years ago, all the restless creativity on the landscape seems to quiet down. In many ways, the Mesolithic Period that followed was relatively unoriginal, culturally speaking. If one judges neurological capacity by its expression in material form, it would appear that the new, biologically superior moderns had somehow de-evolved. This hardly seems likely.

Supporters of the Biological Event scenario also must be willing to spend a lot of time accounting for places in the archaeological record where the neurological capacity for Upper Paleolithic behavior seems to leak backward in time into Middle Paleolithic brains. Alison Brooks's alleged 82,000-year-old bone harpoons are just one extreme example of this paleo-leakage. Mary Stiner and Steven Kuhn's work in Italy is another. If you poke around enough, you can probably find somewhere a Middle Paleolithic version of most of the technological innovations of the Upper Paleolithic. The classic definition of Upper Paleolithic technology centers upon the "three Bs": blades, bone tools, and burins—chisel-like implements made by knocking a spall or two off a flake. But as we have seen, blade tools dominate the Howieson's Poort industry in South Africa 60,000 years before they become the rage in Europe. Around the same time, they are also common in another Middle Paleolithic culture called the Amudian, which flares up and then disappears in the Levant. Another distinguishing feature of the Amudian is an unusual abundance of burins. And even if one can explain away the Katanda harpoons, a few, less carefully crafted bone tools dot the digs of Neandertals and other archaic humans, from southern Africa to northern Europe.

Foreshadowing of the symbolic paraphernalia of the Upper Paleolithic can also be summoned up from before 40,000 years ago, though it may take a bit more effort. While there is no representational art before Vogelherd, scratched lines on an ox rib at the French site of Pech de l'Azé and zigzag markings from various other Neandertal sites *could* be called symbolic. A couple of pierced canine teeth show up here and there in France too, and etched ostrich-egg shell in southern Africa. Ancient stone hand axes have been found in Europe where the knapper seems to have carefully worked around a mollusc fossil, altered the shape of a tool to acknowledge a band of color in the raw stone, or otherwise demonstrated the flickering of an aesthetic sense. There are splashes of ocher here and there, and signs that Middle Stone Age Africans were mining hematite, a silvery, incandescent mineral with the sole useful feature of looking pretty.

Gathering up all these precursors and adding the evidence for Neandertal burials, archaeologists on the opposite extreme from Richard Klein explain the creative explosion by arguing that nothing particularly explosive really occurred. If every novelty flows from some slightly more primitive prototype earlier in time, then the whole business is better seen as just another step in the gradual progress of human culture, requiring neither a better brain, nor any other special biological trigger. "Neandertals may not have been quite capable of doing nuclear physics or calculus," writes Brian Hayden of Simon Fraser University. "However, it is a travesty of the available data to argue that they did not have the full complement of basic human faculties or act in recognizably modern fashions."

This gradualist point of view accounts for the Upper Paleolithic without resorting to an invisible genetic transformation, hauled in like a *deus ex machina* resolution to a hung plot. In the end, however, it too must be swallowed with a heavy dose of denial. No amount of precedent-pointing can explain the astonishing increase in the sheer volume of culture in the Upper Paleolithic and what that increase reflects in the lives of people. A handful of pierced teeth and beads have indeed been found in several late Neandertal sites. But what connection do these isolated curiosities have to the torrent of personal ornaments showered across France and Germany in the Aurignacian, right at the beginning of the Upper Paleolithic? How can they explain the ornamental munificence of Sunghir, a grave site near Moscow about the same age as Dolní Věstonice, where the bodies of three people were festooned with dozens of bracelets, necklaces, painted pendants, and *ten thousand* ivory beads? According to Randy White, an archaeologist at New York University, each of those beads took about an hour to make. That equals ten thousand hours of labor, all to decorate three corpses and lay them in the ground. The Neandertals buried their dead, but they did not devote much time and attention to the act. Sunghir does not represent a little more of the same. It is a different quality of culture altogether.

Another defining feature of Upper Paleolithic culture is its potent infectiousness. Innovations no longer flare up in little pockets and disappear. They metamorphose and diversify and inspire innovations. According to White, the ivory beads made in one site in France 33,000 years ago are exactly the same, in raw material, workmanship, and design, as the ivory beads in another French site two hundred kilometers away. Yet the ones from Germany are utterly different, bespeaking another tradition, another variation on the theme of bead. The Aurignacian industry itself is characterized by an abundance of large, unbeautiful blades, "beaked" burins, and carved-bone projectile points whose bases had been split to accept a shaft. The earliest known Aurignacian sites are in the Balkans, around 43,000 years old. Three thousand years later at the most, the Aurignacian appears across the continent in Spain. Within a few thousand years it covers most of the rest of Europe, picking up regional styles and acquiring new complexions as it goes. This is not simply a little more culture than there was before. For some reason, culture has become an epidemic.

"After two or three hundred thousand years of nothing new," says Berkeley's Tim White, "suddenly, in a tiny segment of time, after this huge gulf of nothing, you've got *everything*. There's one style over here and another one over there; there's trade, there's art, there's differentiation, all of this stuff just blowing up in your face. So you say to yourself, how come?"

Tim White is a demandingly meticulous researcher, one who does not like to waste time with speculations on grand questions. But on this one, he hardly hesitates a moment before answering his own question. "There's only one thing that I can think of that is big-time enough to render such a huge behavioral shift," he says. "It's *got* to be language."

It seems almost too obvious. Take any other innovation—a bone harpoon, for instance—and lay it down on the landscape. Now wrap it up in words. How to make it. How to use it to catch fish. When to expect what sorts of fish to arrive at what time of year, and communicate that information to others with whom you have dealings. *How to have dealings*.

How to fillet fish in long thin strips and dry them on racks, extending their nutritional benefit into a future—a concept uncommunicable without language. How to organize a cooperative fishing strategy and trade fish for other goods. While we are at it, how to name the river god and seek his intervention so that the catch will be plentiful.

Which innovation will travel faster: the naked harpoon, or the one dressed up in language? Language similarly greases the flow of other ideas and inventions—new hunting tactics, new ways of constructing a hearth, preserving meat, or tanning hides. These may have been conceived of before. They may even have been a part of life in one isolated area or another for a thousand years. But language supplies the medium needed to send them zipping across space, human group to human group, brain to brain. It explains the contagion, the pumping up of the cultural volume, even the Neandertals' demise. One has only to imagine two populations, one talking freely among themselves and the other communicating only with grunts and gestures. If they came into competition, would there be any doubt which would survive and which vanish? No wonder so many influential evolutionists from different disciplines ponder the Upper Paleolithic and converge on language as its prime mover: paleoanthropologists like Tim White, archaeologists like Lew Binford, Desmond Clark, Paul Mellars, and Richard Klein, geneticists like Luigi Luca Cavalli-Sforza and Allan Wilson before his death, to name just a few.

Before jumping on this bandwagon, let's look at the evidence. If sophisticated, fully human language came into the world at the beginning of the Upper Paleolithic, all humans who lived before that point must have lacked it. Words and phrases do not fossilize, and I can think of no direct means of ever determining when language entered the human behavioral repertoire. But although language does not turn to stone, two critical areas of human anatomy occasionally do: the bony structures associated with speech production and the imprints of human brains on the insides of skulls. Both of these admittedly indirect witnesses to language capacity suggest that the bandwagon may be headed into muddy ground.

The discovery of the hyoid bone in the "Moshe" skeleton found at Kebara has seriously challenged the notion that Neandertals lacked the anatomy for rapid-fire, fully human speech. For years, Philip Lieberman, Jeffrey Laitman, and their colleagues argued that the uniquely "flexed" skull base of modern humans reflects an adaptation for language, by positioning the larynx lower down in the throat. This means that modern humans, unlike animals with flat skull bases, cannot swallow and breathe at the same time, and thus face a much greater chance of choking to death. This is no small danger: Before the Heimlich maneuver was discovered, choking on food was the sixth leading cause of accidental death in the United States. Most earlier hominid fossils did not have markedly flexed skull bases. Those of Neandertals were particularly flat, suggesting that they were terrific swallowers but poor talkers.

Hyoids were important to this hypothesis because their position in the throat determines the size of the pharynx, which, if it is too small, will be unable to produce certain vowel sounds. The discovery of the first Neandertal hyoid—which turned out to be shaped just like a modern human's—toned down but did not silence the argument. Since then, Lieberman and Laitman have argued convincingly that an isolated Neandertal hyoid won't tell you much, since the position of the bone in the throat is not determined by its shape in the first place. If you want to know whether Moshe could speak, forget his hyoid and try to find his skull base, which is missing along with the rest of his head.

Having fended off Moshe's challenge, however, the Neandertals-can't-talk argument is losing ground on another front. Many of Lieberman and Laitman's arguments were based on the flat skull base of the Neandertal Old Man of La Chapelle, a badly distorted skull that had never been pieced together properly. In 1989, Jean-Louis Heim of the National Museum of Natural History in Paris set out to reconstruct the skull by more meticulous procedures, and lo and behold, his version of the skull base showed a lot more of the flexing seen in modern skulls. Recently David Frayer of the University of Kansas compared Heim's new Old Man to a sampling of modern human specimens from the Upper Paleolithic to the Middle Ages and found that, in the degree of flexing, it held its own with many of them, including a medieval Hungarian skull. "Nobody argues that the medieval Hungarians weren't able to talk," says Frayer. Not surprisingly, Laitman and Lieberman do not put a lot of faith in Heim's new reconstruction. They point to other Neandertal skulls with flat skull bases that continue to buttress their theory.

The debate may not be resolved until a thawed-out Neandertal walks into somebody's lab and asks for directions. But even if the Neandertals did lack a few vowel sounds, one wonders how much it matters. Whether that unfrozen Neandertal says, "Excuse me, I'm lost," or *Eksoose me, I'm loost*, somebody will point him in the right direction. More important, the vowels issuing from his supralaryngeal cavity would have very little to say about the mystery of the Upper Paleolithic. Laitman and Lieberman have never said that *all* Middle Paleolithic hominids lacked modern human vocal anatomy; their focus was on the Neandertals. The skull base of the modern fossil Skhul V is "completely human," according to Lieberman, but the Skhul fossils are associated with Middle Paleolithic tools and are now dated to around 100,000 years old. It is very difficult to see how the addition of some vowel sounds to the human vocal repertoire 100,000 years ago would unleash a "creative explosion" 60,000 years later. Furthermore, if you advance in time all the way to the present, you find plenty of people who communicate freely and expressively without *any* vowels at all: the members of the deaf community. American Sign Language is not a jury-rigged demi-language; it is a rich, fully formed communication system with complex grammar, syntax, and all the other elements of spoken language. It happens to be articulated with hands and fingers instead of hyoids and tongues.

A hominid mouth or hand is going to be only as fluent as the brain behind it. Here, the fossil evidence speaks very faintly, but everything it says points in the same direction. Two areas of the brain have long been known to be associated with language, both of them located in the left hemisphere. The first was discovered by the French anatomist Paul Broca in 1861. Broca had a stroke patient nicknamed "Tan Tan" because *tan* was the only word he could say, even though his comprehension of spoken language was normal. When Tan Tan died, an autopsy revealed an egg-sized patch of damaged brain tissue in the lower back part of the left frontal lobe. A decade after the discovery of "Broca's area," a German neurologist named Carl Wernicke discovered another language center in the left hemisphere, in the upper back part of the temporal lobe. Unlike Tan Tan, Wernicke's patients were verbal in the extreme—but what they said didn't make any sense.

Though they are not the only regions of the brain involved in language production, the vital part played by Broca's area and Wernicke's area has been confirmed hundreds of times. Their uniqueness to our lineage is also well established; no other primates have enlarged Broca's and Wernicke's areas. So it seems fair to say that the neural capacity for language cannot easily be denied to any hominid species whose brain possesses these structures. Brains do not fossilize any better than words or phrases, but sometimes an endocranial cast, or endocast, of a skull's inner surface preserves impressions of some of the grooves and lines outlining specific neural regions. According to Ralph Holloway of Columbia University, the leading authority on ancient hominid brain structure, the markings revealing

Broca's and Wernicke's areas appear millions of years before the Creative Explosion was allegedly triggered by the emergence of language, certainly by the time of *Homo habilis*. Holloway has also shown that *habilis* skulls reveal cerebral asymmetry: a left-hemisphere lopsidedness, which is associated in our species with language. More recently, Terry Deacon of Harvard University has pointed to language-related structures in the prefrontal cortex of the brain that also began to swell beginning with *Homo habilis*.

"This timetable specifically excludes hypotheses that directly correlate the evolutionary appearance of language with the attainment of a 'modern' position of the larynx. . . . " Deacon says, in a pointed reference to the research of Lieberman and Laitman. "It also suggests that the incredible cultural transitions that took place around the same period as the disappearance of the Neanderthals will be best explained in terms of cultural evolution rather than sudden neurologically mediated language evolution."

All this does not, of course, prove that *habilis* or other pre–Upper Paleolithic hominids were language-adept. But it does make it much harder to embrace the notion that they *weren't*, without something tangible to base that notion upon. The last hope for solid support under the wheels of the language bandwagon rests with the record of hominid behavior. Everyone agrees that the art, ornamentation, and sophisticated technology seen in the European Upper Paleolithic and the equivalent Late Stone Age in Africa could not have been accomplished without language. Some experts take this a step further, arguing that the dearth of symbolic expression before the Upper Paleolithic must mean that the early hominids lacked language too. "The Neandertals did not have art," writes physiologist Jared Diamond, "and there was essentially no change in their stone tools for 100,000 years. So they were not inventive. And I cannot believe that people having language could fail to invent for 100,000 years."

Despite isolated puffs of creativity here and there earlier in time, material culture before the "Great Leap Forward," as Diamond calls the Upper Paleolithic, does look impressively unimpressive as it galumphs on through the millennia. But equating language and material inventiveness requires a great leap of another sort. Viewed solely through their stone-tool industries, there are aboriginal societies in Australia and New Guinea that were no more "advanced" than the Neandertals until just a few years ago. Yet these people think and communicate in languages as richly expressive as any others on earth, and use these languages to construct wonderfully inventive myths, stories, and cosmologies. All this highly complex culture would be invisible to an archaeologist 10,000 years from now. Furthermore, Diamond's logic is dizzyingly circular: What defines the Upper Paleolithic is cultural invention, you need language in order to be inventive, therefore language sparked the Upper Paleolithic. How do we know? Because the Upper Paleolithic is defined by invention.

For all its explanatory power, the "Language solution" to the Upper Paleolithic turns out to suffer at bottom from the same inner frailty as the Biological Event solution. It is, in fact, only a more specific rendering of it, with the crucial biological transformation pinned down to the capacity for complex speech. The total lack of evidence for the crossing of *the* language Rubicon at the beginning of the Upper Paleolithic does not mean that such an event could not have happened. It seems more likely, however, that modern language evolved in several smaller steps, scattered through at least the last million years of hominid evolution. Perhaps one of these mini-Rubicons was crossed at the transition to the Upper Paleolithic. But even if an enhancement in language *was* the prime mover behind the Creative Explosion, or the Great Leap Forward, or whatever you want to call the development, what was moving *it*? What could have changed in the landscape and lives of people just beyond the reach of history that led so swiftly to a *need* for a more sophisticated language? What was the mover behind the mover, the secret player in the equation? Having come this far into the mystery of modern origins, I cannot accept language alone as the answer just because it sounds plausible. The answer must emerge from the evidence left behind, in the ground.

This is what makes a place like Dolní Věstonice so alluring. Explain Dolní Věstonice, and you explain humanity.

On a September morning in 1991, I was driving north from Vienna toward the recently opened Czech border. On the seat next to me were directions to the archaeological station in Dolní Věstonice, scrawled on the back of a letter I had received from Jiří Svoboda of the Archaeological Institute of the Czech Academy. Svoboda had been kind enough to answer my list of questions about the triple burial and other archaeological peculiarities of the region with a simple invitation. "My suggestion is that we meet this summer at the base of Dolní Věstonice," his letter read, "where most of the fossils and other materials are stored, and which offers a pleasant surrounding for contemplation."

I reached the border, and the guard grinned me across into the town of Mikulov, dominated by a white castle high on a steep hill. I knew this castle from its tragedy. During World War II the Czechs moved some of their most treasured Neandertals and early modern human specimens here for safekeeping. On the last day of the war, the retreating Germans set fire to the castle, destroying most of the fossils, including an entire collection of over twenty-five skeletons from the 26,000-year-old site of Predmosti. From Mikulov I followed the road north, keeping on my right the Pavlov Hills, a long rocky ridge dotted here and there with ruined fortresses, remnants of the Ottoman wars. Now and then I glimpsed water on the opposite side: the Dyje River, dammed into an artificial lake in the 1980s as part of a massively unsuccessful irrigation project. After a couple of miles I turned off the main road and descended toward the village of Dolní Věstonice. The field station was in the center of town.

Svoboda came out to greet me, blue-jeaned, dark, and slender. There was much on his mind. The recent revolution had not been kind to science, he told me, at least not in the short term. The money to do anything with had simply disappeared. The previous year, a dozen or more busy students and investigators were working out of the field station. Now it was staffed by himself, one student assistant, and a couple of teenage volunteers. One of the volunteers wanted desperately to become an archaeologist, Svoboda said, but his parents insisted that he study engineering. "The way things are now, I told him to listen to his parents."

That day, Svoboda was carrying the added burden of planning a site tour for an international group of archaeologists who would be attending a conference in Bratislava the following week. Originally, the conference organizers had told him to prepare for a dozen or so people. He had just been informed that two hundred had signed up: the lure of Dolní Věstonice. There was no way Svoboda could satisfy five busloads of aggressively curious colleagues, he said. Inevitably, the visit would be a disaster, and his reputation destroyed. On the other hand, it was a beautiful late summer day. Such a day, he declared, should not be wasted with worries.

We left the grim emptiness of the field station and set out on the road to Pavlov. In the 1920s, the construction of this link between the two villages also joined the region to its deep past. A priest who traveled frequently along the road noticed some bones and tools poking out of the newly exposed earth. Archaeologist Karel Absolon was summoned from the Moravian Museum, and the digging did not stop for half a century. First under Absolon's direction, then under Bohuslav Klima's care, the

excavation grew to embrace a cluster of separate sites, representing at least
five distinct periods of occupation between 28,000 and 24,000 years ago.
The dig closed finally in 1979. Svoboda pointed to an inconspicuous stone
rising from a vineyard, all that is left to mark the work.

The most famous feature of the original Dolní Věstonice site is a group
of five dwellings outlined by traces of postholes, limestone blocks, and
mammoth bones and further defined by the density of artifacts on their
floors. They are usually referred to as "huts," although the largest measures fully fifty by thirty feet, which compares favorably to a Cape Cod
cottage. Its interior has five regularly distributed hearths and is littered
with stone, bone, and ivory implements, scattered ornaments, and artwork. Together with the four smaller dwellings, the large hut is surrounded by a framework of mammoth bones and tusks, presumed to be
remnants of a fence erected against animal intruders and Ice Age winds.
In a marshy area just beyond this fence, the excavators also uncovered a
lode of mammoth bones, the remains of at least one hundred animals.

The "village" at Dolní Věstonice, as it is usually called, is utterly unlike
the ephemeral, throwaway encampments of Middle Paleolithic people.
Like the triple burial discovered later, its complexity begs for a narrative.
Klima's version turns on the mammoth bones. He noted the location of
the site, a few hundred meters above the river plain, where a sharp eye
could monitor the passing of migrating herds of mammoth and other
game; he noted the "vast accumulation" of mammoth bones piled up in
the marsh, presumably the leavings of the hunters' meals, and pondered
over other smashed-up mammoth bones on the site, some of which had
been broken to extract their marrow, others bashed apart "in connection
with various magical customs—seemingly to assure success in the hunt."
The story he coaxed from all this became the classic Ice Age tale: a large,
cooperative community, drawn to this spot on the landscape specifically
to hunt "the mighty pachyderm."

And so the people of the time and region—collectively called Pavlovians—became the "mammoth hunters" of the popular romances. But
there is also another way to read these bones. If extinct mammoth behaved like their modern elephant cousins, when they grew old they migrated to marshy areas, where there would be softer vegetation for their
worn teeth to chew. Eventually, the old would die. Over time their bones
would be washed together in these swampy areas by the action of water
flow. Mammoth bones make good building material, and if you can get a
fire hot enough, excellent fuel too. Olga Soffer at the University of Illinois

believes that people were drawn to this spot because there was a natural supply of fuel and building material available. The place may have been a "magnet site," drawing people again and again over the generations.

"If you ask Olga, it was the bones that brought the people here, not the other way around," said Svoboda.

The magnet's pull was remarkably strong. Most of the stone tools at the original Dolní Věstonice site are not made from local raw material. Over 80 percent are flints traceable to distant sources in the north, east, and southwest. It is unlikely that the Pavlovians were relying on trading contacts for such a large proportion of this critical raw material. Instead, apparently, people from far-flung regions were converging at Dolní Věstonice, bringing their flint tools with them. According to Soffer, they weren't coming for mammoth either.

"The site has nothing to do with hunting," she said but cautioned me that no one can know for sure what happened at Dolní Věstonice, because it is impossible to tell how long it took for all the artifacts to accumulate. The site may represent numerous occupations over many generations, even the huts in the alleged village built centuries apart. With this judicious qualification, she stated her "working hypothesis": "If people were getting together, there must have been some kind of relationship among them. So I'd say Dolní Věstonice was some kind of kin aggregation site."

The attraction of the place was the promise of society, an opportunity for related bands to exchange information, find mates, and generally massage relationships. This may explain the plethora of art and ornamentation left behind, and also help illuminate a mystery embedded in the artwork. A little way down the road from the stone marking the excavation, Svoboda led me to an abandoned brick quarry. A solitary wood-frame building stood at the entrance, which he explained was used sporadically as an artists' retreat. Farther on, decades of quarrying had left behind sheer, one-hundred-foot walls of loess—the fine-grained, claylike soil deposited by wind during the last glaciation. Mixed with straw, the loess makes a fairly durable building material, and until a modern cement factory opened nearby, most of the structures in the area, including the archaeological field station, were constructed from loess bricks.

Twenty-six thousand years ago, the Pavlovians were putting the loess to more flamboyant use. In 1951, Klima's team discovered the remains of another structure a few hundred feet up the slope from the ancient village. A circular hut about twenty feet in diameter with posts supporting a roof had been built into the hill on one side. Inside, the center was dominated by the remains of a horseshoe-shaped kiln formed out of earth mixed with limestone. From the blackened floor of the kiln, Klima and his colleagues pulled out more than two thousand bits of fired clay, including numerous little irregularly shaped pellets and fragmented heads and feet of animal figurines, all crafted from loess. Some of the clay pellets still bore the fingerprints of their makers. Similar ceramic objects, including bits of human figures, were found in the village down the slope, as well as in other Pavlovian sites in the vicinity. A second kiln, discovered in 1979 a few dozen yards from the first, was also loaded with splintered ceramics.

There are now more than ten thousand individual clay fragments collected in the area, representing the earliest use of ceramic technology anywhere in the world. If few archaeological textbooks acknowledge the Pavlovians for this epochal invention, the oversight can be traced to the tenacity of preconception. In traditional scenarios of human cultural progress, the emergence of ceramics is linked to its functional use as pottery, which did not appear until the agricultural revolution in the Neolithic period, some 12,000 years after the kilns at Dolní Věstonice had cooled. The Pavlovian hunter-gatherers may not have made pottery simply because they did not need to; they could easily fashion containers out of organic materials like animal skins. As a result, they were never given the credit they deserve for inventing the technology*.

The question, of course, is why they needed clay figurines and oddly shaped pellets. Along with Pamela Vandiver of the Smithsonian Institution, Soffer, Svoboda, and Klima have recently come up with a plausible, if somewhat bizarre explanation. Their initial clue was the dilapidated condition of the figures: Almost all are broken into bits—a fragment of a lion's head, the belted waist of a woman, but never a complete lion or woman. The mystery was resolved in the lens of Vandiver's scanning electron microscope outside Washington, D.C. Under a few hundred degrees of magnification, ceramics that have been broken by handling, trampling, and weathering show a distinctively smooth, simple pattern of breakage. Most of the fragments from the residential areas at Dolní Věstonice showed just such a pattern. In contrast, the bits and pieces pulled out of the kiln areas have rough, branching fractures. The only agent known to

*The Pavlovians may have been the first people to invent a technology with no inkling of its utilitarian future, but they were hardly the last. Native Americans were cold-hammering copper into trinkets long before they used it to make tools. The ancient Greeks developed a workable steam engine for making special effects in their temples, unaware of its potential to move heavy objects through space.

cause such fractures is "thermal shock." Professional potters, especially impatient ones, know that this shattering occurs when inadequately dried clay is placed in a kiln and heated rapidly. The Dolní Věstonice figurines literally exploded in the process of being baked.

It is possible that the Pavlovians left so many fragments around their kilns because they never figured out how to control thermal shock. But could a people be so habitually inept as to litter the ground with their mistakes for six thousand years and leave behind so little evidence of success? Furthermore, when the experimenters fashioned their own clay objects and fired them, the local loess soil proved to be extremely *resistant* to thermal shock. "One would have to try very, very hard to explode objects molded from this stuff," says Pamela Vandiver.

"Either we are dealing with the most incompetent potters the world has ever seen," adds Soffer, "or else these things were shattered on purpose."

We are used to thinking of art as an act of preservation. Apparently the people of Dolní Věstonice were creating art in order to destroy it. The dynamics of thermal shock may provide a clue to their motives. The phenomenon occurs when water trapped in the molded object turns to steam, sizzling at first, then shattering with a loud pop that often sends fragments flying through the air. Perhaps the value of the figures lay not in themselves, but in the dramatic moment of their disintegration, which was the climax to some ritual or performance. The ancient Maya and other more recent people also destroyed pottery and jade ornaments in their rituals. Among all human populations, such ritual behavior is most often associated with large gatherings of people, so Dolní Věstonice looks even more like an aggregation site.

"People tend to engage in ritual more often when there are a number of folks around who do not interact on a daily basis," says Soffer. "It's an integrative mechanism."

I asked Svoboda what he thought about the exploding pottery, but he only mumbled that they had long suspected something of the sort. As we turned to leave, I noticed a modern pottery kiln set up outside the empty artists' retreat. I wondered whether the potters who brought the kiln knew that they were working at the place where ceramics began, 26,000 years ago, among the performance artists of Dolní Věstonice.

I followed Svoboda up a path through farm fields changing into open grassy terrain. Our destination was Dolní Věstonice II, the sequel to the saga. In 1985, a bulldozer quarrying loess for the irrigation project inadvertently flushed out a charcoal- and tool-rich occupation layer about half a mile from the original excavation. The archaeologists moved in and the next year exhumed the three teenagers from their ochered tomb. The following year, Svoboda, who was put in charge of the operation in Klima's absence, unearthed a man in his late forties who had been laid on his right side in a shallow grave, knees flexed, head to the east, feet to the west. A few pierced canine teeth had been thrown in, and the man's head and pelvic area were covered with ocher. The story of his life and death could not begin to be reconstructed from such meager leavings. His skull, however, might speak to the general tone of his existence. Like other skulls from the Moravian Upper Paleolithic, it was pitted and scarred by blows to the head.

We crossed a windy hillside and arrived at what was left of the Dolní Věstonice II site. The real work was over, but the excavation had been continued with whatever funds could be pieced together. Three young workers were busy cleaning up the walls and floor, sprucing it up for the international visitors due in a couple of days. Here and there on the floor, token artifacts were perched on little towers of loess. While Svoboda gave instructions to the workers, I stared across the prospect. A cluster of red-tiled roofs marked the "modern" village—twelfth century on up—and beyond it lay the gray sheen of the artificial lake, prickly with the barren limbs of half-sunken trees. Far out in the middle of the lake, I could see the brilliant sail of a windsurfer tacking around the steeple of a medieval church, the only visible sign of an entire village sacrificed to the reservoir.

"You see what a good view they would have had of the river," Svoboda said. "The site faces north, when it should be on the south slope to catch the sun."

"So they wanted to keep an eye on the river to watch for game?"

"Perhaps. Or other people."

With a stick in the bared earth, he drew a sketch of the site as it had emerged over several seasons of digging. A cluster of ten circles represented hearths, which he surrounded with little dots poked into the earth: small pits probably used for boiling food. Unlike the original Dolní Věstonice, there were no elaborate dwellings here, comparatively little artwork, and nothing suggesting aggregations of folk. "My idea is that people were coming here seasonally," said Svoboda. "It was not a huge gathering, just a modest group that came back each year."

One feature the two site clusters have in common is foreign stone. There is plenty of good chert in the Bohemian massif, a few kilometers to the north, and chert is a good material for stone knapping. Other, possibly contemporaneous Upper Paleolithic people exploited this local resource freely. But for some reason, over 90 percent of the tools from Dolní Věstonice II that Svoboda has examined were made of a type of flint found over two hundred kilometers north, in Poland and northern Moravia. Svoboda thinks the people were migrating seasonally back and forth through the passage between the Bohemian massif to the northwest and the Carpathian mountains to the east, bringing the raw stone for their tools with them. "It's very strange," he said. "They ignored the chert that was here, and imported flint from as far away as southern Poland."

"Why?" I asked. "Was the Polish flint a better raw material?"

"Perhaps a little, but not so much better as to go to such trouble to bring it all this way." As we headed back toward the field station, he suggested a possible explanation: "Maybe they did not use the local cherts because other people were sitting on this resource, claiming it for their own and driving others away." I thought back on the acts of violence preserved in the bones here, the healed head wounds and the injured males in the triple grave.

We arrived at the field station around dusk. Svoboda opened a bottle of wine, and we ate under a dim lamp in the common room downstairs, cluttered with boxes of stone tools and equipment. A row of mammoth skulls cast baleful shadows on the walls. I asked him if I could see some of the archaeological treasure stored here, and he brought out tray after tray—broken clay bears and lion's feet, gleaming polished shell and canine necklaces from the Pavlov site, an ivory lioness ready to spring, a tiny female figure with nubile curves wrought in gorgeous hematite. A thick

Reconstruction of an Upper Paleolithic hut at Dolní Věstonice, around 27,000 years old
DRAWING BY SIMON S. S. DRIVER. REPRODUCED WITH PUBLISHERS' PERMISSIONS FROM STRINGER AND GAMBLE 1993 AND J. WYMER 1982

hunk of stone with a depression in it might have been an oil lamp. There were also bone rings, weird spatulate things, large-eyed needles, smooth batons, and doughnut-shaped objects, ubiquitous in the Gravettian period, that are called "spear straighteners," though no one really has any idea what they were used for. Next came a time-lacquered mammoth tusk, bearing what appeared to be an engraved map. Curving off toward the thick end of the tusk were clusters of herringbone marks that might represent the Pavlov Hills, and below, a series of wavy parallel lines which could well be the river.

Missing from the collection were its most famous treasures, whisked off long ago to the parent museum in Brno. Svoboda showed me replicas: an enigmatic ivory rod with paired swellings part way down its length (male archaeologists see a woman with a long neck and enormous bulging breasts; feminist scholars turn the thing upside down and see male genitalia), two sketchy human faces, one carved in ivory, one molded in clay. Curiously, both faces appear slightly asymmetrical, drooping down on the left side. There are more impressive images of the human face to be found elsewhere from around the same time—the exquisite ivory head of the "Dame of Brassempouy" from France, for example, dated to some 25,000 years ago. The unique significance of these Dolní Věstonice images is revealed only when they are placed in the context of another discovery at the site, a few yards away. In 1949, Klima uncovered the grave of a woman about forty, who had been laid to rest with the usual anointment of ocher. An analysis of the skull revealed that the woman suffered from a bone disease that would have caused her face to droop on the left. Thus the two sketchy images may not be generic depictions of Womanhood, but personal portraits of *this* face, *this* self, rendered 26,000 years before the Mona Lisa composed her smile.

The region's most famous piece of Paleolithic art is the "Venus of Věstonice," the heavy-breasted, swollen-hipped enigma pulled from the ashes of the largest hut's central hearth. The ceramic figurine is all but faceless, with two long slits suggesting eyes. Similar female forms, usually carved in soft stones or ivory, dot the map of Gravettian Europe from southern France to the Urals. Their meaning is mysterious. Interpretations range from the conventional assumption that they represent fertility goddesses to recent suggestions that they are Paleolithic soft porn, carried around by randy hunters to be fondled at will. In a heady mixture of mysticism and Marx, Klima regarded this one as "the symbol of the 'Urmutter,' preserver of the kin group and protectress of the common economic existence."

I asked Svoboda what he thought the Venus meant, what to make of all the other art, and how it came to be in such abundance here. He was cautious. True to the European tradition, he is more interested in the *what* of the archaeology, the carefully made connections that can be traced between details across space and time, than in the *why*. "Why is everyone so interested in origins?" he said. "These are not species, these are *things*. They don't necessarily have a place or a reason why they all begin."

He picked up a curiously pierced reindeer toe bone, brought it to his mouth, and gave a blow. Out came a sharp, clear note.

"A whistle of some kind?" I asked.

"I don't know if it is a whistle," he replied. "I only know that it whistles."

I went to bed shortly after midnight, in a bunk in the attic. My only companions were the three death-linked teenagers and the other ancients of Věstonice, resting in gray metal boxes in a room across the hall. For a while they kept me awake, my mind imagining plots to explain their demise. Earlier in the day, Svoboda had shown me the disarticulated skeletons, the long, cream-white limb bones and the smooth ochered skulls. Some Italian scientists working nearby joined us, and a couple of Czech paleontologists, and for a while we stared into the boxes, fingering the dry mortality. Svoboda brought out a photograph of the burial, and we ogled the queer arrangement of the bodies, the peaceful parallels of the limb bones invaded by the insinuating perpendicular of those arm bones on the left.

Before I came to Dolní Věstonice, the fate of these three young people seemed potent with meaning, as if the pattern of their skeletons on the ground held some preternatural key to the whole tangled mystery, like the tossed bones of a diviner. In the room with their remains, I felt like a voyeur, trying to force a fantasy out of someone else's private affair. As fodder for romance, the burial has actually lost ground recently. Among the Czechs huddled over the bones in the afternoon was Vladimír Novotny of the Czechoslovak Academy of Sciences in Prague. After completing a study of the three skeletons, he has concluded that the sex of the middle one is "enigmatic." "In many ways it looks male," he said, "yet there are numerous morphological points that say it could be female too." Others familiar with the bones, including pelvis-expert Karen Rosenberg of the University of Delaware, are less ambivalent. In spite of the skeleton's small size, its pelvic and lower-limb morphology appears to be that of a young man. His small stature was probably associated with the same pathology that left his legs bent and crippled. A person in his own time, not a protagonist in ours.

I said before that if a site like Dolní Věstonice could be explained, then so could humanity. But this place and these people cannot be explained, because they are already too human. I might as well try to "explain" my wife or my neighborhood. Nevertheless, there was a binding thread running through all that I had seen or heard at Dolní Věstonice. If that thread can be gently lifted up and followed back, it might lead us to the source, the secret player that nudged a species of rare intelligence over the gartel once and for all, into the inexplicable richness of the future.

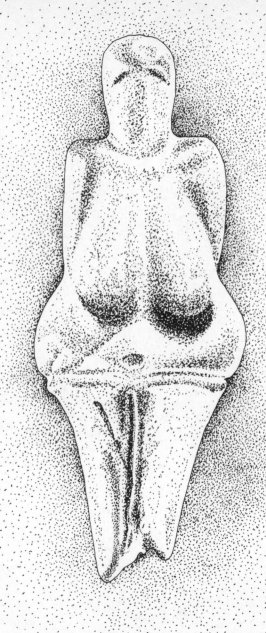

The "Venus" from Dolní Věstonice   DAVID DILWORTH

Chapter Eleven

DOUBLE WISDOM

An Arunta kid is expected to know the genealogy of three hundred people, and depending on their relationship to him, know how to address them. But what happens if he is going over the hill, and sees person number three hundred and one? What does he do then?

—ANTHONY MARKS

Let's begin with the rocks that move. In Bulgaria, a thousand miles east of Dolní Věstonice, there is a cave called Bacho Kiro. Excavated by a team led by the Polish archaeologist Janusz Kozlowski, it is famous for containing the earliest known Aurignacian assemblages, and thus the first formal evidence of an Upper Paleolithic culture, 43,000 years old. The distinctive, blade-based Aurignacian artifacts are utterly different from the Mousterian tools found below them in the cave's deposits, suggesting that they were the handiwork of people who arrived in the area from somewhere else. We do not know where they came from, but we can trace the movements of their rocks. While the Mousterian tools in the cave were hacked out of unwieldy volcanic basalt found in the immediate vicinity, most of the Aurignacian tools in Bacho Kiro were made of high-quality flint imported from outcrops anywhere from fifty to seventy-five miles from the mouth of the cave.

So right from the start of the Upper Paleolithic, stone starts to wander around the terrain. Later, at Dolní Věstonice, distinctive flint arrives from southern Poland, one hundred miles to the north. Slovakian radiolarite of red, yellow, and olive migrates from another place one hundred miles to the east. Even a little obsidian trickles in from outcrops on Tokai Mountain, on the eastern border with Hungary. But by this time, rocks have begun to scoot around all over Europe, wandering hundreds of miles in

Germany and Belgium, Romania and Russia. In most cases, the best-quality flint travels farthest. Later in the Upper Paleolithic, the famous "chocolate flint" of southern Poland is found over a radius of two hundred and fifty miles from its source in the Holy Cross Mountains.

Normally, stone is sedentary; the only time rocks move is when people carry them, or trade them through a chain of exchanges that covers the same distance. The same goes for shells. Molluscs from the Black Sea moved to the Central Russian Plain; others found in Gravettian sites in Germany have been traced to species native to the Mediterranean, four hundred and thirty miles to the south as the crow flies. Pierced marine shells from both the Atlantic and the Mediterranean coasts adorned the necks of Cro-Magnon hunter-gatherers in the Périgord, two or three hundred miles from the sea. They were also making beads out of talc. There is no talc in the Périgord. The nearest outcrops are in the Pyrenees, far to the south.

Rocks traveled across the landscape during the Middle Paleolithic too, but rarely so far and never so much. In the later Middle Paleolithic record of Central Europe, as French archaeologist Jehanne Féblot-Augustins has shown, tools can be found over one hundred miles from the source of the raw material used to make them. But such peripatetic stone was the exception, not the rule; 99 percent of the stone tools in any given site were made on materials found within twelve miles. In the equivalent Middle Stone Age of southern Africa, the percentage of imported flint in some Howieson's Poort sites was close to half, but the distances seldom exceeded twenty-five miles. In short, long-distance procurement of raw material did not seem to be part of the *repeated* pattern of human behavior until the beginning of the Upper Paleolithic. And it is the repeated pattern that reflects what people are really up to. If the raw material was moving more, then so, it would appear, were the people.

For support, look at their legs. One of the hallowed principles of evolutionary biology is Allen's Rule: Legs, arms, ears, and other body protrusions should be shorter in mammals living in cold climates, and longer in mammals of the same type living in the tropics. Having short limbs decreases the body's surface area relative to its volume, which conserves heat, while long limbs dissipate body heat in warm climates. This explains why Eskimos and Laplanders have short legs, relative to the size of their trunks, while most Bantu people have leaner proportions, reaching extremes in open-country tropical people like the Maasai. It also explains why people instinctively curl up into a ball under the covers when it gets cold and stretch out on the sheet on steamy summer nights. Most ancient hominids were equally obedient to the rule. Cold-adapted Neandertals had stubby legs and arms measured against their overall size. The famous *Homo erectus* Turkana boy from Kenya, a million and a half years old, was already a lanky five feet six or so when he died at the age of twelve.

The only ancients who defy Allen's Rule are the modern humans of Europe and the Middle East. The Pavlovian teenagers of Dolní Věstonice, the original Cro-Magnons from the Cro-Magnon rock shelter in Les Eyzies, even the 100,000-year-old skeletons of the Qafzeh and Skhul caves all have longer legs than they "should," judging by the climates they lived in. The deviance of the early Middle Eastern specimens is understandable. Limb length is a conservative trait. You are born with it etched into your genes, and even if you migrate to another, radically different climate, your children and your children's children, and so on for generations, are likely to keep your original body proportions. Witness, for example, the long limbs of African Americans. If the Middle Eastern moderns were relatively recent migrants from Africa, one would expect them to show lanky African proportions, in spite of the more temperate conditions that prevailed in the Levant 100,000 years ago. For the same reason, Moshe, the Neandertal of Kebara Cave, has shorter limbs than the climate would predict, supporting the notion that he was a descendant of migrants—in his case, from the north.

This leaves only the modern humans of the European Upper Paleolithic walking around on legs too long for their bitterly cold habitat, legs that would squander precious body heat in the very climate where you would expect it to be hoarded. Clothing would have helped, but it cannot explain everything—the Eskimos and Lapps use clothing too, and they have much shorter legs. According to Trent Holliday of the University of New Mexico, the lanky limbs of the Cro-Magnons support the notion that they were the descendants of migrants from warmer regions, unrelated to the resident Neandertals. But it seems peculiar that a population of long-limbed people would move from a warm climate into a cold one and remain long-limbed *for a thousand generations*, in spite of the fact that over much of that immense stretch of time the thermometer continued to slide downward. At the glacial maximum, 18,000 years ago, conditions were close to polar in northern Europe. Why was there no lessening of leg? What kind of adaptive trait is it that never adapts? Holliday's interpretation seems to me only part of the story. The Cro-Magnons probably did bring the genes for their long legs in from warmer regions. But once in Ice Age Europe, something else must have put selective pressure on the lower limbs to *stay* long, something important enough to counter the climate's pressure to shrink them down to Neandertal proportions.

There is only one thing you do with long legs besides get rid of heat: You move. You walk or run or trot over long distances, with much less expenditure of energy than short-legged people. What were the Cro-Magnons after, on these restless treks? It must have been something so vital to survival that they would make an unprecedented evolutionary sacrifice in order to obtain it.

I can think of only one resource on earth with so much raw value and dangerous potency that people had to invent art and style and perhaps even new forms of language and consciousness, just to keep it from blowing everything apart. Other people.

They had always been around, of course. Humans are higher primates, and most higher primates are intensely social. Lately, the unusually quick, versatile intelligence of monkeys and apes has been seen as the direct natural product of their fervid sociability, a notion that took seed twenty years ago when a young Cambridge psychologist named Nicholas Humphrey was watching gorillas in the Virunga Mountains of Rwanda. "I could not help being struck by the fact that of all the animals in the forest," said Humphrey, "the gorillas seemed to lead much the simplest existence." With no natural predators other than men, with abundant food available most of the year, the gorillas had little to do but eat, sleep, and play. Yet somehow this easy life-style produced a species that, along with the other great apes, was the most intelligent of all terrestrial animals.

Humphrey speculated that the real pressure driving gorilla intelligence was other gorillas. Simian societies have evolved with relatively long periods of dependence for the young, giving prolonged opportunities for older animals to teach the juveniles essential survival lessons. The increased emphasis on learned behaviors spurs the growth of intelligence. At the same time, the "collegiate community" of old and young, teachers and learners, siblings, cousins, aunts, uncles, and grandparents generates another pressure to "get smart." Though members of a social group tend to behave cooperatively, the bottom line for each individual is the survival of its own genetic legacy. Each competes for food, mates, and other resources with other members of the group, all of whom are armed with an intelligence honed by evolution precisely by and for the "game of social plot and counterplot." "Social primates are required by the very nature of the system they create and maintain to be calculating beings," Humphrey wrote. "They must be able to calculate the consequences of their own behavior, to calculate the likely behavior of others, to calculate the balance of advantage and loss—and all this in a context where the evidence on which their calculations are based is ephemeral, ambiguous and liable to change, not least as a consequence of their own actions. . . . It asks for a level of intelligence which is, I submit, unparalleled in any other sphere of living."

In the last decade, research on various species in the wild has buttressed Humphrey's insight, leading to a *bona fide* revolution in the understanding of primate social behavior. Monkeys and apes have proven uniquely adept in two intellectual spheres: deception and building alliances. For an example of the first facility, there is Paul, a young member of a baboon troop observed in Ethiopia by Andrew Whiten of the University of St. Andrews in Scotland. One day in 1983, Whiten was watching an adult female named Mel dig in the ground for a plant bulb. Paul approached and looked around. There were no other baboons in sight. Suddenly he let out a yell. Within seconds his mother appeared and chased the astonished Mel over a small cliff. Meanwhile, Paul walked over and nabbed the plant bulb.

Deception is rife in the natural world. Stick bugs pretend to be sticks. Harmless snakes resemble deadly poisonous ones. When threatened, blowfish puff themselves up, and cats arch their backs and bristle their hair to seem bigger than they really are. All these animals fool other animals—usually members of other species—into thinking they are something that they patently are not. But they are acting out programmed genetic responses. Their biology leaves them no choice but to dissemble, so their behavior is in fact perfectly honest. *Tactical* deception, the term coined by Whiten and his colleague Richard Byrne, is altogether different. Here an animal has the mental flexibility to take an "honest" behavior and use it in such a way that another animal is misled into thinking that a normal, familiar state of affairs is under way, while in fact something quite different is happening. In the example above, Paul's mother was misled by his scream into believing that Paul was being attacked. The result of her misinterpretation was that Paul was left alone to eat the bulb Mel had carefully extracted—a morsel Paul would not have had the strength to dig out himself.

Byrne and Whiten have collected hundreds of other examples of tactical deception among primates. The most cunning are the chimpanzees, who commit acts of double duplicity, with one chimp outwitting the attempts of another to deceive him. In one case, a captive chimp was alone in a feeding area when a metal box containing bananas was opened electronically. At the same moment, another chimp happened to approach. The first chimp quickly closed the box, walked away, and sat down, looking as if nothing had happened. The second chimp hid behind a tree and peeked back. As soon as the first chimp thought the coast was clear, he opened the box. The second chimp ran out, pushed him aside, and ate the bananas. Byrne and Whiten refer to such calculating mental maneuvers as "Machiavellian intelligence."

"It is good to appear clement, trustworthy, humane, religious, and honest, and also to be so," Machiavelli himself wrote in *The Prince*, "but always with the mind so disposed that, when the occasion arises not to be so, you can become the opposite."

Another talent Machiavelli cultivated in his aspiring prince was the ability to make friends with the right people. Primates understand this too. Far from the old notion that dominance hierarchies in primate troops are strictly determined by a "might makes right" law of the jungle, recent studies show that social relationships among chimpanzees, baboons, and various monkeys are built around a web of alliances that allow individual animals much more power in the group than they could claim from their individual size or strength. A young rhesus macaque soon discovers that she can intimidate members of the group far larger than herself because her high-ranking mother will back her up in any conflict. But alliances need not be between relatives. In some baboon troops, older males have been known to form coalitions against a younger, much stronger male entering the troop. In one long-term study in Amboseli Park in Kenya, Barbara Smuts of the University of Michigan found that baboons of opposite sexes also form tight "friendships" that can last for years. The males often stayed close to their female friends, even when they were not in estrus, grooming and being groomed, sometimes defending them or their young in confrontations with other baboons. In return, the males counted on the females' help in conflicts with strange males. Though paternity is difficult to assess in baboon troops, they were probably favored too when the time came to mate.

Keeping track of alliances in a primate troop would be hard enough if the relationships were unchangeable; in fact, they are in constant flux. Rivals can quickly become allies and enemies, friends, with each shift in alliance sending ripples through the intermingling networks. Such complexity favors the intelligent individual, who by mastering the art of alliance will be more likely to find himself in a position to mate, thus passing on his socially cunning genes. Among the primates, it is once again the chimpanzees who excel at this form of Machiavellian intelligence. Chimps are the only nonhuman species known to have refined the strategy of alliance a crucial step further. In the vast majority of observed cases in primates, alliances are formed between and among individuals within their own social group. The payoff of having allies is thus an ego-specific one. ("If *I* cozy up to so-and-so, grooming him and supporting him in fights with others in the group, then he'll help *me* out too when the time comes.") But chimpanzees seem capable of cooperating on the level of *us* as well: acting cooperatively as a group, usually to the peril of another group.

The most celebrated case of this level of alliance comes from Jane Goodall's research in Tanzania's Gombe National Park. Between 1974 and 1977, the males of a chimpanzee troop Goodall called the Kasakela community systematically hunted down and exterminated the males of the neighboring Kahama community. When the last Kahama male had been eliminated, the Kasakelas annexed their territory and absorbed their females. The gruesomely violent attacks were carried out deliberately, with members of the troop moving stealthily together into the rivals' territory. In subsequent years, the Kasakela community was itself threatened and attacked by a still more powerful group on another border. Similar group aggression has been witnessed by researchers in other areas. After long believing in the intrinsic peacefulness of chimpanzee society, Goodall was forced, to her horror, to concede that the chimpanzee was capable of a level and kind of violence previously witnessed in only one other species on earth. "It seems almost certain that the Kasakela males were making determined attempts, through wounding and battering, to incapacitate the Kahama chimpanzees," she concluded. "If they had had firearms and had been taught to use them, I suspect they would have used them to kill."

Chimpanzees are certainly not the only animals to exhibit aggression between groups; it is to be expected in many species. What is unique to the chimps is the chillingly effective cooperation among one group of animals to defeat another. However unpleasant the evidence, the fact that chimpanzees "cooperate to compete" suggests that this capacity in humans is very ancient. Richard Alexander, a sociobiologist at the University

of Michigan, has even suggested that intense group-against-group competition is the central driving force behind the unique intelligence of the
hominid lineage. As soon as our ancestors became smart enough to
achieve a margin of control over what Charles Darwin called the Hostile
Forces of Nature—predators, drought, food scarcity, and other natural
environmental pressures—an escalating balance of power developed
among human groups sharing the landscape. However much the members
of a particular group might be competing among themselves, the social
unit whose members could cooperate most effectively would have an advantage over neighboring groups. Each ratchet notch of increased social
intelligence would compel another to follow, as competing groups evolved
ever-finer cooperative skills simply to keep pace with their competitors.
The result was a "runaway intellect" fueled by the one "hostile force"
that could never be controlled: *Homo* itself. The end result would be a
species intensely loyal to fellow group members but lethally aggressive to
those on the outside: us and them. "This intergroup competition," says
Alexander, "became increasingly elaborate, direct, and continuous, cul-
minating in the ubiquity with which it is exhibited in modern humans
throughout history and across the entire face of the earth."

Finding evidence for intergroup competition in prehistory is a little
tricky. Judging from traumas on fossil skulls, a number of known *Homo
erectus*, archaic, and early modern human individuals appear to have been
bonked on the head, but there is no way of telling who (or what) did the
bonking, or why. The clearest prehistoric portrait of violent conflict between human groups comes from northeastern Belgium where, 7,000 years
ago, an expanding early Neolithic farming culture battled the Mesolithic
hunter-gatherers who had occupied the territory for centuries. Forager-
farmer relationships in the Neolithic had long been thought to have been
generally peaceful. But when Lawrence Keeley of the University of Illinois
at Chicago and Daniel Cahen of the Royal Institute of Natural Sciences
in Brussels excavated a number of the agriculturalists' settlements in the
1980s, they found that those on the leading edge of the farmers' expanding territory had been fortified with deep ditches, timbered palisades, and
baffled gateways to confound assaults. The farmers were also well supplied
with projectile points, and since they apparently did not hunt, it is hard
to imagine where else they might have projected them except at the bodies of other humans. Among the skeletons in a cemetery from the period
were thirty bodies of men, women, and children who had holes in their
skulls matching the profile of the ax heads used by the intruding culture.

"Clearly these people were occupying a hostile territory," says Keeley.

As Alexander is quick to point out, the historical record for intergroup violence is unambiguous: Virtually all known human societies have shown some tendency for lethal conflict. Today it is hideously evident that no chimpanzee troop can approach the level of group-against-group hostility our species flaunts daily in Bosnia, Rwanda, or a hundred other points of darkness. The ubiquity of the trait suggests that the capacity for it is genetically based and deeply rooted in our evolution.

On the other hand, at any given moment one can find just as many examples of human groups cooperating simply because it makes life easier for everyone involved. *This* is truly extraordinary primate behavior. Chimps patrolling their territorial borders may possess a sense of "we," but they never extend their fellow feeling beyond the vested genetic interest of their individual social unit. Hamadryas baboons sometimes form loose alliances between related bands, but only to defend their common ground against other, unrelated groups. In most cases primate groups can be depended upon to be actively antagonistic to one another. Human groups do not need a direct threat from other groups before they help each other out, nor do they need to be closely related. It is not simply that humans are unique in their ability to cooperate group to group; they are utterly dependent on such cooperation. No collection of human beings can survive without at least occasional assistance from other such human groups. If intergroup violence is rampant in our history, so too is a peculiar amount of mutual reliance.

At some point, our lineage must have developed the most sophisticated Machiavellian tactic of all: the ability to trust. No doubt innate hostility toward strangers remained as potent as ever, perhaps even intensifying as human groups became more complex and more adept at devising schemes to outwit one another. But a wholly original order of relations evolved too, a foil to the xenophobic imperative written into our primate heritage. It was the need to balance *between* cooperation and competition that turned a hominid's social organization into human society. And the evidence shows, I believe, that it happened in the Upper Paleolithic. The explodingly creative culture of the period might be seen as the physical litter of a novel experiment: a species' investment in getting along with itself. The payoff was equally impressive. After four million years of stasis, humankind inherited the earth.

* * *

It all comes down to survival. With their flexible brains, our ancestors may have been better equipped than most species to control the effects of predators and other Hostile Forces of Nature. But like everything else alive, hominids have to eat, and how many hominids can live in a given environment depends on how much food is available. Areas where resources are abundant and available year-round can support higher-density populations, while highly seasonal habitats are more sparsely peopled. If climatic conditions deteriorate, a point may come when a given region is too niggardly and unevenly endowed with resources to support a population. Even if the population does not face immediate starvation, it may become too thinly spread over the landscape for every maturing individual to find a suitable mate. Deaths exceed births, and the population dwindles away.

By the look of the archaeological record of northern and central Europe, that point may have been reached several times during the Middle Paleolithic. The European climate during the last Ice Age was not one long descent of the thermometer, but rather a trend toward increasing cold, punctuated by brief periods of milder temperatures. According to Clive Gamble, evidence of settlement suggests that by intensely exploiting a given territory, Neandertal groups were able to colonize the northern regions during these comparatively balmy respites. When the ice returned, however, they retreated back into the warmer western and southern regions, establishing an "ebb and flow" pattern of occupation that lasted tens of thousands of years. The problem was not simply the cold, but the total collapse of the base of resources it brought about. Less solar radiation, shorter growing seasons, and long snow cover meant meager pickings for gathering and less vegetation for animals to eat, which in turn meant fewer species and smaller herds to hunt. Eventually the environment would leave the Neandertals only two choices: move south again, or die out.

Apparently the Cro-Magnon hunter-gatherers who followed them enjoyed another alternative: They could make friends. Not just friends within the group or among close neighboring groups, but friends in distant places encountered on long treks; friends who, until encountered, might have been total strangers. If you are a hunter-gatherer in a marginal environment, these are the kind of allies you need. They are your insurance policy. In the event of local catastrophe, being part of an alliance is the only way you can be assured of a friendly welcome to the resources of another group over the mountain or across the plain. Ten years later, a drought might give your people a chance to return the favor.

We have already seen one example of this "pooling of risk" among the !Kung Bushmen of the Kalahari desert. Rainfall in the Kalahari is extremely variable. One area can suddenly receive heavy rains, while twenty miles away there is none at all. The !Kung have adapted socially to such conditions by developing an open, mobile society, where individuals move frequently between and among loose-knit foraging groups. Visiting back and forth among groups within an alliance relieves stress on resources in lower yielding areas and at the same time allows people to maintain relationships. When visiting, people present gifts to each other, but it is considered very poor form to keep a gift for long. Instead, on a subsequent visit, you urge it on someone else, who soon gives it to yet another person, the object circulating round and round without ever becoming a permanent possession. The gifts, like the visits, cement trust across vast distances. At the end of their first year of studying the !Kung, Laurence and Lorna Marshall gave cowrie shells to their hosts—one large brown shell and twenty small ones for each woman, intended to be enough for each to make herself a necklace. But when they returned less than a year later, few of the shells remained in the group. They had scattered in ones and twos through a network of exchange over five thousand square miles.

The !Kung's gift network is only one mechanism used to service alliances. Aboriginal hunter-gatherers in the Central Desert of Australia take pains to build social networks by trading or traveling to get foreign raw material even when better quality can be found close to home. Until recently, the Chipewyan of Canada's Hudson Bay depended on caribou herds whose migration routes from their calving areas were often uncertain. Through social alliances established by wife-lending and gift exchanges, local groups were able to move freely into allied territories when the caribou failed to appear on their own horizon. Pastoralists like the Loikop (Samburu) of Kenya also embed networking into the basic structures of their society. They live in territorial clans, but no clan can exist on the resources within its own territory. According to anthropologist Roy Larick, a young Loikop cultivates special relationships with older men in other clans. When the time comes, these "godfathers" assist him through the circumcision ritual and help supply him with the spears he needs but cannot yet afford. Such formalized relationships build interclan connections that would be much harder to establish later in life.

These days, anthropologists are careful not to point to the !Kung, the aborigines, or any other modern human population as a generalized model for what hunter-gatherer behavior must have been like in the past. We cannot assume that Cro-Magnons formed alliances just because modern hunter-gatherers do and it would have been to their advantage to behave the same way. But judging from their disproportionately long legs and the movement of their raw materials, we do know that they were highly mobile. If they were moving greater distances than their archaic predecessors, they must have been encountering other people along the way. Even if the movement of groups across the landscape was random, which is highly unlikely, they were bound to bump into one another more often than more sedentary people like the Neandertals. Given the potentially lethal danger inherent in any contact with strangers, some kind of communication system must have emerged to cope with those bumps, informing people who was who before it was too late to withdraw.

"Whether the message is 'We understand each other, right?' or 'Back off or I'll bash you on the head,' there had to be some kind of codification," says Mary Stiner of Loyola University. "They would have to be investing more in political relations."

If you are an intelligent, preliterate hunter-gatherer, where and how do you invest in political relations? You might begin with your tools. Tools are utilitarian objects, but under certain conditions they can become expressions of social identity too. Martin Wobst, an archaeologist at the University of Massachusetts, believes that material objects begin to carry information about group affiliation, rank, status, and other qualities of identity only when the "social distance" between the user of the object and others who might see it makes this an efficient form of communication. An object that is regularly seen only by one's family and close associates—a kitchen spoon, for instance, or a bed mattress—is not likely to function as a social message, because the people who come within view of such objects already know more about the social identity of the person stirring the pot or sleeping on the mattress than the objects themselves could ever hope to convey. Thus kitchen spoons and mattresses look pretty much the same the world over, with variations in their appearance determined strictly by their functions and the raw materials used to make them. With no target audience, there is no reason for them to "speak" with style. Similarly, material objects do not have much to say to people on the opposite end of the social spectrum: those one is never likely to encounter at all. The bird feathers that a western New Guinean warrior wears into battle may mean a great deal to his allies and enemies. But to me, they mean about as much as a *Go Redskins* T-shirt does to him.

In the social distance between the thoroughly familiar and the never-

met lies the prime target audience for any social information embodied
in the style of a material object: All the strangers and vaguely familiar
people an individual is likely to encounter in the course of a lifetime. By
definition, strangers are stressful presences. The person knocking on your
front door is probably just a delivery man, but he might be a rapist. You
don't know, and that lack of information creates tension. In this situation,
the potential eloquence of material objects suddenly manifests itself.
Looking through the window, you see that the man is wearing a UPS
uniform, and your suspicions are mollified. The outfit is not a perfect
indicator of identity, of course, since the man could still be a rapist in
disguise, but at least you have some information to help you decide
whether or not to open the door.

The UPS man's uniform is even more essential to him than it is to
you, because the announcement of identity it predelivers gets him through
not one but dozens of encounters with strangers every day. Martin Wobst
reasoned that because the prime target of a social message voiced through
material objects is composed of unfamiliar, potentially threatening individuals, the objects most likely to serve this function would be those that
can be seen at a distance—like a uniform—before an encounter becomes
unavoidable. To test this idea, Wobst focused on a region of the modern
world where early information about the identity of someone approaching
could be crucial to survival: the former Yugoslavia. The borders of that
cobbled-together country united three major religions, four major nation-
alities, three major languages, and a potent mix of competing ethnic
groups: Serbs, Slovaks, Croats, Muslims, Czechs, Hungarians, Germans,
Romanians, Albanians, Gypsies, Jews, Greeks, and Italians. All these people were economically dependent on one another but maintained a fierce
allegiance to their own kind. Wobst reasoned that in such a potentially
explosive mix, traditional clothing should clearly announce group affiliation. The most explicit ethnic markers would be any article of clothing
that could be spotted from the other side of a narrow valley or a distance
down the road.

"Being visible over the greatest distance," he wrote, "they are the only
parts of dress which allow one to decipher a stylistic message before one
gets into the gun range of one's enemy."

Wobst homed in on hats as the most conspicuous piece of clothing
worn year-round. If style transmitted social information, traditional headdress in Yugoslavia should be highly distinctive group-to-group, but homogeneous within each ethnic enclave. A survey of the ethnographic

literature on Yugoslavia showed this to be true, at least among the men. Whether a man wore a fez, a skullcap, a wide-brimmed Stetson-like hat, or a flat-topped woolen *sajkaca*, whether his particular headgear was white, red, tasseled, striped, or medallioned, depended entirely upon group affiliation. In Montenegro, before the federal government forced its influence on the population, each isolated mountain tribe wore its own style of headdress. In urban areas, men more often wore hats where competition among ethnic groups was strongest. Sarajevo, a simmering stew of Serbs, Croats, and Muslims, boasted a thriving industry of fancy hatmakers in its bazaar. By contrast, the capitals of ethnically homogeneous Croatia and Slovenia had hardly any millinery at all.*

Hats are, of course, only one kind of artifact that people use to broadcast group identity. Any visible garment can be drafted into service as an information messenger. In Central Brazil, where hats are something of a nuisance, one tribe is distinguished from another by variations in coiffure. The body itself can be painted, shaved, pierced, scarified, or fitted out with lip plugs, penis sheaths, spider-web headbands, beehive coiffeurs, and other elaborations. As Polly Wiessner's study demonstrated, objects like projectile points can also communicate identity group-to-group since they are likely to be left behind on the landscape. Messages borne by clothing and material objects are not limited to simple statements of affiliation. They can express social rank (exotic feathers, mink coats), emotional state (war bonnets, black dress or hair shorn for mourning), and marital status (wedding rings, raised geometric scars in chest). Among the Loikop, the

*In recent decades, before the breakup of the Communist state, distinctions in folk dress in the region had begun to blur. Many costume markers were spreading beyond their group, and a certain amount of "ethnic cross-dressing" was taking place—Croats wearing turbans, Serbs donning the baggy Croatian *dimija* pantaloons, and so on. Dress style among younger people began to signal allegiance to a different identity, typified by blue jeans. ("The major diluting influence on traditional dress would have been not the Communist government, but Calvin Klein," says Lynn Maners, an anthropologist at UCLA who has recently worked in Bosnia.) But true to what Wobst's study would predict, the outbreak of civil war literally brought ethnic clothing out of the closets.

"I thought of Wobst frequently as I watched the Serb nationalist reawakening in the years before the war began," says Staso Forenbaher, an archaeologist from Zagreb. "The dress and the accompanying paraphernalia were coming right from their national guerrillas of World War II. It seemed like a bad dream." The Serbian *sajkaca*, just a few years earlier considered "country bumpkinish," suddenly appeared everywhere, including on the heads of victorious Serbian generals, worn with a swagger of red flashing. Croatian paramilitary troops put on World War II fascistic garb and pinned the distinctive red and white checkerboard emblem of Croatian nationalism to their hats. "Some of them put on the skull and crossbones again from the SS days," says Petar Gumac of the Smithsonian Institution. "It's meant to scare the bejesus out of the population, and it does. There's a lot of symbolism going on in what people wear."

kind of spear a man carries depends upon his stage of life. Only senior warriors, who are responsible for the offense and defense of the community, may carry massive thrusting spears, and they often tote two. From this apex of male potency, the size and quality of spears falls off in two generational directions with the youngest boys and the old toothless men bearing the smallest weapons of all.

From the beginning, the European Upper Paleolithic and the African Late Stone Age are marked by a greater attention to the *shapes* of such finished tools, which become much more uniform within specific geographic regions and periods of time. This is precisely what would be expected if people were using tools as social messages. While before it might have been enough to make a scraper or point in any old shape, so long as it scraped or poked effectively, a tool invested with social information *had* to be rigidly standardized, so that everyone in range of the message got the point. Just as the UPS man cannot occasionally show up for work in a plaid shirt and jeans instead of his familiar brown uniform, an Upper Paleolithic hunter could no longer make tools in any old way that would produce the desired cutting edge.

"A typical Mousterian might make a tool and think, 'As long as this does the job, I don't care what it looks like,' " says Cambridge University's Paul Mellars. "But an Upper Paleolithic fellow says, 'This thing is a burin, I call it a burin, I use it like a burin, and by God, it better look like a burin.' "

The uniformity in Upper Paleolithic artifacts thus reflects a crucial change in the design of human society. If Mellars's "typical Mousterian" couldn't care less what his sharp-thing-that-pokes looked like, it wasn't because he or she lacked the intelligence to invest stone with social information. What was missing was a *target audience of unfamiliar people* to whom that information would have mattered.

"[For Neandertals] there wouldn't be any question that you belong to the resident hominid group," says John Shea, "because you are *there*, the only one in this valley, this cave, this range, this part of the coast, or whatever. It's only when you have lots of contact, lots of people around, that these messages of identity become important."

Using archaeological evidence to calculate population density has always been problematic—does four times as many sites mean four times as many people, or the same number of people moving four times as often? There is no question, however, that the appearance of large, complex sites like Dolní Věstonice shows that "lots of people" were around in particular places. Recent genetic evidence suggests that populations all over the Old World underwent dramatic "explosions" around the beginning of the Upper Paleolithic in Europe and the Late Stone Age in Africa. In 1993, Henry Harpending and his colleagues at Pennsylvania State University used a new method of analyzing mitochondrial DNA data to reconstruct the demographic prehistory of *Homo sapiens*. In almost every population examined, they found statistical evidence for explosive growth around 50,000 to 60,000 years ago.

They suggest that technological innovations may have fueled this increase. While this is possible, the timing is off—in most regions, the cultural changes of the "creative explosion" apparently occurred just *after* the people explosion, so it seems just as probable that the spread of new innovations was the result instead of the cause of population growth. Whatever triggered the burst of new humanity, it amplified the need and the opportunity for more sophisticated political communication. As alliances began to coalesce on the landscape, they introduced the *possibility* that there were people out there, beyond one's circle of relatives and familiars, who might not be something inherently, viciously, lethally hostile. But not every stranger would be friendly. Artifacts thus began to take on style when human social interaction was gripped by a new, energizing quality: ambiguity.

Looked on this way, the first great mystery of the Upper Paleolithic— its suddenness—is no longer so mysterious. The transformation of a largely utilitarian material world into one in which objects are tinged with social meanings *must* happen all at once. Imagine two Paleolithic hunting parties meeting in some border area between their separate territories. Their two clans are knit together by a history of reciprocal exchanges and common traditions, including a particular way of fashioning spearpoints. This common point of reference allows them to meet without hostility: *"Ah, I see by your spear that you are of the River People. Seen any deer?"* Suppose one of the groups later encounters a third hunting party, whose weapons look different. Even if the others never intended their spearpoints to "say" anything when they made them, the weapons will transmit a distinct social message to the River People: *"Weird spear! Not River People!"* and probably, *"Watch out!"* Thus as soon as the River People's spearpoints carry a social message, *all* spearpoints *necessarily* gather meaning, whether or not they were meant to mean anything to begin with. Thereafter, you might find spearpoints that issue garbled or tragically misinterpreted messages—*"Strangers mean us harm! Get them first!"*—but you cannot find a spearpoint that carries no message at all. In Martin Wobst's words, the points have lost their "signaling innocence."

"[This] argues for the sudden appearance of stylistic form in material culture, instead of the gradual incremental evolution often anticipated," Wobst wrote. "A state of no-stylistic-messaging should suddenly be replaced by a state in which stylistic form has pervaded at least one (or more) category of material culture."

The emergence of style in artifact design is only one feature of the Upper Paleolithic. But once utilitarian objects like spearpoints carry social messages, would it take any great leap in imagination or technical know-how to create artifacts that *had* no utilitarian function, whose sole adaptive purpose was to carry social information? And why not wear (or carry) objects that express more about the individual than mere group affiliation, that say "I am a warrior," "I am important," "I am a married woman," or "I believe in the great god Org"? While you're at it, expand your media too. Stone is too rigid for expressing all you have to communicate. Carve bone. Whittle antler. Mold clay. Splash your new, technicolor talents on the dark walls of caves. Creatively explode.

Not every aspect of the Upper Paleolithic can be swept into a single, grand explanation, but it seems to me that an immensely important step occurred when people crossed the barriers of mutual distrust and connected across landscapes, in the process recruiting mere materials to express those connections, to carry information and ideas. After that point the world ceased to be taken at face value and came alive with metaphor, symbol, and layered subtleties of inference and possibility. The archaeologist solemnly pondering a decorated slab or painted wall from the Upper Paleolithic may never know exactly what the thing meant to the person who made it. But there is no doubt that it means *something*. Some artifacts, like the pierced beads and other body ornaments that litter French and German archaeological sites from the beginning of the Aurignacian, 32,000 years ago, are clearly intended to express and elaborate personal identity. Others seem intended to define the communal identity of the alliance. The earliest graphic images in the Aurignacian of southern France, for example, include simple, repeated patterns of Xs, notches, and incisions.

"Such abstract images are found at site after site over broad regions," says archaeologist Randy White of New York University, "suggesting they were signs and symbols shared by members of regional social entities that may have just been coming into existence during the Aurignacian."

The two-dimensional artistic output of these nascent societies included crude engravings and paintings of animals, as well as numerous triangular and oval patterns of lines etched into rocks, long believed to represent vulvas, though they might just as well be mnemonic sketchings of animal tracks. As artists, the Aurignacians were more comfortable in three dimensions. The German site of Hohlenstein-Stadel hosts a lion-man a foot high, carved whole from a mammoth tusk 30,000 years ago, with a hanging penis, dainty feline ears, and the muscular forward lean of a linebacker. Nearby, the delicate animal figures from Vogelherd bear patterns of geometrical incisions that also crop up on pendants, beads, and other portable objects from the region, hinting at a common tradition. Archaeologists have also used various sorts of portable art from later in the Upper Paleolithic to sketch in regional alliances along the coast of the Bay of Biscay in Spain, in the Pyrenees, and in the Moravian sites surrounding Dolní Věstonice.

The most ubiquitous symbols of Paleolithic alliance may be the "Venus figurines" like the clay statuette from Dolní Věstonice. Hundreds of them are now known from the central Russian plain to the Pyrenees, appearing first around 24,000 years ago. In spite of the enormous distances between them, some of the female carvings are remarkably similar in size and design, with faceless heads and sloping shoulders, tapering out to wide hips and tapering in again through corpulent thighs to tiny or nonexistent feet. No one knows whether they were made by men or by women, or in what context they were used. According to Clive Gamble, they may have served a political function as part of a "system of visual display" of a loosely knit, far-flung social alliance, helping to soften and sanction initial contact between groups that might not share even a common language, let alone common genes. With the climate collapsing toward the glacial maximum, the timing was right for such a network to emerge. As resources diminished, the population would be spread thinner and thinner, with individual groups becoming more dependent on contact over increasingly long distances for information, mates, and assistance in times of crisis.

It would, of course, be silly to imagine the social complexion of Europe during any span of the Ice Age as one overarching alliance, a sort of proto-European Economic Community. Among modern hunter-gatherers, alliances are dynamic; they cross and overlap, in some cases encompassing geographical boundaries and divisions of language, in others making mortal enemies of near neighbors who share the same dialect. Small, regional alliances coalesce into larger networks, with allegiances and obligations trickling up through layered levels of reciprocity. Similarly, within, along-side, and perhaps against the Venus network, countless other alliances would have sprung up, expanded, and fallen apart over time, parceling the continent up into a shifting social mosaic, with the fluid borders of each piece defined by the negotiated interactions of people.

Among modern Australian hunter-gatherers, the complexity of the so-cial landscape is expressed in painted, carved sticks called *toas*. Hunters passing through an area leave at waterholes these intricately decorated staffs, which are actually coded instructions, informing those who follow of the group's destination, composition, and activities. But not everyone who finds the toa reads the same text. The arrangement of nicks and swirls pertains to social rituals and discussions, some of which will be known only by other members of the hunter's "estate"—the religious core of the territorial group—while others will be more generally understood. "This gives the message creator immense scope to stratify the levels of meaning," says Gamble. "If the stick is found by a socially distant person, he will be able to decode only the most superficial levels of meaning. However, if discovered by a member of the same estate group, the stick can be interpreted through the shared knowledge of the group rituals, which give meaning to an intricate social landscape that is enacted in physical space."

The toas elegantly exemplify the second great blessing of alliance, over and above the insurance policy of reciprocity: It opens a news network for what is happening in the environment that is of concern to people *of* the environment, who are wholly dependent on what they can glean from their immediate surroundings. Ice Age hunter-gatherers must have been equally hungry for information. The caves of Isturitz in the western Pyrenees have yielded "baguettes" of rounded reindeer antler from the Magdalenian pe-riod, encrusted with squares, spirals, and other perplexing geometrics. To the members of that departed network, their meaning may have shone through as lucidly as the letters on a highway exit sign do for us. In other examples of Paleolithic art, the artist has emphasized and exaggerated the elements of the beast that the hunter needs to know in order to track successfully. In Lascaux and other painted caves, the legs of horses, cows and bison end in hooves in "twisted perspective," displaying not the pro-file of the hoof, but a representation of the track it leaves on the ground. Magdalenian "spearthrowers" from Mas d'Azil, east of Isturitz, display young ibex rendered with exquisite care, rumps up, with great fat turds protruding from their backsides. The people who made these spearthrowers were skilled trackers who understood, viscerally, that every trail of ibex turds must end, eventually, with an ibex.

Information would also have been flowing copiously through alliance networks over nonmaterial channels that do not show up in the archaeological record. Among modern hunter-gatherers, the most effective form of information exchange is the most obvious: People talk. According to Lorna Marshall, conversation in a !Kung camp "is a constant sound like the sound of a brook, and as low and lapping, except for shrieks of laughter." People gossip, criticize each other, make plans and tell in great detail about the comings and goings of friends and relatives. But sooner or later, the talk always turns in one direction.

"Their greatest preoccupation and the subject they talk about most often, I think, is food," writes Marshall. "The men's imaginations turn to hunting. They converse musingly, as though enjoying a sort of daydream together, about past hunts, telling over and over where game was found and who killed it. They wonder where the game is at present, and say what fat bucks they hope to kill."

"I used to think that this idea of hunter-gatherers sitting around talking about the hunt was just romantic nonsense," says John Yellen, who spent several field seasons with the !Kung himself. "But it's real, and no minor matter. Everybody would be sitting around after a meal, and one guy would say, 'By the way, I saw some wildebeest tracks up a ways, only a couple of days old.' And then somebody else says, 'You know, I saw some too! It was a mother with young, just over that hill, and they were only one day old.' These people were constantly ferreting out information, sharing it, passing it along."

No one doubts that Upper Paleolithic and Late Stone Age hunter-gatherers possessed fully human language. The dispute concerns whether earlier hominids had it too, in particular the Neandertals. Personally, I cannot believe that large-brained humans who were emotionally committed to burying their dead were grunting to each other for 100,000 years. But language is not an all-or-nothing proposition, and various elements of modern language may have evolved at different points in time. Linguist Derek Bickerton has suggested that the last major feature of fully modern language to emerge was the concept of time itself—a past and future tense. According to Robert Whallon, an archaeologist at the University of Michigan, this final addition to human language may not have come into play until social changes in the Upper Paleolithic demanded it. So long as foraging groups functioned essentially as integrated, self-contained

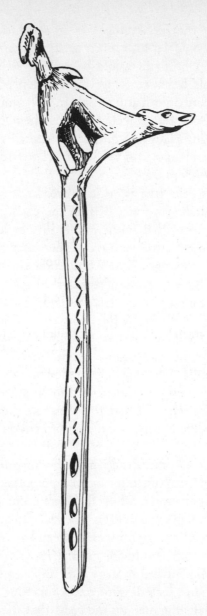

A spearthrower from the Upper Paleolithic site of Mas d'Azil in France. Carved at the top is a young ibex with its rump up, excreting an oversized turd. Once considered to be Stone Age "jokes," such images are now thought to be the adaptive artistry of hunter-gatherers whose survival depended on tracking animals by their droppings and other visual cues. COPYRIGHT © 1982 BY J. WYMER, REPRINTED BY PERMISSION OF ST. MARTIN'S PRESS units, combing the terrain together in territories that excluded other groups, there would be little purpose in discussing the future, because whatever it brought about would be collectively shared or endured. But when individuals began to forage more independently, splitting up and bringing food and information back to the group, or moving between other groups in an interconnected alliance, the need for an efficient means of communicating plans and discussing future events arose. In this context, the shared memories of the !Kung hunters described by Lorna Marshall are hardly idle musings.

"Reference to the past is critical in order to make predictions about the future," Whallon writes. "It is on the basis of generalizations drawn from past experience that future events may be anticipated."

One of the most effective ways to generalize about past experience is to stitch it into narrative. Leah Minc, then an archaeologist at the University of Michigan, found a particularly elegant example of this in the curiously interrelated oral traditions of two separate indigenous groups on the north coast of Alaska. Before Euro-American contact mangled their ancestral ways, the Tareumiut, or "people of the sea," depended on whaling for much of their sustenance, while their landlocked neighbors, the Nunamiut, literally "inland people," relied heavily on the spring migration of caribou. Both groups supplemented these primary resources with other foods, but if the whales failed to appear in the spring, the Tareumiut would starve, while a critical shortage of caribou would bring famine to the Nunamiut.

Using the record of tree rings and other long-term climatic data, Minc determined that the Nunamiut faced a potential crisis in caribou supply about every twenty years, while the Tareumiut could expect a meager spring catch of whales one in every five years. In anticipation of these lean times, members of one group seek and cement trading partnerships with members of the other, through mutual feasts, exchanges of goods, and wife-lending. This survival strategy is so important that it is etched in the myths of creation. According to one Nunamiut myth, the creator-hero Aiyagomahala tells the men on the coast that they should venture inland with seal and whale blubber and skins to exchange with the caribou hunters they find there. "Aiyagomahala told his people that every family . . . should have a partner in each of the other groups and that nobody should trade with anyone else," says the story. "From that time on, the Nunamiut greatly enjoyed giving and attending feasts."

"To attribute the origins of these customs to the time when the people themselves were created," says Minc, "is to establish trade partnerships and the hosting of intercommunity feasts as synonymous with human existence."

Minc's climatic data shows that in addition to the periodic shortages in food, the Nunamiut and Tareumiut could also anticipate disastrous crashes in their prime sources of subsistence every century or so. A century is beyond the reach of any generation's individual memory, so the solution to the crisis could not be handed down through oral instruction. The emergency course of action was instead coded into their cosmology. In its lifetime, every marine animal is believed to have a spirit which has its double in the terrestrial environment inland, and vice versa. Thus killer whales are the spirit doubles of wolves, and mountain sheep find their marine reflections in white beluga whales. When wolves starve on land, they go to their relatives in the sea and turn into killer whales. The myth need not be any more explicit; a starving caribou hunter will know when it is time to take his family and head for the sea.

Suspicious as I am of looking upon language as the prime mover of the Upper Paleolithic, it must have been a potent player in the evolution of intergroup cooperation and alliance. The spoken word opens up the possibility of interaction without emotion. With language, conflicts over territory, food, or mates have the chance of a negotiated solution—a luxury completely beyond the means of the most intelligent animals. "Language," writes the French anthropologist Pierre Clastres, "is the very opposite of violence." Because fully modern language would have been necessary for alliance, it does not follow, however, that the emergence of language was the deciding element that overcame the barrier of intergroup hostility. Negotiations demand language, but language also demands a social climate conducive to cooperative relationships in the first place.

"A capacity for transcending physicality has obviously less to do with the genetic constitution of individuals than with the political/social/sexual situation in which they find themselves," writes British anthropologist Chris Knight. "To the extent that, in any community, issues between individuals or groups are decided purely or primarily physically, language not only cannot evolve—it loses all relevance."

To illustrate his point, Knight offers two chimpanzees, Booee and Bruno, who were trained in American Sign Language under the tutelage of primatologist Roger Fouts, in the 1970s. After their training, Booee and Bruno had no trouble asking for and receiving food from their human companions. Between themselves, however, the free flow of information hit a brick wall: "When one of the two chimpanzees has a desired fruit or drink, the other chimpanzee will sign such combinations as GIMME FRUIT or GIMME DRINK," Fouts reported in 1975. "Generally, when the chimpanzee with the desired food sees this request he runs off with his prized possession."

What we have here is not a failure to communicate. The chimp with the goody knows exactly what the other one is trying to say. But simply adding language to a chimpanzee's behavioral baggage does not create a cooperative social partner. For that, you need a social organization that makes sharing worthwhile. "The problem with Booee and Bruno was not their inadequate linguistic competence or training," says Knight. "It was their lack of involvement in a wider system of cultural meanings. The two animals were not citizens within a chimpanzee republic. . . . Their rights and duties were not codified in the name of a higher authority; neither had they entered into any moral contract regarding the sharing of valuables such as food or sex. It was for these reasons that they lacked a social universe capable of making human language even remotely worth learning."

The citizens of a "Neandertal republic" would not have been as morally depleted as Booee and Bruno. But whatever the archaic human social status quo, I doubt that language fell on it from the sky and altered everything. It is more likely that some enhancement of language capacity evolved out of the increasing complexity of social relations. What was truly revolutionary about the Upper Paleolithic was not language, style, or art, but the opening of the social conduits through which information in all such novel forms could flow. The heated infection of new technologies was also transmitted by alliance. Earlier human populations, both in Europe and elsewhere, might have hit upon a sophisticated technological innovation and passed it down vertically through time, from generation to generation within their own region. But for lack of long-distance connections, the horizontal propagation of innovation from one population to another would peter out. It does not take a new, special kind of brain to make a bone harpoon point, or to learn how to use it. But it does take a chain of brains to keep the idea alive. It was the human need to reach out for other humans, across the landscape, that fueled the first signs of the creative explosion. Twenty thousand years later, the highest expressions of Paleolithic art may have had more to do with shutting them out.

*   *   *

One September afternoon in 1940, four teenage boys set off from the village of Montignac in southwestern France and slipped quietly into the woods. Their destination was a hole in the ground that one of them had discovered in the forest a few days before, exposed beneath a newly up-rooted tree. Like all children of the region, the boys had heard the legend of a mysterious passageway under the Vézère River that linked the ancient castles of Montignac and Lascaux. That afternoon, they brought along rope and a crude oil lamp, widened the hole, dropped some stones down to test the depth, and squeezed through. One by one, they disappeared into the darkness. Sliding down a steep tunnel of wet clay, they landed on a level platform and, fumbling with matches, lit the lamp.

One can imagine the depth of their wonder at what the lamp's light revealed. On the wall of rock above them was a fantastic being, a beast with a dappled coat, the hind quarters of a bison, the distended stomach of a cow, two long delicate horns, and what appeared to be a bearded face. Before this creature flew a headless horse, its foresection lost where a piece of the wall face had peeled away. Out of the scarred portion of wall emerged another horse, running faster, then a third, and a fourth leaping ahead. Above them floated the most splendid equine of all with a sweeping black tail and tawny coat. The violent commotion of its mane narrowed toward the head, then burst out again and became the back of an enormous aurochs, a wild cow with speckled shoulders and a winking, friendly eye. Behind her, another horse delicately cantered off into the shadows, without the aid of legs.

In all, there are twenty-odd animal figures painted in the first, largest chamber of the famous Lascaux Cave, known as the Hall of the Bulls. When my wife and I were in among them, half a century later, stopping to count their number seemed not only irrelevant but physically danger-ous. They circled around us, horse leaping out of cow, bull chasing deer, a boil of antlers, horns, and galloping hooves. Farther into the cave, the wall bulged with the anger of two ibex charging each other with lowered horns. A stag roared. A herd of horses toppled headlong over a cliff, their hooves futilely scrabbling the slippery rock face. In another passage, branching off to the right, the scenes were more tranquil: a quartet of stags with magnificent antlers and gentle benevolent eyes floated by. Op-posite them, a cow with dainty horns and pointed hooves seemed to sigh against the weight of a great pregnant belly dragging on the ground. No wonder she was frustrated: She had been pregnant for over 17,000 years.

Staring in utter awe at that cow, it struck me how naive it would be to apply the word "primitive" to such a creation, no matter its age, or to a creator who could so effortlessly take the heft of that gravid belly. Lascaux is probably the most spectacular example of an artistic phenomenon that developed in the later Upper Paleolithic, and was by and large confined to southwestern France and Cantabrian Spain. Hundreds of other caves are known to have painted, etched, or decorated walls, and more still are being found. In 1991, a French diving instructor discovered a cave two hundred yards up a submerged tunnel in a limestone cliff on the Mediterranean coast near Marseilles, which rivals Lascaux in its beauty and the abundance of its art. While the site would have been inland in the Ice Age, today the only access is through an opening 121 feet underwater.

The earliest interpreters of such cave paintings viewed them as idle doodlings, a sort of "art for art's sake." This is highly improbable. In the 1960s, André Leroi-Gourhan showed that the art had definite structure and recurrent patterns. Three animal species—horse, bison, and aurochs—account for 60 percent of all the images in the caves, with half a dozen others making up the bulk of the rest. Human images are rare, as are carnivorous animals and plants, and the placement of the images reveals deliberate purpose. Many engravings and paintings can be seen only by squeezing through terrifyingly tight passages; others are discovered after walking miles deep into labyrinthine cave systems. In a recent study of three caves in the foothills of the French Pyrenees, Légor Reznikoff and Michel Dauvois demonstrated that images had been painted precisely where sounds in the cave resonate the most—hardly what one would expect from a bunch of Paleolithic *dilettanti* with time on their hands.

Cave art is probably best viewed as part of carefully planned ritual experience, which may have relied on incantatory chanting and rhythm as well. No doubt the meaning and purpose of the rituals varied considerably—perhaps some were indoctrination ceremonies for the young, as John Pfeiffer believes, while others were performed as symbolic enactments of the hunt, a view long put forth by the French archaeologist and cleric Henri Breuil. Arguing from analogy to the rock art of South African Bushmen, archaeologist David Lewis-Williams believes many of the images sprang directly from the hallucinations of shamans. Transported into a trance state, the shaman produced the art as an attempt to fix the hallucinated image and give it permanence.

Considering that hundreds of caves were painted over several thousand years, there is no reason to believe that all of the art sprang from a single ritual purpose. Modern ritual behavior is as diverse and complex as modern societies themselves, but no matter what the specific form and content of the ritual—a circumcision ceremony, a mass, an animal sacrifice, a homecoming game—the experience reinforces the solidarity of whatever tribe, clan, or other social entity formalized the ritual in the first place. What was happening in Upper Paleolithic society that would unleash such an intense, unprecedented *need* for ritualized behavior, in this one region, at this one point in human history?

Significantly, the apogee of Franco-Cantabrian cave art corresponds to the maximum reach of the glaciers. The already frigid climate of northern and central Europe had begun to decline even more around 25,000 years ago, and by 18,000 years before the present, much of the northern region of the continent was uninhabitable polar desert. People were able to survive in the cold, dry shrub tundra skirting the polar no-man's-lands, but with increasing difficulty and in decreasing numbers as the glaciers continued their slow advance. "Around the glacial maximum," says Olga Soffer, "people were forced in two directions. One was into southern France, the other was out onto the Central Russian Plain. Nobody lived up there between the Scandinavian glacier and the Alpine glacier. That corridor was pure hell."

In contrast, the climate of southwestern France remained reasonably hospitable. Pollen studies show that trees grew in sheltered valleys, and the faunal record reveals a remarkable variety of large herbivores alive at the time, including mammoth, woolly rhinoceros, horse, bison, aurochs, and numerous species of deer and antelope. The river valleys of the Central Russian Plain were equally rich in resources. As the ice reached its full extent 20,000 to 18,000 years ago, these comparatively balmy "refuge areas" became the only places where humans were able to survive on a permanent basis.

Having eluded the worst environmental stress, the Franco-Cantabrians were faced with what was in some ways a more difficult challenge. They had to learn to cope with other Franco-Cantabrians. As we have seen, in a resource-poor environment like the Australian outback or the Arctic tundra, it pays to have far-flung friends. But if people crowd together in areas where food is available but not unlimited, what were once potential allies can quickly turn into potent threats. Communication networks would have played an essential role in the people-packed refuges of the late Upper Paleolithic. But more effort would have been put into marking the borders of alliance than into opening them up to all comers. Territoriality would become increasingly enforced. According to Clive Gamble, the exchange of exotic goods would become less an investment in "insurance premiums," and more a payment of "membership dues" to establish one's status as a member of the larger territorial group.

The art of the late Upper Paleolithic reflects this switch to more closed societies. In the Solutrean Period, 20,000 to 17,000 years ago, portable art grew increasingly scarce. The cave art that came to predominate in Franco-Cantabria may be beautiful, but it is anything but mobile. Painting on walls attaches meaning and importance to fixed geographic space: *our* magic, *our* cave, *our* valley. According to Michael Jochim of the University of California at Santa Barbara, population pressures forced people to rely less on reindeer migrations and other resources that might or might not appear within their territory in any given year, and instead focus on more predictable food sources. Salmon in particular became a major staple. During the Ice Age they were abundant in the rivers of southern France and northern Spain that drained into the Atlantic, and while salmon, too, are migratory, their routes can be predicted by the course of the rivers. By decorating caves, Jochim believes, individual groups were, in effect, laying ritual claim to a stretch of river and its pink-fleshed harvest.

"It is not the case that [Franco-Cantabria] formed a Garden of Eden with leaping salmon and running deer, thus allowing [late Upper Paleolithic] populations to lie back in their rock shelters and pass the time by painting them," says Clive Gamble. "On the contrary, the cave art serves as a measure of the intensity of competition between these populations and their physical and social environments."

In the eastern refuge area on the Russian Plain, territory could not be claimed by ritually painted caves, because there were no caves to paint. According to Olga Soffer, individual groups established their presence on the landscape instead by the style they used in the construction of their mammoth-bone dwellings. "There is one dwelling where you find left scapulae arranged vertically on the left side, an upturned skull and right scapulae on the right. Nearby you have other redundant patterns. Of course, this is deliberate. On another hut, you get a sort of woven herringbone effect. What I think they are doing is marking *place*, much the same thing as the painted caves."

Ritualistic activity bolstered group identity against a more dangerous threat than any posed by a competing group on the landscape: the powerful ancient imperatives of the individual and its blood-rooted loyalty to immediate kin. Any alliance system generally works to the advantage of its members, but not everyone within the alliance benefits equally all of the time. Even under the best of circumstances, the vagaries of the environment make it unlikely that everyone can enjoy the same access to resources. When too many people are forced to compete for a limited supply of food, or simply interact with each other over a long period of time, self-interest threatens alliances from within. Packing people together on the landscape also causes competing alliances to overlap. If members on the periphery of one group are unsatisfied, they may be tempted to desert the coalition for the advantages inherent in forming bonds with other groups instead. Conflicts are bound to arise at the same time that the most efficient, painless way to resolve social disputes disappears.

"When the !Kung have a real disagreement, somebody just packs up and leaves the group for a while," says Soffer. "But you can't do that if you are stuck in one place, or if you risk leaving behind your meals for the next six months. Under those circumstances, everybody had better be taking their nice pills."

The best way to persuade people to swallow that medicine is through ritual. By taking part in public social acts, people signal their acceptance of the higher moral order of the group, which transcends their status as individuals. Ritual can be expected as soon as alliances develop—remember the exploded ceramics at Dolní Věstonice 28,000 years ago. But the importance and complexity of rituals increases along with the potential for social and economic imbalance latent in more settled societies. According to Soffer, the change over time in the placement of food-storage pits in open-air sites on the Russian Plain traces this growing imbalance. Early in the Upper Paleolithic, such pits were usually dug in the middle of a group of dwellings, suggesting an egalitarian sharing of resources. Later, each dwelling had its own set of pits. At some sites, after the glacial maximum, only one or two dwellings had control over most of the stored goods. These are the same dwellings that contain the most portable art and body ornaments. Apparently some members of the group were becoming more important than others. If society was crystallizing into some kind of hierarchy, those on the top would have to have some means to placate the have-nots.

"It's hard to tell archaeologically, but after the glacial maximum, I think there were people who were running the irrational—leaders in charge of ritual," Soffer says. "Sacred information is, after all, the easiest to control, because it can't be checked. If I tell you that there are reindeer over the next hill, you can climb up and see for yourself. But if I tell you that I speak to God and He speaks to me, how are you going to prove me wrong? In the ethnographic record, wherever you get inequality, it is justified by invoking the sacred. Inequality is very costly. Some people are going to be unhappy. To keep them pacified, you supply them with sacred justification and ritual—keep them dancing and waving purple feathers. Wherever you go, whatever big shot is running the show, he or his brother is also the guy in charge of the rituals."

Even a decade ago, most archaeologists would have agreed that true social complexity originated with the emergence of agriculture, some 10,000 years ago. Influenced by the inordinate research attention given to the !Kung, hunter-gatherer societies of the Upper Paleolithic were seen as fairly simple affairs. It has become increasingly evident that late Upper Paleolithic society was anything but simple. Groups who were more or less sedentary 20,000 years ago were developing sophisticated solutions to the social problems of living together in a limited space. The mere *possibility* of cooperation and alliance with non-kin, which had emerged even earlier, represents an enormous enrichment in the complexity of society over anything evident in the previous four million years of hominid presence on the planet.

Such a fundamental transformation in the human social universe must have left its mark on the interior landscape of the human mind as well. Human beings all share a conscious sense of self, the "inner eye," as Nicholas Humphrey punningly puts it, that is able to peer in on its own behavior, moods, needs, and volitions. The running monologue of this inner voice is so integral to what we think of ourselves that it is impossible to imagine what life would be like without it. But why is it there in the first place? Why did it evolve? A minute ago, the cat was clawing at the back door, so I got up and let him in. I could just as well have heard the cat and responded unconsciously, with the same net result—the irritating noise would cease, and I could continue to write undisturbed. Why do I need to *know* that I am irritated by something? Why not go through life possessed of all the cleverness, learning acumen, and sheer technical expertise that we humans are so famous for, but remain blissfully oblivious to it all?

The growth of human consciousness must be an adaptation for processing the kind of information our ancestors were trying to make sense of in the first place. According to Humphrey and others, by far the most challenging information was social. The "collegiate community" of primate social life spurs intelligence from two directions: by providing the individual with a rich context for learning, and by surrounding him or her with other, equally clever individuals who are cooperating for the good of the group, but simultaneously, and probably with greater attention, looking out for themselves. This means that at any point in our evolution, the extent to which hominid social life was founded on mutual aid and cooperation was precisely the extent to which it would be riddled with cunning and deceit, as individuals pursued their own interests in the "game of plot and counterplot" I referred to earlier. In such a complex social world, a time might come when mere intelligence no longer sufficed. The advantage would go to the player who was not only a keen observer of others' behavior, but who also could peek inside their minds and anticipate their next moves. No one has yet evolved a nervous system capable of directly reading the minds of others. But we have evolved the means to read our own. By providing awareness of the motivations and consequences of our own actions, consciousness grants us insight into the actions of others as well.

"The trick which nature came up with was *introspection*," writes Humphrey. "It proved possible for an individual to develop a model of the behavior of others by reasoning by analogy from his own case, the facts of his own case being revealed to him through examination of the contents of consciousness."

It is certain that consciousness evolved. Pinpointing *when* is purely speculation. Many investigators are reluctant to deny a sense of self to primates, dogs, or even less intelligent animals, which would mean either that consciousness emerged before the hominid split from the apes some six million years ago or that the faculty evolved separately in different lineages. Others place its origin much more recently; Julian Jaynes of Princeton University, to take the most extreme case, does not believe consciousness emerged until less than a millennium before the Greek philosophers. Looking at hominid brain casts—or even whole brains, if we had them—could not possibly resolve this question, because consciousness has no fixed address in the brain. But if consciousness is the evolutionary product of complex social interaction, it would seem to make sense that it emerged when society suddenly became much more complex.

I do not know if the inner voice that represents ourselves to ourselves is a legacy of the Upper Paleolithic revolution, but I cannot imagine a change more cataclysmic, more stressful to the individual nervous system, more in need of an introspective self, than the eruption of social information that attended that event. Hard as it might have been to read the motives and intentions of those within one's immediate group, how much more difficult it would be to anticipate the actions of total strangers, to calculate one's own interests in the face of a hundred ambiguous situations, to position oneself in relation to a social hierarchy, a layered system of alliance, a higher moral order, that by its very nature demands a more articulated awareness of the boundary between the individual and society.

While the birth of self-consciousness is invisible to archaeology, the first clear trace of that border between self and society is not. For the past several years, Randy White of New York University has been studying the abundance of beads and other body ornaments that suddenly appears in Aurignacian sites in France, Belgium, and Germany. By now he has examined some 18,500 body ornaments from before 28,000 years ago, only a handful of which predate the Upper Paleolithic. What impresses White most about the Aurignacian beads is the amount of time and care put into making them. Most are manufactured on materials imported from elsewhere. People with access to shells did not wear shells but instead traveled or traded to obtain mammoth ivory, while those who lived near mammoth disdained ivory and chose exotic shells. Rarity enhanced value in an ornamental object, just as it does today. White even has examples of ivory carved to look like shell, and deer teeth faked from limestone. Once the prized raw material was procured, it was shaped, polished and drilled, using highly standardized production techniques to ensure uniformity in design. The ornaments were complicated not only technically but conceptually. While many kinds of animal teeth were available, in most places only those of carnivores were pierced and worn as pendants, perhaps as a way of expressing the individual's identity as a hunter.

Archaics like the Neandertals may have rubbed their bodies with ocher, worn hawk feathers in their hair, or otherwise adorned their persons. But there is no evidence of such a wholesale, formalized effort to transform elements of the material world into social messages. The universal human habit of ornamenting the body to communicate gender, social status, group affiliation, and other information about the wearer appears to have sprung into being at the beginning of the Upper Paleolithic.

"I don't think the beads represent the beginnings of self, but rather the earliest evidence for the creation of *selves*," says Randy White. "A lot of primates have a sense of themselves as individuals. But this sense remains relatively unmanipulative. Modern humans are able to fabricate multiple identities, depending on whomever you're dealing with at the moment. You do not present the same version of yourself to a job recruiter that you do to your spouse. You don't even talk to your father-in-law the same way you talk to your mother-in-law, even if they're standing ten feet apart. We aren't born with our complete social identity, it's a constructed phenomenon. I view the use of beads and pendants as one of the ways early Upper Paleolithic people were constructing their social identities."

If complex, multiversioned self-identities did not exist before the Upper Paleolithic, perhaps it was because nobody needed them. Neandertal social relations were far from primitive—vastly more sophisticated than the interactions of chimpanzees, which are hardly simple either. But as long as the "game of plot and counterplot" was played out on the local level, with familiar players and little ambiguity in the rules, why construct complicated psychological strategies to cope with it? To the extent that a Neandertal's social world was more homogeneous and straightforward than ours, so, I believe, was a Neandertal's sense of self.

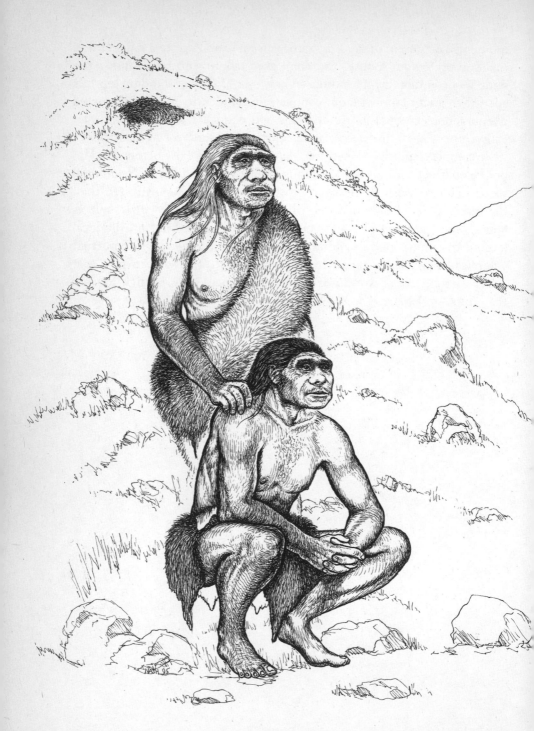

STEPHEN NASH

Chapter Twelve

BUT WHY?

The true paradises are paradises we have lost.

—MARCEL PROUST

A few years ago, I clipped out of a magazine an advertisement for *U.S. News & World Report*. The ad is a comic take on the human evolutionary time line, consisting of three sepia-toned, dog-eared, ancient cartoon frames. In the first, a naked, hairy, stooped australopithecine lopes along left to right. His expression is stubbornly aimless. The same figure appears in frame two, but he has stopped to inspect an object he has found. The last frame reveals the result of his scrutiny. The demi-ape has evolved abruptly into Modern Man: handsomely upright, resolute, full of clear-eyed, iron-jawed purpose.

The enlightening object back in frame two was, of course, a copy of *U.S. News & World Report*. The Modern Guy at the end of the line is not necessarily any smarter than his primitive predecessor; he is just better informed. Leaving aside the comically telescoped perspective, the gist of the ad strikes me as surprisingly apt. Substitute an archaic *sapiens* for the australopithecine, let the copy of the magazine represent the information riding the social conduits opening up on the landscape, and you have a fair rendering of what might have been happening in Europe in the Upper Paleolithic. Much the same was going on in the African Late Stone Age— or perhaps even sooner—and although there is not enough research yet to tell, the links were probably developing in other inhabited parts of the world too. These early information highways delivered vital news of the environment to the doorstep of the individual group, and by their very nature created new realms of social data to be processed: who was who, their status and degree of relatedness and alliance, what ancient obligations bound people together, and what deep-remembered rift in the past blackened their dealings in the future.

At the heart of it all was the uniquely original capacity to engage other people. The payoff from that experiment would be astonishing: Humankind literally inherited the earth. Bolstered by social networks, human populations were able to penetrate and take hold in the inhospitable regions of northern Europe and Asia. But they did not stop there. Since the 1930s, anthropologists have taken it as fact that the first humans to cross the Bering land bridge and colonize the New World were the Clovis people, dating back some 11,500 years. New evidence suggests that the initial migration may have been much earlier. Artifacts from isolated sites in New Mexico, Chile, and Brazil have been tentatively dated to around 30,000 years old. Genetic studies of mitochondrial DNA variation among American Indians suggest that at least some populations trace their occupation of the Americas back to between 42,000 and 21,000 years ago. More recently, a comparative study of grammatical structures in different language families suggested that humans reached the New World not long after the Cro-Magnons appeared in Western Europe. All of these contentions are hotly disputed. But if the estimates of an earlier arrival prove true, one could add to the moderns' résumé the lightning-quick conquest of two new continents.

"Whether the Middle to Upper Paleolithic transition is a revolution, you judge by its outcome," says Ofer Bar-Yosef. "The best evidence that a revolution occurred is the colonization of new habitats, the spread and dispersal itself."

It isn't difficult to understand the adaptive edge afforded by long-term sharing of help and information, which is implicit in alliances forged across the landscape. But there is still a Gordian knot to unravel. What set this system of exchange going in the first place? Who made the first move? Why help other people when such reciprocity between unrelated people still had no precedent?

Speculating on original causes is a frustrating temptation. No matter how deep you go, there will always be another *"but why?"* generated by the act of answering, something akin to what children discover to their solemn delight at around the age of four, usually on rainy afternoons when there is nothing to do but bother a parent. When my daughter used to play this game, I sometimes tried to shortcut the next *but why?* by calmly declaring, "Because that's the way it happened, and that's all there is to it." One could similarly dismiss the enigma at the beginning of the Upper

Paleolithic—it had to happen sometime, and it happened then, and that's all there is to it. To which my daughter would have replied, after waiting a teasing moment, "But *why* did it happen then?"

Why didn't alliances develop earlier, among archaic populations like the Neandertals? They, too, suffered through intense cold phases in Eurasia. With their big brains, they, too, were well equipped, so far as we know, to process large amounts of information, social or otherwise. Why did the revolution begin forty thousand years ago, and not seventy? Why us, and not them?

Those who are tired of mysteries can simply grab on to the old standby explanation: Modern human populations succeeded because they were inherently smarter or more "modern" than their predecessors. Never mind that Neandertals had big brains; there must have been something missing in their neurological makeup that kept them from processing information with the sophistication needed to evolve a new social order, or to compete with those who could. If you want to be fashionable, call this missing element language, or fully modern language, or "language-as-we-know-it." While there is no evidence of this superiority other than the evident replacement itself, let that suffice. Cro-Magnon Man stands at the end of the time line not by some evolutionary whim but because he *belongs* there.

But why? Why were the moderns better able to form those chains of connection, group to group, brain to brain? If the answer is language, why did one kind of human develop it, and the other not? I don't have the answers, but I can suggest a better way of approaching the questions. When trying to account for the origin of a new adaptation, people often fall into the trap of assuming that a particular anatomical trait or behavior

Evolution through information   COURTESY *U.S. NEWS & WORLD REPORT*

arose because of the benefit that would accrue to the species, once the trait was established. But natural selection works in the here and now. Species do not know the future and so cannot evolve to adapt to it. For a trait to be selected in a population, it must serve some immediate function that grants an edge to those who possess it, not to their descendants yet to be born. If by chance an adaptation eventually proves to have an additional side effect that is also advantageous, so much the better for those who bear it and their offspring. But that future utility cannot be the reason the trait evolved in the first place. Evolutionary biologists call the initial utility of the trait a *preadaptation*. Harvard's Stephen Jay Gould cites the evolution of feathers in the bird lineage. "Feathers work beautifully in flight," he writes, "but the ancestors of birds must have evolved them for another reason—probably thermoregulation—since a few feathers on the arm of a small running reptile will not induce takeoff."

If this is true, feathers-for-warmth were a preadaptation to feathers-for-flight, a trait that could evolve only in the scantily feathered reptile's future. The adaptive behavior whose origin we are trying to understand is intergroup cooperation and trust. Is there some trait that the ancestors of moderns possessed that the Neandertals did not, predisposing them to a greater mobility on the landscape and the attendant increase in contact with other groups?

One possibility immediately comes to mind—the moderns' long legs. The lanky body proportions of Upper Paleolithic humans might have evolved first in Africa south of the Sahara, or some other tropical region, to help dissipate body heat, and they only later proved doubly useful by allowing for greater mobility in cold-country climates. Erik Trinkaus's studies on hominid limb-bone development suggest that early moderns were not only born with limb lengths suitable for linear, long-distance walking, they were *using* those legs for precisely that purpose. This is as good a guess as any, but there are some problems. For one, the timing seems off. Judging from the Kenyan Turkana Boy, African *Homo erectus* was already long-limbed 1.5 million years ago, but there is no indication of greater mobility among African hominid populations until the Late Stone Age. And if Trinkaus's studies suggest what he thinks they do, at least some of the early moderns from Skhul and Qafzeh populations were accustomed to prolonged, goal-directed walking some 60,000 years before the archaeological evidence for alliance pops up farther north.

More important, it does not seem likely that merely moving around more would overcome the principal obstacle to alliance in the first place:

the wall of suspicion when unrelated, unfamiliar people met. Before any precedent for reciprocal aid or other cooperation between unrelated groups existed, such a meeting would not go well. Using living primates as a guide, at best they would be cautiously indifferent to each other, if a glut of resources were available. At worst they would be lethally aggressive. Most of the time they would try to avoid meeting in the first place. Greater mobility might set the stage for increasing cooperation by multiplying the number of group-to-group confrontations and thus putting selective pressure on any behavior that might soften their impact. But it could not work the transformation by itself.

There is an added element to consider, a means not only of maintaining alliances but of jump-starting the whole system. Art, ritual, and regional style might have worked as "social glue" to cement far-flung, fragile networks, but such cultural practices would never initiate and maintain alliances on their own. Contemporary societies use an even stronger adhesive; groups are bound together by the exchange of mates through marriage. Such exchanges can be direct—two groups regularly serve as a source of mates for each other—or indirect—one group supplies mates to another in the expectation of later receiving mates in turn from other groups in the alliance. In 1889, Edward Tylor, a founding father of anthropology, recognized this as a fundamental principle of human social organization. "Among tribes of low culture," he wrote, "there is but one means known of keeping up permanent alliance, and that means is intermarriage." Faced with "the practical alternative between marrying-out and being killed out," these level-headed savages naturally chose the former.

The social behavior that Tylor saw as essential to alliance building is called exogamy. In an exogamous system, young men or women leave their natal group on reaching marriageable age and transfer to the community of their spouses, ensuring that unions take place between groups, rather than within them. Later theorists like Leslie White and Claude Lévi-Strauss picked up and retooled this theme into the "incest taboo," which they saw as the marker for the beginning of human society. "Cooperation *between* families cannot be established if parent marries child; and brother, sister," wrote White in 1949. "A way must be found to overcome this centripetal tendency with a centrifugal force. This way was found in the definition and prohibition of incest. If persons were forbidden to marry their parents or siblings they would be compelled to marry into some other family group—or remain celibate, which is contrary to the nature of primates. The leap was taken; a way was found to unite families with one another, and social evolution as a *human* affair was launched upon its career."

It should be noted that exogamy and incest prohibitions are not really the same thing—the first has to do with marriage, the second with sex. One can, of course, have sex without marriage, and in some societies, such as the Nayar of India's Malabar coast, one can be married without any expectation of sex with one's spouse. When they declared exogamy and incest taboos to be the basis of human social organization, Tylor, White, and Lévi-Strauss were not aware of the patterns of exogamy that have since been discovered to guide the affairs of nonhuman primates. Among apes and monkeys, one sex or the other leaves its natal group on reaching sexual maturity. In most circumstances this renders incest a moot point, since an animal's immediate kin are not around to mate with. Even when sexually mature individuals reside with their parents or siblings, they tend to shun incestuous intercourse instinctively. Judging by the total lack of sexual interest displayed between men and women who are brought up together in Israeli kibbutzim and similar environments, human beings share this instinctive avoidance.

As social anthropologist Lars Rodseth of the University of Michigan and his colleagues have recently pointed out, there are nevertheless telling differences between the human pattern of exogamy and that of all other primates. First, humans do not simply wander off from their natal group to join another one. The young man or woman (usually the latter) who transfers to another group forms a stable, more-or-less exclusive mating relationship with one or more of its members. (We call these relationships "marriages," but what is important here is the behavior, not the cultural name for it.) Second, in most societies, *after the transfer, the new bride or husband continues to maintain relationships with his or her blood relatives in the natal group.* These two human patterns combine to make the marriage a living link between the two groups, a bond sealed in blood as soon as any offspring appear. The parents of a new bride do not lose a daughter, they gain a son—and an alliance, through affinity with his parents, brothers, cousins, and so on. In contrast, the dispersal of nonhuman primates tends to weaken any social networks that might already exist.

"[The human] contribution is not the invention of kinship," writes anthropologist Robin Fox in *The Red Lamp of Incest,* "but the invention of in-laws."

One other characteristic of human mate exchange sets the stage for alliance. Marriages do not normally occur at random between individuals from any two groups; marriage systems in human societies instead ensure that mates are exchanged between particular groups, presumably with the hope that both will benefit. Among some aboriginal tribes in Australia, the prohibition against incest goes far beyond one's immediate family. Since practically everyone in a local community is off-limits, a young man or woman is forced to search for a spouse over a wide geographic area, a system tailor-made to promote the open-ended, long-distance ties of reciprocity needed for a population to survive in the resource-poor deserts of inland Australia. But the potential threat need not be the environment. As Rodseth and his colleagues point out, if most maturing individuals in a population simply disperse into whatever other group happens to be close by, without regard to the political value of the union, the population could "marry out" and still face the probability of being "killed out" by more politically savvy groups, who better understood the game of cooperating-to-compete.*

Which brings us back to the Gordian knot at the beginning of the Upper Paleolithic. While it makes no adaptive sense to help complete strangers, it makes extraordinarily good sense to help groups in which you have a genetic stake through marriage. After alliances-through-kinship become established and begin to confer the advantages of cooperation, does it matter if the shared gene pool becomes more and more diluted as the size of the alliance expands? Once human populations began "creating kin" by ties between and among in-laws, all that might have been needed to underscore the initial, genetic basis for the alliance would be some cultural cue that a special relationship existed among its members. When a Bushman meets someone whose name he has never heard, one of the first steps toward reducing apprehension and establishing a positive relationship is to rename him, using a more familiar kinship term, perhaps the name of the Bushman's own brother or father. For examples of "fictive

---

*This does not mean that swapping mates with another group ensures peaceful, mutually beneficial relations, or that there is some simple one-to-one correspondence between social alliances and mating networks. Human societies are far more complicated and unpredictable than that— not surprisingly, since they are composed of such complicated and unpredictable beings. It is quite common for groups to form mutually beneficial trading relationships with people they would rarely look to for mating opportunities. According to Keith Otterbein of the State University of New York at Buffalo, groups can just as commonly be united by mate exchanges and still feud and fight. Perhaps some precautions are taken to ensure that in-laws do not come to grief at each other's hands—Otterbein cites one Nigerian tribe that rearranges its battle lines so that kinsmen do not directly shoot at each other—but the battle lines are drawn just the same.

kinship" closer to home consider a college "fraternity," or a Christian prayer meeting of "brothers and sisters."

"A consanguine [blood relative] is someone who is defined *by the society* as a consanguine," says Robin Fox, "and 'blood' relationship in a genetic sense has not necessarily anything to do with it."

It seems safe to assume that in the Upper Paleolithic, as in all later periods, mate exchanges were being arranged specifically for the political and economic advantages inherent in the unions. The currency of alliance was not just reciprocal aid, but common blood; not just Venus figurines and other art, but young women and the reproductive potential they represented. You cannot see this genetic currency in the archaeological record. But it would *have* to have been there for the alliances to germinate in the first place. Aiding in the survival of one's genes in a neighboring group is the same as helping those in one's own group, only less so. Helping those whom you only *perceive* to be related still works out to your benefit, if in fact they do reciprocate with help for your group later on. Eventually, as alliances expand, all the time connected by the flow of genes through the system and the restrictions against mate exchanges outside of it, you end up with people who may not ever have met one another, but who have much in common biologically and culturally, including shared physical traits and an unswerving loyalty to each other, especially in confrontations with outsiders. We call them ethnic groups. Give these groups separate, mutually unintelligible languages, and history, for better or worse, is ready to begin.

Mate exchange loosens the Gordian knot at the beginning of the Upper Paleolithic, but it does not untangle it. If mate exchange developed alliances, what prevented the archaics from taking the same step? Remember the hint of an answer left behind in the French cave of Combe-Grenal. Lewis Binford speculates that the pattern of bones and tools in that cave reveals a curious and very un-like-us aspect of Neandertal social organization: the presence of two separate but contemporaneous *kinds* of living area, one suggesting a life lived very locally, the other showing the comings and goings of more mobile folk. Binford calls the first type "the nest," and believes its occupants were females and their dependent children, foraging in the immediate vicinity. The second kind—smaller, more peripheral to the cave itself—he calls "scraper sites," because, unlike the nests, they are full of scrapers and other more carefully crafted tools, made from raw materials found a few miles away. These sites, he suggests, were produced by males. There is a connection between the two types, but a small one. Binford sees some provisioning of the young on the part of the males, but only in a haphazard sort of way.

This dual pattern, which occurs again and again through the 75,000 Mousterian years in Combe-Grenal, is utterly unlike anything produced by modern hunter-gatherers. Modern men and women live together and divide the labor of food procurement and preparation between them. The Neandertal sexes may have lived apart. "I'm not arguing that males and females weren't interacting," Binford says. "Rather, that it's an interaction we don't commonly see in modern humans. There's independent food preparation, different land-use patterns, different uses of technology. In modern humans the relationships are more integrated. But the Neandertals are separate, yet they're interacting. That's the important point. That's what makes them different."

Even if the patterns at Combe-Grenal mean what Binford suggests, the idea will have to be confirmed at other sites, too, before it is taken seriously. But let's suppose for a moment that Combe-Grenal means what Binford thinks it does. Then the reason Neandertals were not developing alliances through mate exchange could be that they had no mates to exchange in the first place. Of course they were mating. But the lives of sexually mature males and females were not bound together by the long-term, reasonably stable sexual and economic relationships that exist between the sexes in all known human communities today. Without these relationships, "mates" do not exist. If they do not exist, they cannot be exchanged.

It used to be assumed that the basis for human society was the nuclear family, a lifelong, sexually exclusive pair bond between a man and a woman jointly responsible for raising their children to maturity. As a human universal, this permanently monogamous pair-bonded family lives only in the minds of certain former American vice presidents. Western cultures are among a small minority of known societies that actually prescribe monogamy; almost 85 percent permit the husband to have more than one wife, and in a very few others the wife is permitted to take more than one husband. In practice, most men and women do have only one spouse at one time. But divorce is common in most cultures, especially in young marriages, and extramarital sex is even more so.

Some patterns of modern human sexual and reproductive behavior seem, however, to be nearly universal. In virtually all human societies, the female lives with her dependent young and is usually associated with a male formally recognized by society as her mate and father of her children.

Unlike other primate males, he regularly provisions them and their mother with food. In return for this economic assistance, he enjoys more or less continual and exclusive sexual access to the female, who, unlike the females of other primates, remains sexually receptive throughout her reproductive cycle. Her continual receptivity has obvious biological underpinnings. A human female does not display swollen genitalia or other sexual advertisements when she is fertile, and ovulation is carefully concealed, both behaviorally and physiologically. Not only do the female's mate and the other males remain uninformed of this crucial biological event, so does the woman herself, suggesting that it is somehow to her evolutionary advantage not to know.

This mating arrangement is the essence of what anthropologist Helen Fisher calls "the sex contract." British anthropologist Chris Knight more bluntly dubs it "sex-for-meat": "All over the world, wherever hunting was part of the traditional way of life, women treated marriage as an economic-and-sexual relationship, claiming for themselves the meat which their spouses obtained. . . . Marital relations (in contradistinction to mere 'sexual relations') were the means by which women, supported by their kin, achieved something that no primate females ever achieved. They were the means by which they secured for themselves and their offspring the continuous *economic* services of the opposite sex."

How this uniquely human pattern evolved has been a matter of much debate. Most experts agree that the two sexes have very different reproductive agendas, which are often at odds. Males can potentially produce a new offspring with every copulation; it would thus seem to be in their genetic interest to mate frequently with as many different fertile females as possible, and move on looking for more. Traditional theories, usually thought up by traditional males, approach the question by asking why males instead choose to settle down with a single female. Concealed ovulation is seen to have evolved to reduce competition for estrus females among the males in the group, making cooperative hunting possible, or else it is seen to have promoted the human family by keeping men happy at home with sex whenever they wanted it.

More recent theories seek to explain how the female's sexual strategies work to her advantage, rather than to that of her male or the group as a whole. A woman's reproduction dictates an arduous nine months of gestation followed by years of nutritional dependency by the young. With so much invested, it would seem to be in the woman's interest to secure the father's help in getting the child through to adulthood. Concealed ovulation, some investigators say, compelled the male to stick around through her whole reproductive cycle to ensure that his sperm, and nobody else's, fertilized her egg, spending his time providing for his sure-bet offspring rather than casting his seed all over the place to produce others who would not benefit from his aid. Another theory has it that concealed ovulation did not assure the male's knowledge of paternity but rather confused it, to the female's advantage, allowing her to mate on the sly with several potentially infanticidal males, all of whom would be less reluctant to kill her offspring if it might be their own. Yet another theory suggests that if the female's ovulation were not concealed even from herself, she would assiduously *avoid* sex when she knew she was fertile, in order to escape the rigors of childbirth and the burdens of childcare.

There are many such interesting notions about *how* the sex contract may have evolved, but surprisingly little argument about *when*. Something so intrinsically, universally human is assumed by virtually everyone to be very, very ancient. Some accounts take it all the way back to the primeval African forests of the Miocene, when bipedalism arose to enable early hominid fathers to carry food back to "home bases" occupied by their mates and children. More popular is the notion that the sex-for-meat exchange developed some two million years ago, when these ancestors left the forest for the savanna and were forced by its relative unpredictability to supplement their diet with hunted or scavenged game. But archaeological evidence for food sharing among early hominids remains controversial, at least until the end of the Middle Paleolithic.

Contemporary hunter-gatherer males do, in fact, hunt medium and large-sized game and bring it back to share with their kin and others in the group, a behavior traditionally seen as the essence of the Pleistocene hominid life. But in cultures whose environments most resemble the Pleistocene savanna, the hunters' contribution actually accounts for as little as 15 percent of the calories consumed by the group. Dry-country African groups like the !Kung and the Hadza of Tanzania make up the bulk of their diet with nuts, roots, tubers, small game, insects, honey, and other goods gathered by the females. Hadza men are especially focused on large game—and bring home nothing, ninety-seven days out of one hundred. No matter, says the early food-sharing theory—though the protein supplied by provisioning males may have been paltry, it was enough to tip the evolutionary balance toward individuals who adopted the sex-for-meat reproductive pattern. Their women were able to bear more children and raise them to maturity, so the behavior became the human norm.

The jury is still out on the archaeological evidence, but it seems to me more likely that sex-for-meat would have evolved first in an environment where the hunting of game animals was absolutely critical to the female's and young's survival. Today, there are hunter-gatherer groups where in some months the males provide virtually 100 percent of their family's food in the form of hunted meat. These are Eskimo and other high-latitude peoples, and the reason their women do not pitch in more with gathered plant foods is that there are none to be gathered, at least in winter. And when did our ancestors first occupy such environments? At the beginning of the Upper Paleolithic. According to Olga Soffer, Neandertals of the Middle Paleolithic occupied some northerly regions, but only places where the proximity of mountain ranges and foothills softened the climate, offering "more complex, diverse, and productive biotic communities." In other words, where there were more kinds of food available both to hunt *and* to gather. Not until the Upper Paleolithic did people settle permanently on the cold forbidding steppes, where the major source of nourishment came in big, hunted packages like reindeer and mammoth. It is certainly plausible, then, that sex-for-meat did not develop as a survival strategy until the Upper Paleolithic.

Soffer, like Binford, challenges the idea that Neandertal males brought home food to the women and children. She believes that additional evidence against this conventional wisdom can be found in the remains of the women and children themselves. Citing studies by David Frayer, she points out that while *both* Neandertal males and females were very strong and robustly built, the skeletons of early modern human females are much more slender and lightly constructed compared to their males. This pattern of females-first gracilization would be expected if human males were beginning to assume some of the females' foraging burdens. "Once the men were helping with provisioning," she says, "there would be less selection for these linebacker women."

The bones of the children may also bear witness to a major change in social organization. Without help from provisioning males, juveniles would be particularly vulnerable to food shortage and disease. Studies on juvenile Neandertal teeth show a higher incidence of nutritional stress than their modern human counterparts in the Upper Paleolithic. Many, apparently, did not make it through adolescence. Less than 30 percent of the early modern humans in Soffer's fossil sample were sub-adults, as opposed to 43 percent of the Neandertals. Like Binford, Soffer admits that her evidence is still meager. But combined with more established patterns—such as lack of evidence for a division of labor in Neandertal sites, which in modern groups usually falls along gender lines—she sees the burden of proof lying with those who unquestioningly *assume* that Neandertal homelife was like ours.

"Although each data set I presented is problematic," she says, "their patterning suggests that the time has come for us to abandon our latent but unproven assumptions that biparental provisioning of the young, division of labor, and food sharing . . . go back to the australopithecines."

To carry this line of thought one step further, if the sex-for-meat contract had not evolved in Neandertals, then the never-ceasing machinery of love we moderns use to reproduce ourselves had not yet evolved either. The hungered emotions, the physical responses, and all the physiological underpinnings of our uniquely incessant sexuality may have been circumscribed by the need to *advertise* fertility rather than conceal it. I suspect that Neandertal women were not very coy. They may have shown visible signs of estrus, or aggressively pursued or presented to the males during their fertile time. The males may have been attentive to sex only when the women's estrus inspired their interest. For the rest of their days, both genders may have been wholly indifferent to each other's physical allure. This is a far cry from the sexual behavior of modern humans today. Such a profound difference in sexual behavior would clearly bolster the notion discussed earlier, that Neandertals and moderns were separate on a species level because of their distinct mate-recognition systems.

In this scenario, the transition to the Upper Paleolithic involved a critical biological transformation, but it had little to do with native intelligence or a heightened capacity for language. It had to do with sex. The Neandertals and the moderns of the Upper Paleolithic took two different paths to solve the same ecological problem: how to survive in harsh northern latitudes. The early modern way leaned on a major change in social and reproductive strategy, one that allowed the females the chance to raise their young without having to forage so much on their own. The solution was an evolutionary pact that sent the males out onto the frigid plain looking for meat—and more important, brought them back again. Once out there, the moderns found the plain inhabited by more than just mammoth, bison, and reindeer. They found other people, and now they had something to offer them: through mate exchange, the possibility of social connections across the landscape. Neandertal groups, on the other hand, could not build alliances through the exchange of mates because their society was essentially mateless. Their alternative solution was to stick tight to those areas that provided a reasonably abundant and diverse resource base close at hand.

Eventually, the Neandertal solution ran out of time, or room, or luck. Or maybe all three.

In northern Burgundy, about one hundred miles southeast of Paris, there is a tranquil little river called the Cure. Near the hamlet of Arcy, the Cure whispers past a solemn series of limestone caves hunched along the left bank. Archaeologists have named the caves after animals long extinct in these parts: Cave of the Reindeer, Cave of the Hyena, the Bison, the Lion, the Bear. Perhaps as recently as 31,000 years ago, the caves sheltered another forgotten European. The Cave of the Reindeer in particular was found to be full of artifacts from a tool industry known as the Châtelperronian, an odd blending of the flake-based Mousterian tradition and the blade-based Aurignacian. Until recently, most investigators believed the Châtelperronian was the handiwork of the earliest Cro-Magnon immigrants into Europe. But in 1979, in another French cave called Saint-Césaire, a Neandertal partial skeleton turned up in deposits full of Châtelperronian artifacts. Abruptly, the significance of the Châtelperronian was turned upside down. Instead of the first sign of the new Europeans, it is now believed to be the last trace of the old.

With the Châtelperronians now identified as Neandertals, it becomes much more difficult simply to shoo them offstage so that the great modern Upper Paleolithic show can begin. They linger around, complicating the plot in Western Europe for fully 10,000 years *after* the arrival of Aurignacians into the region, refusing to play either one of the two roles scripted for them by the traditional theories. Replacement advocates who see the Neandertals as doomed by their innate limitations have to acknowledge that they were fully capable of emulating an Upper Paleolithic standard of technology, one that includes not only a heavy reliance on blade technology but also respectably elaborate bone tools at several sites. Châtelperronian deposits in the Grotte du Renne itself have also surrendered ivory beads, pierced animal teeth, and evidence of dwelling structures supported by mammoth tusks. Such cultural sophistication makes the continuity believers happy. They would like to see the industry as a technological transition between the Mousterian and the Aurignacian, perhaps paralleling the anatomical evolution of moderns out of Neandertal stock.

But this traditional scenario is no more tenable than the other. As we have seen, the Aurignacian emerged first in *eastern* Europe, sweeping westward across the continent to reach northern Spain thousands of years before the Châtelperronian made its brief, restricted, tenuous appearance on earth. It is hard to understand how the Châtelperronian could be transitional to something older than itself. And where the two industries touch in time, there is no sense of one gradually merging into the other. On the contrary, in several French and Spanish caves, beginning 35,000 years ago, Châtelperronian and Aurignacian deposits distinctly overlap, first one, then the other, then the first again, like the fingers of two hands interlaced. This does not look like behavioral evolution in a single human lineage. As in the Levant earlier in time, but with much greater clarity, this appears to be two distinct human lineages sharing the same region for thousands of years.

If you did not know the eventual end of the story, you could not say which was doomed, and which was destined. But we do know the end of the story. We can read the separate fates of the Neandertals and the moderns on the map, a little later in time. After millennia of coexistence, the Aurignacian expands north to dominate the region, as it dominates all of Europe. At the same time, the Châtelperronian dwindles away into isolated flashes. A thousand years later, the map shows nothing left of it but one lonely trace: the Grotte du Renne at Arcy-sur-Cure. Then it disappears from there, too. With it, presumably, go the Neandertals.

The currently fashionable interpretation of the Châtelperronian is that it represents an example of acculturation—the Neandertals' attempts to "get a grip," as Paul Mellars puts it, on techniques not of their own devising. As the modern Aurignacians penetrated into Western Europe, the story goes, the resident Mousterians picked up some of their tricks and technologies, adapting them to their own purposes in much the same way that American Plains Indians welcomed the horse from the invading Europeans and adapted it to their traditional hunting life-style. In the end, however, the cultural graft did not take. Unable to compete with the intruders, the Neandertals succumbed. Judging by their long period of coexistence, the adaptive difference between the two populations was not pronounced. In evolution, however, you do not have to fail to go extinct. You have only to succeed a little less often than someone else. The Cro-Magnons did not have to massacre the Neandertal groups they came across, or chase them away from river fords or flint outcrops, or infest them with new and deadly viruses that they themselves were immune to.

They had only to produce a few more babies every year than the beetle-browed others they occasionally met, and after a couple of thousand years, the job was done.

"The Neanderthals," says Chris Stringer, "probably went out with a whimper, not a bang."

Extinction is the ultimate private act. It buries both hope and memory, disconnecting the one from the many, the species moment from the continuum of time. Life goes on, but not *this* life, and not *these* deaths, with their promises of rebirth. There is nothing left behind to grieve for, because the grievers, too, are silenced. The only witnesses to Neandertal extinction are biased: the ancient Cro-Magnons, whose intrusion into their territories at least helped seal their fate; and today, the Cro-Magnons' even more intrusive descendants, who will not be content to leave the issue alone until every rock in every cave is turned over, analyzed, and argued about. I think there is something a little smug, a little prurient, about our century-long probing into the Neandertals' fate, as if the only element of their humanity that mattered was the single secret imperfection that kept them from becoming just like us. Perhaps we feel that once that dark failing is revealed, we can reclaim our sense of special destiny, in spite of Charles Darwin and his uncomfortable legacy, and get back to the way we had it in the old days, before morphology replaced Moses and genetics, Genesis.

But there is a price, and the Neandertals are the ones who have to pay. For every ounce of human uniqueness we posit in the moderns, we dehumanize the hominid closest to us by precisely that amount. It is a zero-sum game: Our rise is their fall. They had everything *except* the one special trick, be it the requisite tools, adequate planning depth, fully modern language, or whatever other shortcoming we use to explain their fade from sight. I have tried hard to avoid such assumed deficiencies, because I do not think that the evidence supports them. But the social scenario I have outlined also sells them short, as it must. By adding our mating patterns, our talent for alliance, or our politicized consciousness to the transformations occurring during the Upper Paleolithic, I have taken all these away from the Neandertals, making it more difficult, by just that much, to understand them as fully qualified to be imagined as human.

Unless, of course, we are not the only imaginable humanity.

"Anybody can visualize an earlier hominid with a slightly more rudimentary language, or a little less intelligence or whatever, developing into us," says University of Pennsylvania archaeologist Harold Dibble. "But try imagining a very smart creature who *isn't* like us! Now *there's* the challenge. Now *that's* exciting."

Before leaving the Neandertals to the private peace of their oblivion, let's try envisioning their version of humanity. If they had not gone extinct, I suspect their world today would not be greatly different from their world 50,000 years ago, not because Neandertals and other archaics lacked the power to innovate, but because, to them, innovation was anathema. Primitive societies in our own world can be described as conservative, with deep, superstitious attachments to the ways of the past, the ways of the ancestors. But modern hunter-gatherers are fad-driven lightweights compared to the Neandertals. Their conservatism was profound and immense, as voluminous as the great cranial vaults in which they stored the remembered minutiae of their days. Novelty, deviation, surprise—strangeness in all its forms was assiduously avoided. Their distrust of the new did not hold them back except in the elaboration of more newness, which they did not need in the first place. We think of our fantastically elaborate material culture as liberty from labor, which, for us, it is. But to a Neandertal, more adept in physical space, all these extra *things* might define a kind of prison.

"All of our decisions are made for us already, by the culture we have invented," muses Dibble. "If you walk into a room and there is a bunch of chairs, you simply sit down. You don't think about it, you just do. You don't sit facing away from the speaker, you don't sit on the back of the chair . . . where and how you act has been predetermined by the cultural creation. We don't see those patterns in the Mousterian."

Neandertal society would be new-averse, circumscribed by the known. Without the bridges of alliance extending over space, each separate breeding population would enclose in on itself, uncurious about the unfamiliars beyond its borders. But it does not necessarily follow that relationships between such populations would be hostile and defensive. The absence of alliance might also mean the absence of alliance *against*. The raw spectacle of chimp intergroup aggression notwithstanding, territorial animals do not normally seek out each other and attack; instead they take pains to meet as little as possible. I imagine Neandertals regarding strangers not as an anxious, imminent threat, but as a sort of superfluous blur on the very fringe of awareness. Strangers do not matter.

Familiar people, on the other hand, matter for everything. To be known at all would be to be implicitly, deeply understood. Naturally, Neandertals would be guided by self-interest just as much as any other

species, human or otherwise. But in the society I imagine, where all ac-
quaintances are intimates, some of the most successful ways *we* pursue
self-interest would be utterly futile. I discussed earlier the "tactical
deception" guiding the affairs of higher primates. By far the most adept
deceivers are chimpanzees. Curiously, the chimps' closest relative, the
gorilla, rarely shows a taste or talent for deception in the wild. This does
not necessarily imply that chimps are inherently more intelligent; the difference may derive from the two species' contrasting social organizations.
Chimpanzee society is commonly referred to as "fission-fusion." Members
of a troop meet, drift apart into smaller foraging groups, and reconnect
as food distribution, weather, and whim decide. Fission-fusion provides
an ideal setting for deception, because the shifting swirl of relationships
ensures that knowledge—of the environment, and of each other—is unevenly distributed within the troop. In contrast, gorillas live in stable
groups that forage together day after day. What is known to one is known
to all. The hardest individuals to outwit are the ones who know you best,
and in a social organization like the gorilla's, perhaps deception does not
thrive simply because the effort to deceive would seldom pay off.

I am not arguing that Neandertals lived like gorillas. But I imagine
them as far more familiar with the other individuals they regularly came
in contact with than were their contemporary early moderns, with their
wide-ranging pursuits. Neandertal social circumstances would render
Neandertals politically transparent, more gullible. But more honest too.
The transparency I imagine is not the innocence of the primitive, which
depends on the existence of the cynic. What I see is a world that man-
dates directness. The purpose of knowledge, to a Neandertal, would not
be to gain control, but to increase intimacy, not just between individuals
but between the individual mind and whatever it sees, touches, smells,
and remembers.

To this end, I imagine Neandertals possessing a different sort of self
and a different kind of consciousness. The plurality of "selves" we invent
to negotiate our guarded social encounters would be a waste of psychic
energy for Neandertals. Instead, let's give them a single but infinitely
graded ego, an analog self, as opposed to our own digitized identities. The
borders between the Neandertal and the Neandertal world are fuzzy. For
us, consciousness seems like an inner "I" resting somewhere deep in the
mind, eavesdropping on our stream of thoughts and perceptions. This, of
course, is a neuro-fiction; there is no special center of the brain where
consciousness resides. I would give the Neandertals a fictive inner voice,

too, but move it out, away from the center, so that it speaks from nearer to that fuzzy border with the world. A Neandertal thought would be much harder to abstract from the thing or circumstance that the thought is *about*. The perception of a tree in a Neandertal mind feels like the tree; grief over a lost companion *is* the absence and the loss. Neandertal psyche floats on the moment, where the metaphor of consciousness as a moving stream is perfect, the motion serene and unimpeded by countercurrents of re-think, counter-think, and double-think. I picture two Neandertals sitting side by side, their intimacy so exact that their interior voices cross and coalesce, like two streams merging into a river, their waters indistinguishable.

Finally, I imagine a Neandertal ethos and aesthetic. If the Upper Paleolithic people created art, ritual, symbol, even "invented meaning," this would appear to dump the Neandertals and other archaics into a starkly unenlightened, functional frame of mind. Maybe they did live their yawning millennia in a valueless stupor. But maybe not. The absence of ritual, for example, does not mean the absence of religion. I think the Neandertals had natural spirits like those of modern peoples who also live tight against nature. But where the modern's gods might inhabit the eland, the buffalo, or the blade of grass, the Neandertals' spirit *was* the animal or the grass blade, the thing and its soul perceived as a single vital force, with no need to distinguish them with separate names. Similarly, the absence of artistic expression does not preclude the apprehension of what is artful about the world. Neandertals did not paint their caves with the images of animals. But perhaps they had no need to distill life into representations, because its essences were already revealed to their senses. The sight of a running herd was enough to inspire a surging sense of beauty. They had no drums or bone flutes, but they could listen to the booming rhythms of the wind, the earth, and each other's heartbeats, and be transported.

Compared to our own, the Neandertal world would be more comfortable, but infinitely less spectacular. There would be no sciences, arts, media, none of the splendid diversity of human type and human accomplishment, and none of the vastly intricate economic and political systems we have devised to make it all hang together. On the other hand, there would be none of the heated, sustained hatred and aggression of war, no oppression of one folk by another, no contamination of the one earth by all.

This is not the Neandertals, of course, but merely the imaginings of a long-obsessed modern human. I should leave them now; but one final fantasy holds me. In 1990, Jean-Jacques Hublin introduced me to my first Neandertal fossil in a Parisian café. At the time, Hublin suspected that the mandible he showed me from Zafarraya cave in southern Spain was around 30,000 years old. This would make Zafarraya, rather than the Grotte du Renne, the last known Neandertal address. Recently, Hublin told me that he has a new date for the Zafarraya specimen, which now seems to be around 28,000 years old. So their moment of extinction creeps even closer to the present.

Hublin suggests that southern Spain may have been a *cul-de-sac* where Neandertals hung on a little longer than elsewhere in Europe, keeping to their old Mousterian traditions, having little or no contact with the Cro-Magnons just to the north. If this was possible twenty-eight thousand years ago, why not, say, twenty thousand? Or fifteen? Is it wholly impossible that evidence of their extended survival might turn up in some other isolated mountain pocket in Europe, perhaps in the Caucasus Mountains, the Zagros or Urals? What if they were still alive, unknown and unknowing, when the first agriculturalists planted their seeds twelve thousand years ago? Or when the Greeks destroyed Troy, in 1230 B.C.?

Take this one step further, and picture one last enclave somewhere, left alive today. Suppose we could get our curious hands on one of these Neandertals and actually perform the famous hypothetical experiment suggested in the late 1950s: give him a bath, a shave, and a new set of clothes, and take him for a ride on a New York subway train in rush hour. The original purpose of this thought experiment was to imagine what the other commuters would think of him. Would he go unnoticed, or stand out as grotesque? It is still an interesting question. But there is a much more intriguing one left unasked. I see the Neandertal clutching on to a pole in that shaking metal cage, staring at the commotion of diverse and unfamiliar faces which do not look back, amazed by the unholy smells and sounds that everyone else is so strangely ignoring, and I wonder.

What would he think of us?

Suggested Reading and Bibliography

SUGGESTED READING

For an historical account of the Neandertal debate, see *The Neandertals: Changing the Image of Mankind* by Erik Trinkaus and Pat Shipman. Also recommended is *In Search of the Neanderthals* by Chris Stringer and Clive Gamble. Good recent books on human evolution generally include *Lucy's Child: The Discovery of a Human Ancestor* by Donald Johanson and James Shreeve and *Origins Reconsidered: In Search of What Makes Us Human* by Richard Leakey and Roger Lewin. For reference, aficionados of human evolution should have close at hand *The Human Career* by Richard Klein and/ or *The Encyclopedia of Human Evolution and Prehistory*, edited by Ian Tattersall, Eric Delson, and John Van Couvering.

BIBLIOGRAPHY

Aiello, L. C. "The Fossil Evidence for Modern Human Origins in Africa: A Revised View." *American Anthropologist* 95 (1 1993).

————, and R.I.M. Dunbar. "Neocortex Size, Group Size, and the Evolution of Language." *Current Anthropology* 34 (2 1993):184–193.

Alexander, R. D. "Evolution of the Human Psyche." In *The Human Revolution: Behavioural and Biological Perspectives on the Origins of Modern Humans*, ed. P. A. Mellars and C. B. Stringer. Princeton N.J.: Princeton University Press, 1989.

Ambrose, S. H., and K. G. Lorenz. "Social and Ecological Models for the Middle Stone Age in Southern Africa." In *The Emergence of Modern Humans: Biocultural Adaptations in the Later Pleistocene*, ed. P. Mellars. Edinburgh: Edinburgh University Press, 1990.

Arensburg, B., et al. "A Middle Palaeolithic Human Hyoid Bone." *Nature* 338 (1989): 758–760.

————. "A Reappraisal of the Anatomical Basis for Speech in Middle Palaeolithic Hominids." *American Journal of Physical Anthropology* 83 (1990):137–146.

Auel, J. *The Clan of the Cave Bear.* New York: Bantam Books, 1980.

Avise, J. C. "Nature's Family Archives." *Natural History* 3 (1989):24–27.

Bahn, P. G., and J. Vertut. *Images of the Ice Age.* London: Windward, 1988.

Barinaga, M. " 'African Eve' Backers Beat a Retreat." *Science* 255 (1992):686–687.

Bar-Yosef, O. "The Role of Western Asia in Modern Human Origins." *Philosophical Transactions of the Royal Society of London B* 337 (1992):193–200.

———, et al. "New Data on the Origin of Modern Man in the Levant." *Current Anthropology* 27 (1986):63–64.

Bickerton, D. *Language and Species*. Chicago: University of Chicago Press, 1990.

Binford, L. R. *Faunal Remains from Klasies River Mouth*. Orlando, Fla.: Academic Press, Inc., 1984.

———. "Human Ancestors: Changing Views of Their Behavior." *Journal of Anthropological Archaeology* 4 (1985):292–327.

———. "Isolating the Transition to Cultural Adaptations: An Organizational Approach." In *The Emergence of Modern Humans: Biocultural Adaptations in the Later Pleistocene*, ed. E. Trinkaus. Cambridge, England: Cambridge University Press, 1989.

Binford, S. R. "A Structural Comparison of Disposal of the Dead in the Mousterian and the Upper Paleolithic." *Southwestern Journal of Anthropology* 24 (1968):139–154.

Bordes, F. *A Tale of Two Caves*. New York: Harper and Row, 1972.

Bowler, P. J. *Evolution: The History of an Idea*. Berkeley, Calif: University of California Press, 1984.

Brace, C. L. "A Consideration of Hominid Catastrophism." *Current Anthropology* 5 (1 1964):3–43.

Brain, C. K. *The Hunters or the Hunted?* Chicago: Chicago University Press, 1981.

Bräuer, G. "The Evolution of Modern Humans: A Comparison of the African and Non-African Evidence." In *The Human Revolution: Behavioural and Biological Perspectives on the Origins of Modern Humans*, ed. P. A. Mellars and C. B. Stringer. Princeton, N.J.: Princeton University Press, 1989.

———, and F. H. Smith, eds. *Continuity or Replacement: Controversies in Homo sapiens Evolution*. Rotterdam: Balkema, 1992.

Brooks, A. S., and C. C. Smith. "Ishango Revisited: New Age Determinations and Cultural Interpretations." *The African Archaeological Review* 5 (1987):65–78.

Brooks, A. S., et al. "Dating Pleistocene Archaeological Sites by Protein Diagenesis in Ostrich Eggshell." *Science* 248 (1990):60–64.

Burling, R. "Primate Calls, Human Language, and Nonverbal Communication." *Current Anthropology* 34 (1 1993):25–54.

Byrne, R., and A. Whiten. "The Thinking Primate's Guide to Deception." *New Scientist*, 3 December 1987, 54–57.

Cann, R. L. "DNA and Human Origins." *Ann. Rev. Anthropol.* (17 1988):127–143.

———. "In Search of Eve." *The Sciences*, September/October 1987, 30–37.

———, M. Stoneking, and A. C. Wilson. "Mitochondrial DNA and Human Evolution." *Nature* 325 (1987):31–36.

Cavalli-Sforza, L. L. "Genes, Peoples and Languages." *Scientific American*, November 1991, 104–110.

———, et al. "DNA Markers and Genetic Variation in the Human Species." *Cold Spring Harbor Symposia on Quantitative Biology* LI (1986):411–417.

———. "Reconstruction of Human Evolution: Bringing Together Genetic, Archaeological, and Linguistic Data." *Proceedings of the National Academy of the Sciences* 85 (1988):6002–6006.

Chagnon, N. A. *Yanomamö: The Fierce People*. New York: Holt, Rinehart and Winston, 1968.

Chase, P. G. "How Different was Middle Paleolithic Subsistence?" In *The Human Revolution: Behavioural and Biological Perspectives on the Origins of Modern Humans*. ed. P. A. Mellars and C. B. Stringer. Princeton, N. J.: Princeton University Press, 1989.

———. "Scavenging and Hunting in the Middle Paleolithic: The Evidence from Europe." In *Upper Pleistocene Prehistory of Western Eurasia*, ed. H. Dibble and L. Montet-White. Philadelphia: The University of Pennsylvania Museum, 1988.

———, and H. Dibble. "Middle Paleolithic Symbolism: A Review of Current Evidence and Interpretations." *Journal of Anthropological Archaeology* 6 (1987):263–296.

Cheney, D., R. Seyfarth, and B. Smuts. "Social Relationships and Social Cognition in Nonhuman Primates." *Science* 234 (1986):1361–1366.

Clark, G. A. "Continuity or Replacement?—Putting Modern Human Origins in an Evolutionary Context." In *The Mousterian and Its Aftermath*, ed. H. Dibble and P. Mellars. Philadelphia: The University of Pennsylvania Museum, 1991.

———, and J. M. Lindly. "The Case for Continuity: Observations on the Biocultural Transition in Europe and Western Asia." In *The Human Revolution: Behavioural and Biological Perspectives on the Origins of Modern Humans*, ed. P. A. Mellars and C. B. Stringer. Princeton, N.J.: Princeton University Press, 1989.

———. "Modern Human Origins in the Levant and Western Asia: The Fossil and Archeological Evidence." *American Anthropologist* 91 (4 1989):962–985.

Clarke, R. J. "The Ndutu Cranium and the Origin of *Homo sapiens*." *Journal of Human Evolution* 19 (1990):699–736.

Coon, C. S. *The Origin of Races*. New York: Alfred A. Knopf, Inc., 1962.

Cosgrove, R. "Thirty Thousand Years of Human Colonization in Tasmania: New Pleistocene Dates." *Science* 243 (1989):1706–1708.

Cosmides, L., and J. Tooby. "From Evolution to Behavior: Evolutionary Psychology as the Missing Link." In *The Latest on the Best: Essays on Evolution and Optimality*, ed. J. Dupré. Cambridge, Mass.: The MIT Press, 1987.

Davidson, I., and W. Noble. "The Archaeology of Perception: Traces of Depiction and Language." *Current Anthropology* 30 (2 1989):125–155.

Deacon, H. J. "Origins of Modern Man—Ancestors of Us All." *The Phoenix: Magazine of the Albany Museum* 2 (1 1989):9–16.

———. "Southern Africa and Modern Human Origins." *Philosophical Transactions of the Royal Society of London* B 337 (1992):177–183.

Diamond, J. "The Great Leap Forward." *Discover*, May 1989, 50–60.

———. *The Third Chimpanzee: The Evolution and Future of the Human Animal*. New York: HarperCollins Publishers, 1992.

Dibble, H. "The Implications of Stone Tool Types for the Presence of Language During the Lower and Middle Paleolithic." In *The Human Revolution: Behavioural and Biological Perspectives on the Origins of Modern Humans*, ed. P. A. Mellars and C. B. Stringer. Princeton, N.J.: Princeton University Press, 1989.

———. "The Interpretation of Middle Paleolithic Scraper Morphology." *American Antiquity* 52 (1987):111–117.

Di Rienzo, A., and A. C. Wilson. "Branching Pattern in the Evolutionary Tree for

Human Mitochondrial DNA." *Proceedings of the National Academy of the Sciences USA* 88 (1991):1597–1601.

Dupont, E. *Bulletins de l'Academie Royale de Belgique,* 2d series, 22 (1866).

Fagan, B. M. *The Journey from Eden: The Peopling of Our World.* London: Thames and Hudson, Ltd., 1990.

Farizy, C. "The Transition from Middle to Upper Paleolithic at Arcy-sur-Cure (Yonne, France):Technological, Economic and Social Aspects." In *The Emergence of Modern Humans: Biocultural Adaptations in the Later Pleistocene,* ed. P. Mellars. Edinburgh: Edinburgh University Press, 1990.

Féblot-Augustins, J. "Mobility Strategies in the Late Middle Palaeolithic of Central Europe and Western Europe: Elements of Stability and Variability." *Journal of Anthropological Archaeology* 12 (1993):211–265.

Fischman, J. "Hard Evidence." *Discover,* February 1992, 44–51.

Fisher, A. "On the Emergence of Humans." *Mosaic* 19 (1 1988):34–45.

Fisher, H. E. *Anatomy of Love: The Natural History of Monogamy, Adultery, and Divorce.* New York: W. W. Norton & Company, Inc., 1992.

Foley, R. *Another Unique Species: Patterns in Human Evolutionary Ecology.* New York: Longman, Scientific & Technical, 1987.

―――. "The Ecological Conditions of Speciation: A Comparative Approach to the Origins of Anatomically Modern Humans." In *The Human Revolution: Behavioural and Biological Perspectives on the Origins of Modern Humans,* ed. P. A. Mellars and C. B. Stringer. Princeton, N.J.: Princeton University Press, 1989.

Fox, R. *Kinship and Marriage: An Anthropological Perspective.* New York: Penguin Books, Ltd., 1967.

―――. *The Red Lamp of Incest.* Notre Dame, Ind.: University of Notre Dame Press, 1980.

Frayer, D. W. "On Neanderthal Crania and Speech: Response to Lieberman." *Current Anthropology* 34 (5 1993):721.

Gamble, C. S. "Culture and Society in the Upper Paleolithic of Europe." In *Hunter-Gatherer Economy in Prehistory: A European Perspective,* ed. G. Bailey. Cambridge, England: Cambridge University Press, 1983.

―――. "Exchange, Foraging and Local Hominid Networks." In *Trade and Exchange in Prehistoric Europe in University of Bristol,* ed. Chris Scarre and Frances Healy. Oxbow Books, 35-44, Year.

―――. "Interaction and Alliance in Palaeolithic Society." *Man* 17 (1982):92–107.

―――. *The Palaeolithic Settlement of Europe.* Cambridge, England: Cambridge University Press, 1986.

―――. "The Social Context for European Palaeolithic Art." *Proceedings of the Prehistoric Society* 57 (1991):3–15.

―――, and O. Soffer, eds. *The World at 18,000 BP.* Vol. 2: *Low Latitudes.* London: Unwin Hyman, 1990.

Gargett, R. H. "Grave Shortcomings: The Evidence for Neandertal Burial." *Current Anthropology* 30 (2 1989):157–190.

Geneste, J.-M. "Systèmes d'approvisionnement en matières au paléolithique moyen et au paléolithique supérieur en Aquitaine." *L'Homme de Néandertal* 8 (1988):61–70.

Gibbons, A. "Pleistocene Population Explosions." *Science* 262 (1993):27–28.

Golding, W. *The Inheritors.* London: Faber and Faber, Ltd., 1955.

Goodall, J. *The Chimpanzees of Gombe*. Cambridge, Mass.: The Belknap Press of Harvard University Press, 1986.

Gould, S. J. "Grimm's Greatest Tale." *Natural History* 2 (1989):20–28.

———. "A Novel Notion of Neanderthal." *Natural History* 6 (1988):16–21.

Graves, P. "New Models and Metaphors for the Neanderthal Debate." *Current Anthropology* 32 (5 1991):513–541.

Grün, R., P. B. Beaumont, and C. B. Stringer. "ESR Dating Evidence for Early Modern Humans at Border Cave in South Africa." *Nature* 344 (1990):537–539.

Grün, R., C. B. Stringer, and Henry P. Schwarcz. "ESR Dating of Teeth from Garrod's Tabun Cave Collection." *Journal of Human Evolution* 20 (1991):231–248.

Hammond, M. "Anthropology as a Weapon of Social Combat in Late-19th-Century France." *Journal of the History of the Behavioral Sciences* 16 (1980):118–132.

———. "The Expulsion of the Neanderthals from Human Ancestry: Marcellin Boule and the Social Context of Scientific Research." *Social Studies of Science* 12 (1982):1–36.

———. "The Shadow Man Paradigm in Paleoanthropology 1911–1945." In *Bones, Bodies, Behavior: Essays on Biological Anthropology*, ed. G. W. Stocking. Madison, Wis.: University of Wisconsin Press, 1985.

Harpending, H. C., et al. "The Genetic Structure of Ancient Human Populations." *Current Anthropology* 34 (1 1993):483–496.

Hayden, Brian. "The Cultural Capacities of Neandertals: A Review and Re-evaluation." *Journal of Human Evolution* 24 (1993):113–146.

Hedges, S. B., et al. "Human Origins and Analysis of Mitochondrial DNA Sequences." *Science* 255 (1991):737–739.

Holloway, R. L. "Human Paleontological Evidence Relevant to Language Behavior." *Human Neurobiology* 2 (1983):105–114.

Hublin, J. J. "Human Fossils from the North African Middle Pleistocene and the Origin of *Homo sapiens*." In *Ancestors: The Hard Evidence*. ed. E. Delson. New York: Alan R. Liss, Inc., 1985, pp. 283–288.

———. "Recent Human Evolution in Northwestern Africa." *Philosophical Transactions of the Royal Society of London* B 337 (1992):185–191.

Humphrey, N. *Consciousness Regained: Chapters in the Development of Mind*. Oxford, England: Oxford University Press, 1983.

———. *The Inner Eye*. London: Faber and Faber, Ltd., 1986.

Huxley, Thomas. *Man's Place in Nature*. New York: D. Appleton and Company, 1896.

Jelinek, A. J. "The Tabun Cave and Paleolithic Man in the Levant." *Science* 216 (1982):1369–1375.

Jochim, M. "Paleolithic Cave Art in Ecological Perspective." In *Hunter-Gather Economy in Prehistory: A European Perspective*, ed. G. Bailey. Cambridge, England: Cambridge University Press, 1983.

Johanson, D. C., and M. Edey. *Lucy: The Beginnings of Humankind*. New York: Simon and Schuster, 1981.

Johanson, D. C., and J. Shreeve. *Lucy's Child: The Discovery of a Human Ancestor*. New York: William Morrow and Company, 1989.

Jones, J. S., and S. Rouhani. "How Small Was the Bottleneck?" *Nature* 319 (1986): 449–450.

Kendon, A. "Some Considerations for a Theory of Language Origins." *Man* 26 (1991): 199–221.

Kimbel, W. H. "Species, Species Concepts and Hominid Evolution." *Journal of Human Evolution* 20 (1991):355–371.

Klein, R. G. "The Archeology of Modern Human Origins." *Evolutionary Anthropology* 1 (1 1992):5–14.

———. "Biological and Behavioural Perspectives on Modern Human Origins in Southern Africa." In *The Human Revolution: Behavioural and Biological Perspectives on the Origins of Modern Humans,* ed. P. A. Mellars and C. B. Stringer. Princeton, N.J.: Princeton University Press, 1989.

———. *The Human Career: Human Biological and Cultural Origins.* Chicago: University of Chicago Press, 1989.

Klima, B. "A Triple Burial from the Upper Paleolithic of Dolní Věstonice, Czechoslovakia." *Journal of Human Evolution* 16 (1988):831–835.

Knight, C. *Blood Relations: Menstruation and the Origins of Culture.* New Haven, Conn.: Yale University Press, 1991.

Kocher, T. D., and A. C. Wilson. "Sequence Evolution of Mitochondrial DNA in Humans and Chimpanzees: Control Region and a Protein-Coding Region." In *Evolution of Life—Fossils, Molecules and Culture,* ed. S. Osawa and T. Honjo. Tokyo: Springer-Verlag, 1991, pp. 391–413.

Larick, R. "Age Grading and Ethnicity in the Style of Loikop (Samburu) Spears." *World Archaeology* 18 (2 1986):269–283.

Lartet, Edouard, and Henry Christy. *Reliquiae aquitanicae.* London: Williams and Norgate, 1875.

Leakey, R., and R. Lewin. *Origins Reconsidered: In Search of What Makes Us Human.* New York: Doubleday, 1992.

Lewin, R. "Africa: Cradle of Modern Humans." *Science* 237 (1987):1292–1295.

———. "The Biochemical Route to Human Origins." *Mosaic* 22 (3 1991):46–55.

———. "Species Questions in Modern Human Origins." *Science* 243 (1989):1666–1667.

Lewis-Williams, J. D., and T. A. Dowson. "Entoptic Phenomena in Upper Palaeolithic Art." *Current Anthropology* 29 (2 1988):201–245.

———. *Images of Power.* Johannesburg: Southern Book Publishers, 1989.

Lieberman, D. E., and J. J. Shea. "Behavioral Differences Between Archaic and Modern Humans in the Levantine Mousterian." *American Anthropologist* 96 (2 1994):300–332.

Lieberman, P. *The Biology and Evolution of Language.* Cambridge, Mass.: Harvard University Press, 1984.

———. "On the Kebara KMH 2 Hyoid and Neanderthal Speech." *Current Anthropology* 34 (2 1993):172–175.

———. "The Origins of Some Aspects of Human Language and Cognition." In *The Human Revolution: Behavioural and Biological Perspectives on the Origins of Modern Humans,* ed. P. A. Mellars and C. B. Stringer. Princeton, N.J.: Princeton University Press, 1989.

Lindly, J. M., and G. A. Clark. "Symbolism and Modern Human Origins." *Current Anthropology* 31 (3 1990):133–261.

Long, J. C., et al. "Phylogeny of Human Beta-Globin Haplotypes and Its Implications for Recent Human Evolution." *American Journal of Physical Anthropology* 81 (1990):113–130.

Manson, J. H., and R. W. Wrangham. "Intergroup Aggression in Chimpanzees and Humans." *Current Anthropology* 32 (4 1991):369–390.

Marks, A. E. "Early Mousterian Settlement Patterns in the Central Negev, Israel: Their Social and Economic Implications." In *L'Homme de Néandertal*. Vol. 6 Liège: ERAUL, 1989, pp. 115–126.

———. "The Middle to Upper Paleolithic Transition in the Southern Levant: Technological Change as an Adaptation to Increasing Mobility." In *L'Homme de Néandertal*. Vol. 8. Liège: ERAUL, 1988, pp. 109–123.

Marshack, A. "Evolution of the Human Capacity: The Symbolic Evidence." *Yearbook of Physical Anthropology* 32 (1989):1–34.

Marshall, L. *The !Kung of Nyae Nyae*. Cambridge, Mass.: Harvard University Press, 1976.

Masters, J. C. "Primates and Paradigms: Problems with the Identification of Species." In *Species, Species Concepts, and Primate Evolution*, ed. W. H. Kimbel and L. B. Martin. New York: Plenum Press, 1993.

Maugh, T. H., II. "We're All Children of African 'Eve,' Scientist Says." *Los Angeles Times*, October 5, 1989, 3, 34.

Mellars, P. A. *The Emergence of Modern Humans*. Edinburgh: Edinburgh University Press, 1990.

———. "Major Issues in the Emergence of Modern Humans." *Current Anthropology* 30 (3 1989):349–385.

———. "A New Chronology for the French Mousterian Period." *Nature* 322 (1986): 410–411.

———, and C. B. Stringer, eds. *The Human Revolution: Behavioural and Biological Perspectives on the Origins of Modern Humans*. Princeton, N.J.: Princeton University Press, 1989.

Mercier, N., et al. "Thermoluminescence Dating of the Late Neanderthal Remains from Saint-Césaire." *Nature* 351 (1991):737–739.

Minc, L. D. "Scarcity and Survival: The Role of Oral Tradition in Mediating Subsistence Crises." *Journal of Anthropological Archaeology* 5 (1986):39–113.

Mullis, K. B. "The Unusual Origin of the Polymerase Chain Reaction." *Scientific American*, April 1990, 36–43.

Nei, M., and G. Livshits. "Genetic Relationships of Europeans, Asians and Africans and the Origin of Modern *Homo sapiens*." *Human Heredity* 39 (1989):276–281.

Paterson, H.E.H. "The Recognition Concept of Species." In *Species and Speciation*, ed. E. S. Vrba. Transvaal Museum Monograph No. 4. Pretoria: Transvaal Museum, 1985, pp. 21–29.

Pfeiffer, J. E. *The Creative Explosion: An Inquiry into the Origins of Art and Religion*. Ithaca, N.Y.: Cornell University Press, 1982.

———. *The Emergence of Humankind*. 4th ed. New York: Harper & Row, 1985.

Pope, G. G. "Evolution of the Zygomaticomaxillary Region in the Genus *Homo* and Its Relevance to the Origin of Modern Humans." *Journal of Human Evolution* 21 (1991):189–213.

———. "Recent Advances in Far Eastern Paleoanthropology." *Annual Review of Anthropology* 17 (1988):43–77.

Protsch, R. "The Absolute Dating of Upper Pleistocene Sub-Saharan Fossil Hominids and Their Place in Human Evolution." (1975):297–322.

Rak, Y. "The Neanderthal: A New Look at an Old Face." *Journal of Human Evolution* 15 (1986):151–164.

———. "On the Differences Between Two Pelvises of Mousterian Context from the Qafzeh and Kebara Caves, Israel." *American Journal of Physical Anthropology* 81 (1990):323–332.

Rak, Y., and B. Arensburg. "Kebara 2 Neanderthal Pelvis: First Look at a Complete Inlet." *American Journal of Physical Anthropology* 73 (1987):227–231.

Rak, Y., and W. H. Kimbel. "A Neandertal Infant from Amud Cave, Israel." *Journal of Human Evolution* 26 (1994):313–322.

Ramaswamy, V. "Explosive Start to Last Ice Age." *Nature* 359 (1992):14.

Rampino, M. R., and S. Self. "Volcanic Winter and Accelerated Glaciation Following the Toba Super-Eruption." *Nature* 359 (1992):50–52.

Restak, R. *The Brain.* New York: Bantam Books, 1984.

Reynolds, P. C. *On the Evolution of Human Behavior: The Argument from Animals to Man.* Berkeley, Calif.: University of California Press, 1981.

Rightmire, G. P. *The Evolution of Homo Erectus.* Cambridge, England: Cambridge University Press, 1990.

———. "The Relationship of *Homo erectus* to Later Middle Pleistocene Hominids." *Courier Forschungs-Institut Senckenberg* XXX (1994):323–329.

Roberts, R. G., R. Jones, and M. A. Smith. "Thermoluminescence Dating of a 50,000-Year-Old Human Occupation Site in Northern Australia." *Nature* 345 (1990):153–155.

Rodseth, L., et al. "The Human Community as a Primate Society." *Current Anthropology* 32 (3 1991):221–254.

Rosenberg, K. R. "The Evolution of Modern Human Childbirth." *Yearbook of Physical Anthropology* 35 (1992):89–124.

Rouhani, S. "Molecular Genetics and the Pattern of Human Evolution." In *The Human Revolution: Behavioural and Biological Perspectives on the Origins of Modern Humans,* ed. P. A. Mellars and C. B. Stringer. Princeton, N.J.: Princeton University Press, 1989.

Scarre, C. "Painting by Resonance." *Nature* 338 (1989):382.

Schepartz, L. A. "Language and Modern Human Origins." *Yearbook of Physical Anthropology* 36 (1993):91–121.

Schwarcz, H. P., et al. "ESR Dates for the Hominid Burial Site of Qafzeh in Israel." *Journal of Human Evolution* 17 (1988):733–737.

Seyfarth, R. M. "Talking with Monkeys and Great Apes." *International Wildlife* 12 (1982):12–18.

———, and D. L. Cheney. "Meaning and Mind in Monkeys." *Scientific American,* December 1992, 122–128.

Shea, J. J. "A Functional Study of the Lithic Industries Associated with Hominid Fossils in the Kebara and Qafzeh Caves, Israel." In *The Human Revolution: Behavioural and Biological Perspectives on the Origins of Modern Humans,* ed. P. A. Mellars and C. B. Stringer. Princeton, N.J.: Princeton University Press, 1989.

———. "A New Perspective on Neandertals from the Levantine Mousterian." *AnthroQuest,* 1990, 14–18.

———. "A Preliminary Functional Analysis of the Kebara Mousterian." In *Investigations in South Levantine Prehistory,* ed. O. Bar-Yosef and B. Vandermeersch. B.A.R. International Series, 1989, pp. 185–201.

Shevoroshkin, V. "The Mother Tongue: How Linguists Have Reconstructed the Ancestor of All Living Languages." *The Sciences*, May/June 1990, 20–27.

Shreeve, J. "Argument over a Woman." *Discover*, August 1990, 52–59.

———. "The Dating Game." *Discover*, September 1992.

———. "*Erectus* Rising." *Discover*, September 1994, 80–89.

———. "Machiavellian Monkeys." *Discover*, June 1991, 68–73.

Singer, R., and J. Wymer. *The Middle Stone Age at Klasies River Mouth in South Africa.* Chicago: University of Chicago Press, 1982.

Smith, F. H. "Upper Pleistocene Hominid Evolution in South-Central Europe: A Review of the Evidence and Analysis of Trends." *Current Anthropology* 23 (6 1982): 667–687.

Smith, F. H., A. B. Falsetti, and S. M. Donnelly. "Modern Human Origins." *Yearbook of Physical Anthropology* 32 (1989):35–68.

Smith, F. H., and F. Spencer, eds. *The Origins of Modern Humans.* New York: Alan R. Liss, Inc., 1984.

Soffer, O. "Ancestral Lifeways in Eurasia—the Middle and Upper Paleolithic Records." In *Origins of Anatomically Modern Humans*, ed. M. H. Nitecki and D. V. Nitecki. New York: Plenum Press, 1994, pp. 101–119.

———. "The Middle to Upper Paleolithic Transition on the Russian Plain." In *The Human Revolution: Behavioural and Biological Perspectives on the Origins of Modern Humans*. ed. P. A. Mellars and C. B. Stringer, Princeton, N.J.: Princeton University Press, 1989.

———. *The Upper Paleolithic of the Central Russian Plain.* Orlando, Fla.: Academic Press, Inc., 1985.

Solecki, R. S. *Shanidar: The First Flower People.* New York: Alfred A. Knopf, Inc., 1971.

Spencer, F. *Piltdown: A Scientific Forgery.* Oxford, England: Oxford University Press, 1990.

Stiner, M., and S. L. Kuhn, "Subsistence, Technology, and Adaptive Variation in Middle Paleolithic Italy." *American Anthropologist* 94 (1992):306–339.

Stoneking, M., K. Bhatia, and A. C. Wilson. "Rate of Sequence Divergence Estimated from Restriction Maps of Mitochondrial DNAs from Papua New Guinea." *Cold Spring Harbor Symposia on Quantitative Biology* LI (1986):433–439.

Stoneking, M., and R. L. Cann. "African Origin of Human Mitochondrial DNA." In *The Human Revolution: Behavioural and Biological Perspectives on the Origins of Modern Humans*, ed. P. A. Mellars and C. B. Stringer. Princeton, N.J.: Princeton University Press, 1989.

Stoneking, M., et. al. "New Approaches to Dating Suggest a Recent Age for the Human mtDNA Ancestor." *Philosophical Transactions of the Royal Society of London* B 337 (1992):167–175.

Straus, L. G. "Age of Modern Europeans." *Nature* 342 (1989):476–477.

Straus, W. L., and A.J.E. Cave. "Pathology and Posture of Neanderthal Man." *Quarterly Review of Biology* 32 (1957):348–363.

Stringer, C. B. "The Dates of Eden." *Nature* 331 (1988):565–566.

———. "The Emergence of Modern Humans." *Scientific American*, December 1990, 98–104.

———. "Out of Africa: A Personal History." In *Origins of Anatomically Modern Humans*, ed. M. H. Nitecki and D. V. Nitecki. New York: Plenum Press, 1994.

Stringer, C. B., and P. Andrews. "Genetic and Fossil Evidence for the Origin of Modern Humans." *Science*, 1988, 1263–1268.

Stringer, C. B., and C. Gamble. *In Search of the Neanderthals*. New York: Thames and Hudson Ltd., 1993.

Stringer, C. B., and R. Grün. "Time for the Last Neanderthals." *Nature* 351 (1991): 701–702.

Stringer, C. B., et al. "ESR Dates for the Hominid Burial Site of Es Skhul in Israel." *Nature* 338 (1989):756–758.

Svoboda, J. "A New Male Burial from Dolní Věstonice." *Journal of Human Evolution* 16 (1988):827–830.

Tattersall, I. "Species Recognition in Human Paleontology." *Journal of Human Evolution* 15 (1986):165–175.

———, E. Delson, and J. Van Couvering, eds. *Encyclopedia of Human Evolution and Prehistory*. New York: Garland Publishing, 1988.

Templeton, A. R. "The 'Eve' Hypothesis: A Genetic Critique and Reanalysis." *American Anthropologist* 95 (1 1993):51.

Thorne, A. G., and M. H. Wolpoff. "The Multiregional Evolution of Humans." *Scientific American*, April 1992, 76–83.

Trinkaus, E. "Morphological Contrasts Between the Near Eastern Qafzeh-Skhul and Late Archaic Human Samples: Grounds for a Behavioral Difference?" In *The Evolution and Dispersal of Modern Humans in Asia*, ed. T. Akazawa, K. Aoki, and T. Kimura. Tokyo: Hokusen-sha, 1992.

———. "The Neandertals and Modern Human Origins." *Annual Review of Anthropology* 15 (1986):193–218.

———. *The Shanidar Neandertals*. New York: Academic Press, 1983.

Trinkaus, E., ed. *The Emergence of Modern Humans: Biocultural Adaptations in the Later Pleistocene*. Cambridge, England: Cambridge University Press, 1989.

Trinkaus, E., and F. H. Smith. "The Fate of the Neandertals." In *Ancestors: The Hard Evidence*, ed. Eric Delson. New York: Alan R. Liss, Inc., 1985.

Trinkaus, E., and P. Shipman. *The Neandertals: Changing the Image of Mankind*. New York: Alfred A. Knopf, Inc., 1992.

Turner, A. "The Concept of the Ecological Niche and Its Application to Studies of Hominid Evolution." *Archaeozoologia* III (1,2 1989):111–120.

———. "Hominids and Fellow Travellers: Human Migration into High Latitudes as Part of a Large Mammal Community." In *Human Evolution and Community Ecology*, ed. Robert Foley. London: Academic Press, 1984.

———. "The Recognition Concept of Species in Palaeontology, with Special Consideration of Some Issues in Hominid Evolution." In *Species and Speciation*, ed. E. S. Vrba. Pretoria: Transvaal Museum Monograph No. 4, 1985.

Valladas, H. "Thermoluminescence Dating of Flint." *Quaternary Science Reviews* 11 (1992):1–5.

———, et al. "Thermoluminescence Dates for the Neanderthal Burial Site at Kebara in Israel." *Nature* 330 (1987):159–160.

———. "Thermoluminescence Dating of Mousterian 'Proto-Cro-Magnon' Remains from Israel and the Origin of Modern Man." *Nature* 331 (1988):614–616.

Vandiver, P. B., et al. "The Origins of Ceramic Technology at Dolní Věstonice, Czechoslovakia." *Science* 246 (1989):1002–1008.

Vigilant, L., et al. "African Populations and the Evolution of Human Mitochondrial DNA." *Science*, 1991, 1503–1507.

———. "Mitochondrial DNA Sequences in Single Hairs from a Southern African Population." *Proceedings of the National Academy of Sciences USA* 86 (1989):9350–9354.

Vogel, J. C., and P. B. Beaumont. "Revised Radiocarbon Chronology for the Stone Age in South Africa." *Nature* 237 (1972):50–51.

Wainscoat, J. S., et al. "Evolutionary Relationships of Human Populations from an Analysis of Nuclear DNA Polymorphisms." *Nature* 319 (6 1986):491–493.

Weidenreich, Franz. "Facts and Speculations Concerning the Origin of *Homo sapiens*."*American Anthropologist* 49, No. 2 (1947):187–203:189.

Weiss, K. M. "On the Number of Members of the Genus *Homo* Who Have Ever Lived, and Some Evolutionary Implications." *Human Biology* 56 (4 1984):637–649.

Whallon, R. "Elements of Cultural Change in the Later Paleolithic." In *The Human Revolution: Behavioural and Biological Perspectives on the Origins of Modern Humans*, ed. P. A. Mellars and C. B. Stringer. Princeton, N.J.: Princeton University Press, 1989.

White, R. "Production Complexity and Standardization in Early Aurignacian Bead and Pendant Manufacture: Evolutionary Implications." In *The Human Revolution: Behavioural and Biological Perspectives on the Origins of Modern Humans*, ed. P. A. Mellars and C. B. Stringer. Princeton, N.J.: Princeton University Press, 1989.

———. "Rethinking the Middle/Upper Paleolithic Transition." *Current Anthropology* 2 (1982):169–192.

———. "Thoughts on Social Relationships and Language in Hominid Evolution." *Journal of Social and Personal Relationships* 2 (1985):95–115.

———. "Visual Thinking in the Ice Age." *Scientific American*, July 1989, 92–99.

White, T. D. "Cannibalism at Klasies?" *Sagittarius* 2 (1987):6–9.

———, and N. Toth. "The Question of Ritual Cannibalism at Grotta Guattari." *Current Anthropology* 32 (1991):118–138.

Whiten, A., and R. W. Byrne. "Tactical Deception in Primates." *Behavioral and Brain Sciences* 11 (2 1988):233–273.

Wiessner, P. "Style and Social Information in Kalahari San Projectile Points." *American Antiquity* 48 (2 1983):253–276.

Wilson, A. C., and R. L. Cann. "Where Did Modern Humans Originate? The Recent African Genesis of Humans." *Scientific American*, April 1992, 66–73.

Wobst, H. M. "The Demography of Finite Populations and the Origins of the Incest Taboo." In *Population Studies in Archaeology and Biological Anthropology*, ed. A. C. Swedlund. Society for American Archaeology Memoir 30, 1975, pp. 75–81.

———. "Locational Relationships in Paleolithic Society." *Journal of Human Evolution* 5 (1976):49–58.

———. "Stylistic Behavior and Information Exchange." In *Papers for the Director: Research Essays in Honor of James B. Griffin*, ed. C. E. Cleland. Anthropological Papers, Museum of Anthropology, University of Michigan, No. 61, Ann Arbor: 1977.

Wolpoff, M. H. "Describing Anatomically Modern *Homo sapiens*: A Distinction Without a Definable Difference." *Anthropos* 23 (1986):41–53.

————. "Multiregional Evolution: The Fossil Alternative to Eden." In *The Human Revolution: Behavioural and Biological Perspectives on the Origins of Modern Humans*, ed. P. A. Mellars and C. B. Stringer. Princeton, N.J.: Princeton University Press, 1989.

Wolpoff, M. H., and A. Thorne. "The Case Against Eve." *New Scientist*, June 22, 1991, 37–41.

Wolpoff, M. H., et al. "Letters: Modern Human Origins." *Science* 241 (1988):772.

Wymer, John. *The Paleolithic Age*. London: St. Martin's Press, 1982.

Yellen, J. E., et al. "A Middle Stone Age Worked Bone Industry from Katanda, Upper Semliki River (Kivu), Zaire." *Science* 268 (April 28, 1995).

Zubrow, E. "The Demographic Modelling of Neanderthal Extinction." In *The Human Revolution: Behavioural and Biological Perspectives on the Origins of Modern Humans*, ed. P. A. Mellars and C. B. Stringer. Princeton, N.J.: Princeton University Press, 1989.

Index

Page numbers in *italics* refer to illustrations.